The Application of Endoscopy in Gastrointestinal Cancers

The Application of Endoscopy in Gastrointestinal Cancers

Editors

Marcello Maida
Alberto Murino
Alan Moss

Basel • Beijing • Wuhan • Barcelona • Belgrade • Novi Sad • Cluj • Manchester

Editors

Marcello Maida
University of Enna "Kore"
Enna
Italy

Alberto Murino
University College London (UCL)
London
United Kingdom

Alan Moss
University of Melbourne
Melbourne
Australia

Editorial Office
MDPI AG
Grosspeteranlage 5
4052 Basel, Switzerland

This is a reprint of articles from the Special Issue published online in the open access journal *Cancers* (ISSN 2072-6694) (available at: https://www.mdpi.com/journal/cancers/special_issues/9J035C31PS).

For citation purposes, cite each article independently as indicated on the article page online and as indicated below:

Lastname, A.A.; Lastname, B.B. Article Title. *Journal Name* **Year**, *Volume Number*, Page Range.

ISBN 978-3-7258-1721-4 (Hbk)
ISBN 978-3-7258-1722-1 (PDF)
doi.org/10.3390/books978-3-7258-1722-1

© 2024 by the authors. Articles in this book are Open Access and distributed under the Creative Commons Attribution (CC BY) license. The book as a whole is distributed by MDPI under the terms and conditions of the Creative Commons Attribution-NonCommercial-NoDerivs (CC BY-NC-ND) license.

Contents

About the Editors vii

**Bruno Costa Martins, Renata Nobre Moura, Angelo So Taa Kum,
Carolina Ogawa Matsubayashi, Sergio Barbosa Marques and Adriana Vaz Safatle-Ribeiro**
Endoscopic Imaging for the Diagnosis of Neoplastic and Pre-Neoplastic Conditions of the
Stomach
Reprinted from: *Cancers* **2023**, *15*, 2445, doi:10.3390/cancers15092445 1

Ana Clara Vasconcelos, Mário Dinis-Ribeiro and Diogo Libânio
Endoscopic Resection of Early Gastric Cancer and Pre-Malignant Gastric Lesions
Reprinted from: *Cancers* **2023**, *15*, 3084, doi:10.3390/cancers15123084 16

Julia Chaves, Michael Fernandez Y Viesca and Marianna Arvanitakis
Using Endoscopy in the Diagnosis of Pancreato-Biliary Cancers
Reprinted from: *Cancers* **2023**, *15*, 3385, doi:10.3390/cancers15133385 35

Oliver Cronin and Michael J. Bourke
Endoscopic Management of Large Non-Pedunculated Colorectal Polyps
Reprinted from: *Cancers* **2023**, *15*, 3805, doi:10.3390/cancers15153805 48

**Leonardo Henry Eusebi, Andrea Telese, Chiara Castellana, Rengin Melis Engin,
Benjamin Norton, Apostolis Papaefthymiou, et al.**
Endoscopic Management of Dysplastic Barrett's Oesophagus and Early Oesophageal
Adenocarcinoma
Reprinted from: *Cancers* **2023**, *15*, 4776, doi:10.3390/cancers15194776 59

Edward Young, Louisa Edwards and Rajvinder Singh
The Role of Artificial Intelligence in Colorectal Cancer Screening: Lesion Detection and Lesion
Characterization
Reprinted from: *Cancers* **2023**, *15*, 5126, doi:10.3390/cancers15215126 78

**Giacomo Emanuele Maria Rizzo, Lucio Carrozza, Gabriele Rancatore, Cecilia Binda,
Carlo Fabbri, Andrea Anderloni and Ilaria Tarantino**
The Role of Endoscopy in the Palliation of Pancreatico-Biliary Cancers: Biliary Drainage,
Management of Gastrointestinal Obstruction, and Role in Relief of Oncologic Pain
Reprinted from: *Cancers* **2023**, *15*, 5367, doi:10.3390/cancers15225367 94

**Rocio Chacchi-Cahuin, Edward J. Despott, Nikolaos Lazaridis, Alessandro Rimondi,
Giuseppe Kito Fusai, Dalvinder Mandair, et al.**
Endoscopic Management of Gastro-Entero-Pancreatic Neuroendocrine Tumours: An Overview
of Proposed Resection and Ablation Techniques
Reprinted from: *Cancers* **2024**, *16*, 352, doi:10.3390/cancers16020352 114

**Aurelio Mauro, Davide Scalvini, Sabrina Borgetto, Paola Fugazzola, Stefano Mazza,
Ilaria Perretti, et al.**
Malignant Acute Colonic Obstruction: Multidisciplinary Approach for Endoscopic
Management
Reprinted from: *Cancers* **2024**, *16*, 821, doi:10.3390/cancers16040821 128

Tomonori Yano and Hironori Yamamoto
Endoscopic Diagnosis of Small Bowel Tumor
Reprinted from: *Cancers* **2024**, *16*, 1704, doi:10.3390/cancers16091704 145

About the Editors

Marcello Maida

Professor Marcello Maida is an Associate Professor of Gastroenterology at the Department of Medicine and Surgery, Kore University of Enna, Italy. He is also member of the steering committee of the Italian Society of Gastroenterology (SIGE) and member of the Guidelines Committee of the European Society of Gastrointestinal Endoscopy (ESGE). His research interests include several areas of gastroenterology, including endoscopy, gastrointestinal cancers, and liver diseases. He has authored more than one hundred scientific papers published in impact factor journals (indexed on PubMed and Scopus) and is an editorial board member of many journals focused on gastroenterology.

Alberto Murino

Dr. Alberto Murino is a Consultant Gastroenterologist and Advanced Endoscopist with a distinguished British and international background. He currently practices at the Royal Free Hospital NHS Trust and Cleveland Clinic London. Additionally, he serves as an Honorary Associate Professor at University College London (UCL), within the Institute for Liver and Digestive Health.

Dr. Murino has authored over 70 peer-reviewed publications and serves as a reviewer for several leading endoscopy journals. He is an active member of the ESGE guidelines committee, the WEO educational committee, and the ASGE international committee.

He is also the director of the International Advanced Endoscopy Masterclass in London.

Dr. Murino earned his medical degree in Italy, specializing in gastroenterology and gastrointestinal endoscopy in 2009, graduating with the highest honors (50/50 cum laude). From 2010 to 2014, he completed three prestigious endoscopy fellowships at internationally renowned institutions: St. Mark's Hospital in London, Erasmus Hospital in Brussels, and Nagoya University Hospital in Japan. Following his fellowships, he worked as a consultant in Italy for two years before relocating to the UK.

Alan Moss

Professor Alan Moss is the Head of Unit Endoscopic Services and the Head of Unit Gastroenterology at Western Health, a large academic tertiary level hospital network in Melbourne, Australia. He is a Clinical Professor at the University of Melbourne, an editorial board member of the *Endoscopy* journal, and an International Editorial Board member of the *Gastrointestinal Endoscopy* journal. He trained in interventional gastrointestinal endoscopy at Westmead Hospital, Sydney, and at St Michael's Hospital, Toronto. He has special interests (clinical and research) in colonic Endoscopic Mucosal Resection (EMR) for large or complex polyps. He performs a range of endoscopic procedures, including gastroscopy, colonoscopy, EMR, ERCP, EUS, and GI luminal stenting. He is a member of the World Endoscopy Organisation (WEO) Emerging Stars program. Alan's formal qualifications include an MBBS (Hons), MD, FRACP, FASGE, MHM, and GAICD.

Review

Endoscopic Imaging for the Diagnosis of Neoplastic and Pre-Neoplastic Conditions of the Stomach

Bruno Costa Martins [1,2,*], Renata Nobre Moura [1,2], Angelo So Taa Kum [3], Carolina Ogawa Matsubayashi [3], Sergio Barbosa Marques [2,3] and Adriana Vaz Safatle-Ribeiro [1]

1. Endoscopy Unit, Instituto do Cancer do Estado de São Paulo, University of São Paulo, São Paulo 01246-000, Brazil
2. Fleury Medicina e Saude, São Paulo 01333-010, Brazil
3. Endoscopy Unit, Hospital das Clínicas da Faculdade de Medicina da Universidade de São Paulo, University of São Paulo, São Paulo 05403-010, Brazil
* Correspondence: bruno.costa@hc.fm.usp.br

Simple Summary: Gastric cancer has a poor prognosis when diagnosed in advanced stages, but curative treatment is possible if an early diagnosis is made. Endoscopy represents an essential tool for the detection of early neoplastic and pre-neoplastic gastric lesions and for surveillance. Many endoscopy imaging technologies have been developed to increase diagnostic accuracy. In this review, we summarize these endoscopy technologies.

Abstract: Gastric cancer is an aggressive disease with low long-term survival rates. An early diagnosis is essential to offer a better prognosis and curative treatment. Upper gastrointestinal endoscopy is the main tool for the screening and diagnosis of patients with gastric pre-neoplastic conditions and early lesions. Image-enhanced techniques such as conventional chromoendoscopy, virtual chromoendoscopy, magnifying imaging, and artificial intelligence improve the diagnosis and the characterization of early neoplastic lesions. In this review, we provide a summary of the currently available recommendations for the screening, surveillance, and diagnosis of gastric cancer, focusing on novel endoscopy imaging technologies.

Keywords: gastric cancer; endoscopy; artificial intelligence; early detection of cancer; early diagnosis

1. Introduction

Gastric cancer (GC) is the fourth most common cause of cancer deaths worldwide, with 768,793 cases and an estimated incidence of 10,839,103 new cases yearly, according to The International Agency for Research on Cancer. Eastern Asia has the highest incidence, accounting for 60% of the cases [1]. According to gender, both the incidence and death rate in males were more than twice that in females. Most GC cases occur in patients older than 45 years of age, with a higher incidence at 65 to 70 years.

Although advanced GC is associated with a poor prognosis and high mortality, early detection and treatment have a good prognosis, with 5-year survival rates higher than 95% [2]. Unfortunately, more than 70% of GCs in Western countries are diagnosed in advanced stages [3].

Histologically, GC is classified as diffuse (composed of non-cohesive cells) or intestinal types (gland-forming), with different epidemiological patterns [4]. *Helicobacter pylori* (*H. pylori*) is considered a risk factor for non-cardia GC, with nearly 89% of cancers associated with this pathogen infection [5], mainly intestinal-type cancer. According to Pelayo Correa's model of carcinogenesis, a cascade of events beginning with active chronic inflammation may progress to multifocal atrophic gastritis (AG), intestinal metaplasia (IM), dysplasia, and carcinoma [6]. The eradication of *H. pylori* infection has been associated with a significant reduction in the incidence and mortality of GC, especially among patients younger than 50 years of age [7]. In a meta-analysis, a pooled incidence rate ratio of 0.53 (95% CI 0.44–0.64) was observed comparing individuals who received *H. pylori* eradication with individuals who did not receive eradication therapy [8].

Esophagogastroduodenoscopy (EGD) is the gold-standard exam for the diagnosis of neoplastic and pre-neoplastic gastric conditions, such as AG and IM. However, in a meta-analysis of 22 studies, the authors demonstrated that nearly 10% of GCs were potentially missed during white-light endoscopy (WLE), mainly adenocarcinomas located at the gastric body. Predictive factors for diagnostic failure were a younger age (<55 years), female, advanced atrophy, adenoma, ulcer lesions, and an insufficient number of biopsies [9].

Image-Enhanced Endoscopy (IEE) has been used to overcome the diagnostic limitations of standard endoscopy. IEE refers to various methods, such as dye chromoscopy, high-resolution imaging, virtual chromoscopy, and artificial intelligence. IEE provides a better assessment of the mucosal surface, increasing the detection of subtle changes and improving the diagnosis of pre-neoplastic lesions [10–12]. In this article, we will discuss the technical measures and imaging technologies that can be adopted to increase endoscopy diagnostic yield.

2. Indications for Endoscopic Screening of Gastric Cancer

In Eastern countries, radiographic screening programs for GC diagnosis were implemented in the 1960s, reducing its mortality [13]. Nowadays, screening methods include radiography and EGD. A cohort study comparing the two methods showed that subjects screened by EGD had a 67% reduction in GC mortality compared with subjects screened by radiography (RR 0.327; 95% CI, 0.118–0.908) [14]. In South Korea, GC screening by EGD every 2 years was shown to be associated with a significant ($\geq$80%) reduction in GC mortality, while for those undergoing radiographic examinations every 2 years, the reduction in mortality was only 20% [7]. In a metanalysis, EGD screening detection of GC (0.55%, 95% CI 0.39–0.75%) and early-GC (EGC) (0.48%, 95% CI 0.34–0.65%) was superior to radiography screening (GC 0.19%, 95% CI 0.10–0.31%; EGC 0.08%, 95% CI 0.04–0.13%) [15].

In Western countries, where the incidence of GC is lower than in Eastern countries, screening focuses on high-risk patients with AG or IM. The management of epithelial precancerous conditions and lesions in the stomach (MAPS) guideline recommends the use of a staging system in patients with AG and/or IM, such as the Operative Link on Gastritis Assessment (OLGA) or the Operative Link on Gastritis Assessment based on Intestinal Metaplasia (OLGIM) systems [16,17]. Patients with extensive atrophy and/or IM i.e., affecting both antral and corpus mucosa, should be identified and sampled as they are considered to be at higher risk for GC [17]. These patients should be followed-up with a high-quality EGD every 3 years. In patients with a family history of GC, close follow-up is suggested (e.g., every 1–2 years after diagnosis). The association of high-risk OLGA stages (III/IV) with GC was demonstrated in a meta-analysis including 6 case–control studies and 2 cohort studies (2.64 and 27.7 times higher chance of GC than lower OLGA stages respectively). Among patients with OLGIM III/IV, the risk of GC was 3.99 times higher (95% CI 3.05–5.21) [18].

The American Gastroenterological Association (AGA) suggested EGD surveillance every 3–5 years in patients at high risk for GC, including those with advanced AG, extensive and incomplete IM, a family history of GC, and new immigrants from areas of high risk, such as East Asia or South America. The AGA also recommended testing and the eradication of *H. pylori* infection among individuals with IM [19,20]. Similarly to the AGA, the American Society of Gastrointestinal Endoscopy (ASGE) recommended surveillance for high-risk individuals with IM, as well as testing and treating *H. pylori* infection [21].

The British Society of Gastroenterology emphasized the importance of IEE for the diagnosis of pre-neoplastic lesions. The society advised a baseline endoscopy among individuals with laboratory evidence of pernicious anemia, such as vitamin B12 deficiency and positive gastric parietal cell or intrinsic factor antibodies [22].

3. White-Light Endoscopy (WLE)

EGD has the possibility of identifying pre-neoplastic mucosal changes, detecting early-stage GC, and reducing cancer-related mortality by diagnosing and treating gastric mucosal alterations and/or EGC. Under WLE, EGC should be suspected in the presence of mucosal surface irregularity and/or mucosal coloration changes. Spontaneous bleeding, pallor coloration, and alterations in the mucosa surface and in light reflection should raise a concern about neoplastic lesions, as shown in Figure 1.

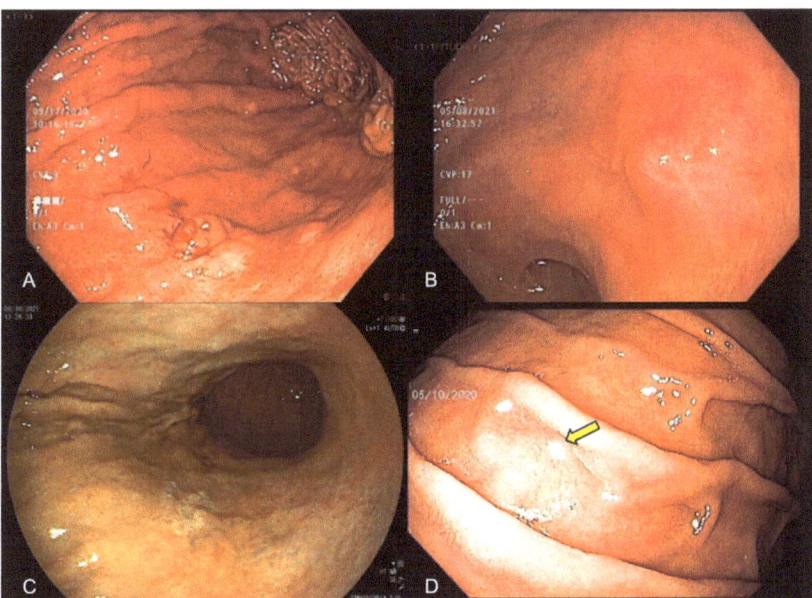

Figure 1. Endoscopic appearance of early gastric cancer (**A**) white-light endoscopy (WLE) showing spontaneous bleeding, (**B**) WLE showing irregular surface, showing a slightly elevated lesion with a central depression, (**C**) pallor color change in the posterior wall of the stomach enhanced by FICE (Flexible spectral imaging color enhancement), (**D**) change in light reflection (arrow) under WLE.

Early gastric cancer is defined as a lesion confined to the mucosa and submucosa (T1), regardless of lymph node involvement. These lesions usually manifest as superficial lesions (type 0), which can be subclassified into polypoid (Type 0-I), flat (Type 0-II), and excavated (Type 0-III). Flat lesions are categorized as slightly elevated (0-IIa), completely flat (0-IIb), or slightly depressed (0-IIc). Superficial tumors with two or more components should have all the components described (e.g., 0-IIa + IIc) [23] (Figure 2).

A high-quality level of endoscopic examination is imperative to make a proper diagnosis of early neoplastic lesions. The use of pre-endoscopy medications (mucolytic and defoaming agents), high-definition endoscopes, adequate inspection time, obtaining index images, and the application of the MAPS biopsy protocol when AG and chronic inflammation are suspected have been recommended by the European Society of Gastrointestinal Endoscopy (ESGE) guidelines [24,25].

3.1. Mucolytic and Defoaming Agents

To achieve a proper mucosal evaluation during EGD, the cleansing of mucus, bubbles, and foam is important. The most common mucolytic agents used worldwide are Pronase® and n-acetylcysteine, and both are used to eliminate gastric mucus using non-osmotic solutions. Simethicone (activated dimethicone) is a commonly used defoaming agent that decreases the surface tension of gas bubbles without significant adverse interactions. It can improve the mucosal observation when used 20 min before the procedure [26].

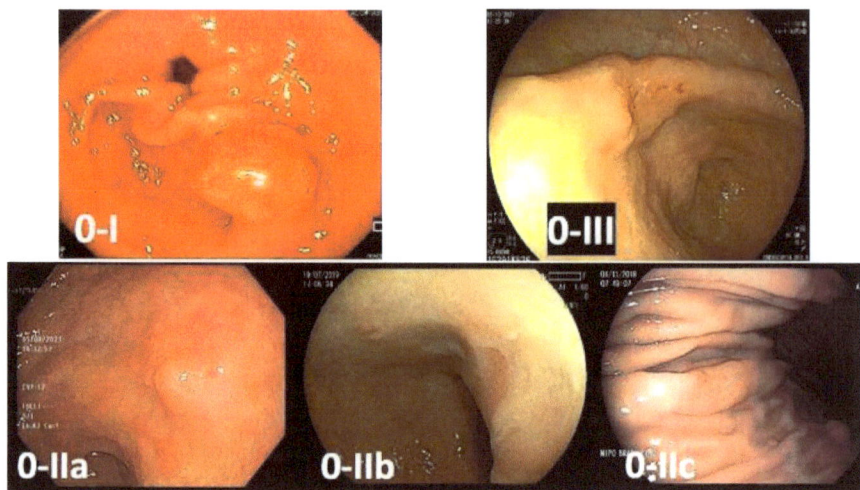

Figure 2. Paris classification of gastric superficial lesions—0-I: polypoid lesion; 0-IIa: flat and slightly elevated lesion; 0-IIb: completely flat lesion; 0-IIc: flat and slightly depressed lesion; 0-III excavated lesion.

3.2. Antispasmodic Agents

Antispasmodic agents, such as cimetropium bromide, scopolamine, and hyoscine N butyl bromide, can reduce peristalsis and may be used during EGD [27]. However, there is no scientific evidence supporting the benefits of antispasmodic agents in the detection rate of upper gastrointestinal (GI) neoplasia. In addition, patients may have adverse reactions to these drugs, such as arrhythmia, benign prostate hypertrophy, and glaucoma. Thus, its use should be selective and at the discretion of the endoscopist.

3.3. Inspection Time

Limited consensus exists about the optimum inspection time for EGD compared to the colonoscopy quality protocol recommendation (withdrawal time of 6 min or more). In a retrospective Korean study [28], endoscopists who dedicated at least 3 min to evaluate the gastric mucosa detected more gastric adenomas or cancers than faster endoscopists (0.28% versus 0.20%, respectively; $p < 0.01$). A Japanese study showed that faster endoscopists, with a mean inspection time below 5 min, may overlook neoplastic lesions in the upper gastrointestinal tract. In this study, endoscopists were classified into fast (<5 min examination), moderate (between 5–7 min), and slow (>7 min) groups. The odds ratio of diagnosing neoplastic lesions was 1.90 for the moderate group and 1.89 for the slow group compared to the fast group ($p = 0.03$ and $p = 0.06$, respectively) [29].

3.4. High-Resolution Endoscopes

The resolution of an image depends on the pixel density, which is directly associated with the capacity to distinguish two adjacent points. Higher pixel density endoscopes provide higher imaging resolution. Standard endoscopes produce a signal image with a resolution of 100,000 to 400,000 pixels. High-resolution or high-definition endoscopes generate images with up to 1,000,000 pixels [30]. The development of high-resolution endoscopes enabled to distinguish subtle mucosal surface details compared to standard endoscopes, allowing more accurate suspicious diagnoses and targeted biopsies [31].

3.5. Obtain Index Images

The requirements of minimum photo documentation vary from each endoscopy society recommendation. While USA guidelines do not specify the minimum number of

EGD images to report, the ESGE suggested at least ten images indicating the anatomical index [32]. In Japan, a systematic screening protocol for the stomach included 22 images, with the rationale to study the full gastric mucosa and avoid blind spots [33]. The World Endoscopy Organization proposed a total of 28 image areas, including the hypopharynx [34]. Adherence to a standardized photo protocol may increase the EGD neoplastic detection rate because the endoscopist must examine the entire stomach, including potential blind spots.

3.6. Target Biopsies of Suspicious Lesions

Target biopsies improve the diagnosis of GC. The ESGE recommends at least six biopsies of suspected advanced GC [35]. For EGC, only one to two targeted biopsies are recommended to avoid scars or submucosal fibrosis, as it is best staged and treated by endoscopic resection [36].

Moreover, some conditions, including chronic AG or IM, carry a higher risk for GC. The updated Sydney system recommended at least five biopsies: two from the antrum (greater and lesser curvature, 3 cm from the pylorus), one from the incisura, and two from the gastric body (from lesser curvature and greater curvature) [37]. The MAPS I and II guidelines question the necessity of sampling the incisura, as it yielded minimal additional diagnostic information, with more costs [17]. The inclusion of an incisura biopsy, however, increased the proportion of patients classified as high-risk stages (OLGA III/IV or OLGIM III/IV) [38,39]. It is also important to label the biopsies from different sites, and to apply a validated staging system, such as OLGA or OLGIM.

4. Chromoendoscopy

Chromoendoscopy (CE) is an IEE modality consisting of spraying dyes on the mucosal surface to improve visualization of the lesions under investigation. The use of CE in the screening of malignant and premalignant lesions may increase the detection rate and provide a better understanding of the lesion boundaries and microsurface, which helps differentiate benign or inflammatory from suspected malignant conditions and to determine the adequate region for biopsy. CE has a relatively low cost and it can be used in any endoscopy unit. CE with acetic acid, methylene blue, and indigo carmine are the main dye spray techniques to improve the diagnosis of GC. CE had higher accuracy compared to SD-WLE for the diagnosis of EGC ($p = 0.005$) and premalignant lesions ($p = 0.001$) [10].

4.1. Acetic Acid

The acetic acid solution is used in a concentration of 1.5 to 3%. After spraying it into the gastric mucosa, an acetowhite reaction is seen immediately: the structure of the cellular protein is reversibly altered and the superficial pH is lowered in the mucosa, causing a white reflection. The intensity of the whitening differs for normal mucosa and IM, as well as the duration for cancerous and non-cancerous tissue, disappearing earlier in the carcinoma. This creates a contrast between the regular mucosa versus IM and the pinkish cancer lesion surrounded by non-neoplastic tissue [40,41], as seen in Figure 3.

4.2. Methylene Blue

Ida et al. first described the methylene blue staining in endoscopy to improve the diagnosis of EGC. Methylene blue is absorbed by the small bowel and colon mucosa and is not absorbed by the stratified squamous epithelium of the esophagus or normal gastric mucosa [42]. The effectiveness of methylene blue, used as a 5% solution, is to highlight subtle mucosal changes through the staining of IM within the stomach. This vital staining dye improves the accurate delineation of the anatomical extent of histological abnormalities in the stomach. This strategy improves the accuracy of mapping IM and guides target biopsies.

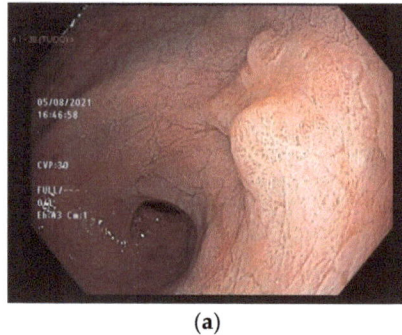 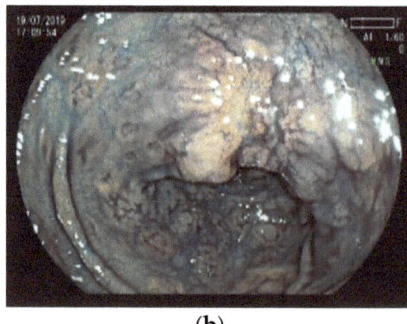

(**a**)   (**b**)

Figure 3. (**a**) White-light endoscopy with acetic acid chromoendoscopy showing an elevated lesion with central depression. The lesion is slightly reddish compared to the surrounding mucosa, suggesting a neoplastic lesion. Biopsy revealed tubular adenocarcinoma; (**b**) chromoendoscopy with indigo carmine dye delineating gastric lesion's edges (Paris 0-IIa + IIc).

4.3. Indigo Carmine

CE with indigo carmine dye was first described by Tada et al. [43] in 1976. This contrast staining dye is not absorbed by the mucosa. It is used to enhance crevices and valleys, defining irregularities in the mucosal architecture more accurately (Figure 4). Indigo carmine is used in a concentration of 0.2–0.5% and is an important modality to classify and delimit gastric lesions (e.g., Figure 4).

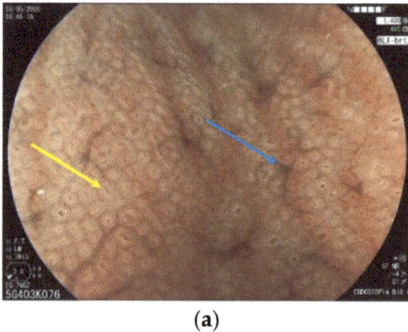 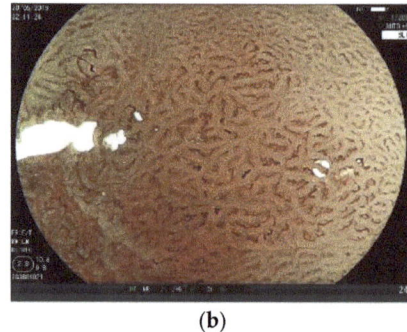

(**a**)   (**b**)

Figure 4. (**a**) High-definition magnifying view of normal fundic mucosa. Note the round pits surrounded by a white-colored structure (marginal crypt epithelium—yellow arrow). The regular arrangement of collecting venules is easily seen (blue arrow); (**b**) normal pyloric gland mucosa shows a reticular aspect with grooves.

5. Virtual Chromoendoscopy

Virtual or electronic chromoendoscopy are image-processing techniques capable of enhancing mucosal surface patterns. It is easy to use because it is activated by pressing a button in the endoscope, being less time-consuming than conventional dyes. Currently there are several types of electronic IEE, which can be basically separated into two methods post-processed images (ex., FICE and i-SCAN), which electronically select filters and reconstruct images, and pre-processed images, which use optical filters (ex., NBI, LCI and BLI).

The basic principle consists of different tissue wavelengths absorption according to the depth of penetration. Narrow-band imaging (NBI) developed by Olympus (Olympus Medical Systems Co., Ltd., Tokyo, Japan) relies on a bandwidth light filter resulting in an increased contrast of the mucosal microsurface and microvessels. Several studies and meta-analyses have demonstrated the efficacy of NBI on the detection, characterization, differentiation, and margin delineation of GC [12,44–47].

FICE (Flexible spectral imaging color enhancement), developed by Fujifilm (Fujifilm Co., Kanagawa, Japan), consists of a post-selection of a wide range of spectral combinations, improving the resolution and enhancement (Figure 1C). Afterward, Fujifilm developed blue laser imaging (BLI), which uses two different monochromatic lasers to produce a narrow blue band to intensify changes in the mucosal surface. Similarly to NBI, BLI also showed high diagnostic accuracy for the detection and characterization of GC [44,48]. The linked color imaging (LCI), also by Fujifilm, emphasizes the contrast of hemoglobin by expanding the color redness and, therefore, producing bright images that can improve the visibility of lesions [49,50]. In a multicenter randomized trial, the percentage of patients with one or more neoplastic lesions diagnosed with LCI was higher than with WLE (8.0%; 95% CI, 6.2–10.2% vs. 4.8%; 95% CI, 3.4–6.6%), and the proportion of patients with overlooked neoplasms was lower in the LCI group than in the WLE group (0.67%; 95% CI, 0.2–1.6% vs. 3.5%; 95% CI, 2.3–5.0%). The I-SCAN method, developed by PENTAX (Tokyo, Japan), consists of various types of post-processing images to enhance the surface, contrast, and tone. In a comparison trial, i-SCAN showed similar results compared with dye CE (acetic acid and indigo carmine) to delineate the margins of gastric lesions [51].

6. Magnifying Endoscopy

Image-enhanced magnifying endoscopy is a tool to better characterize the detected lesion and to correlate it with pathology, helping the endoscopist to differentiate benign from malignant or pre-neoplastic conditions. In a meta-analysis study including 1724 patients and 2153 lesions, the pooled sensitivity, specificity, and area under the curve for the diagnosis of EGC using WLE were 48%, 67%, and 62%, respectively. The use of magnifying endoscopy with NBI improved these rates to 83%, 96%, and 96%, respectively [12].

To diagnose subtle mucosal changes, it is necessary to recognize the normal magnifying features of gastric mucosa. Histologically, there are two different types of gastric glandular epithelium: fundic and pyloric. Normal fundic mucosa is present in the gastric body and fundus of patients without pathological changes, such as inflammation or AG secondary to *H. pylori* infection. Chromoendoscopy with the magnification of normal fundic mucosa is characterized by an epithelial surface where the crypt opening is seen as oval or round surrounded by a white-colored structure (marginal crypt epithelium). Collecting venules have a regular arrangement (RAC) and a greenish (cyan) color. In normal pyloric mucosa, the glands open obliquely and not perpendicularly as in the fundic pattern, resulting in a reticular aspect with grooves. The capillaries form spiral or spring loops (Figure 4):

As stated above, IM is considered a risk factor for the development of intestinal GC. Even though the definitive diagnosis of IM relies on histopathologic evaluation, magnifying endoscopy has also been shown to be an important tool for diagnosis, characterization, and guiding target biopsies. A fine blue-white line on the epithelial surface called the "light blue crest" was described by Uedo et al. and is a good predictor of IM [52].

Yao et al. first described the vessel plus surface (VS) classification, and later, Muto et al. reported the magnifying endoscopy simple diagnostic algorithm for early gastric cancer (MESDA-G) for the differentiation between neoplastic and non-neoplastic lesions [53,54]. The MESDA-G classification relies on the concept that neoplastic changes in gastric mucosa are followed by a clear demarcation line between the normal and the altered tissue. This line is defined as an abrupt change in the surface pattern, either in vascular or glandular structures, as shown in Figure 5.

According to the vs. classification, microvascular and microsurface patterns are classified into three categories: regular (normal), irregular, and absent. The irregular vascular pattern shows a variety of different capillaries or shapes irregularly distributed. When no microvessels are seen, the V pattern is considered absent. This occurs especially when a white opaque substance (WOS) is present in the superficial mucosal layer and prevents the visibility of the epithelium vasculature [55]. In such cases, instead of evaluating the vascular pattern, the distribution type of the WOS is classified as homogeneous or heterogeneous to make a differential diagnosis of cancer and adenoma (Figure 6). The surface pattern (S) is characterized by mucosal glandular structures. When asymmetrically distributed or in various morphologies, it is classified as irregular. If no glandular structure is seen, it is classified as absent. In this case, the V pattern is used for differentiation between neoplastic and non-neoplastic lesions (Figure 7).

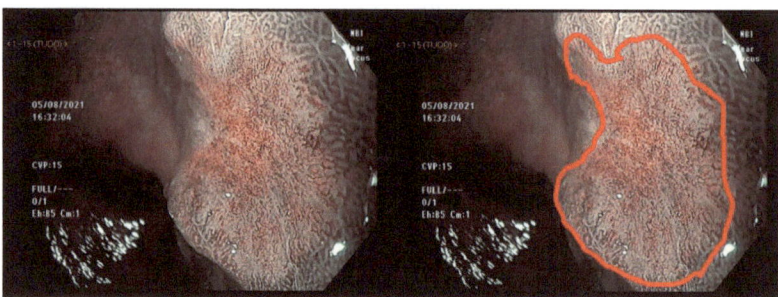

Figure 5. NBI with near-focus magnification showing the demarcation line of a gastric cancer NBI = narrow-band imaging.

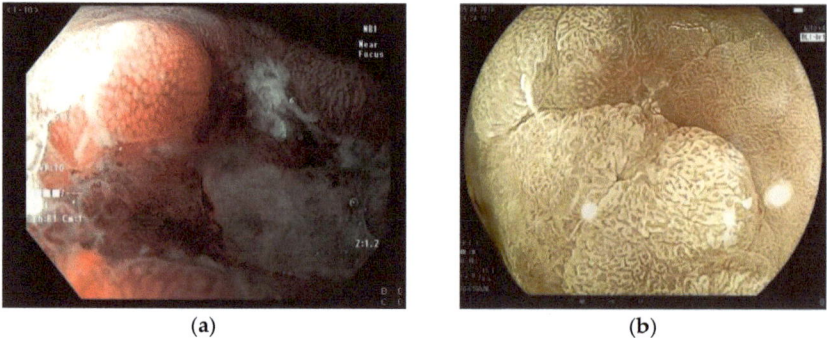

Figure 6. (a) Irregular vascular pattern seen with NBI and near-focus magnification; (b) absent vascular pattern due to the presence of a white opaque substance inside the demarcation line (BLI—bright view plus magnification). It has a homogeneous distribution, which leads to classifying this lesion as non-cancerous (adenoma). NBI = narrow-band imaging; BLI = blue laser imaging.

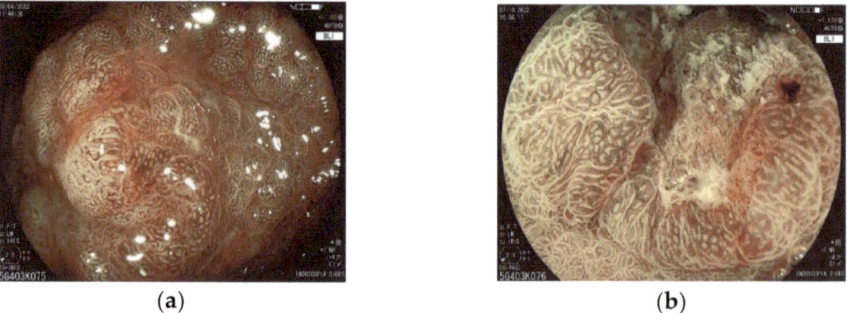

Figure 7. (a) Irregular surface pattern and irregular vascular pattern; (b) absent surface pattern with a clear demarcation line. BLI with magnification view. BLI = blue laser imaging.

The vs. classification showed high accuracy (97%), a positive predictive value (79%) and a negative predictive value (99%) for the diagnosis of intestinal-type EGC [56]. Never theless, diffuse- or undifferentiated-type carcinoma characterization may be a limitation for endoscopic diagnosis. This occurs because neoplastic involvement occurs horizontally in the deeper layers of the mucosa without glandular formation, and during the early phase there may be no visible endoscopic alterations, and even biopsies may be negative because the cancer cells are usually deeper and more widespread. This difficulty in detection contributes to a late diagnosis and, consequently, a worse prognosis. In addition, it is also difficult to determine the depth of invasion and distinguish the lateral limits of undifferentiated lesions.

Under WLE, the typical presentation of undifferentiated tumors is a flat or depressed lesion with a pale color. According to Yao, this is due to the reduction in hemoglobin levels [57]. Magnifying endoscopy is important for the evaluation of undifferentiated GC. Usually, in this type, there is no glandular formation, and, consequently, the microsurface pattern is classified as absent. In this scenario, only microvessels can be seen [58] and typically present as a corkscrew pattern with a loop-opened look (Figure 8).

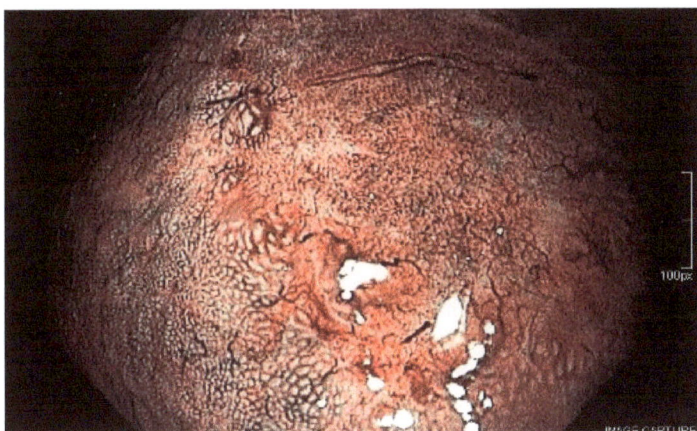

Figure 8. Corkscrew microvessel pattern present on an undifferentiated-type early gastric cancer. Magnified view with NBI.

7. Artificial Intelligence

Artificial intelligence (AI) has been used to support physicians interpreting medical images. The related applied terminology is computer-aided detection (CADe) and computer-aided diagnosis (CADx), when algorithms are developed to detect and differentiate pathologies, respectively [59]. For this purpose, the most used system is deep learning due to the capacity to be trained with a large number of images and to extract specific clinical features to further predict or classify new images [60,61].

As already discussed, physicians' ability to adequately diagnose upper GI lesions varies greatly, with a missing rate of EGC of 9.4% (95% CI 5.7–13.1%) [9]. In contrast to colorectal lesions, GC is often subtle (e.g., flat or slightly depressed) and hard to recognize due to the inflammation environment of chronic gastritis [59].

After the publication of the first deep learning-based AI system, which showed promising results with a sensitivity of 92.2% for detecting GC [62], other studies have been conducted, and the commercialization of the first software may not take long. A tandem randomized prospective trial showed a significantly lower miss rate of EGC in the AI-first group than in the routine-first group (6.1%, 95% CI 1.6–17.9% vs. 27.3%, 95% CI 15.5–43%) [63]. Additionally, it was shown that AI was able to predict the superficial cancer invasion depth with a specificity of 78–95% [64]. Consequently, it will help a more homogeneous diagnosis across physicians, improving the decision of referring patients to endoscopic or surgical resection.

In a meta-analysis, the overall sensitivity, specificity, and area under the curve of AI were 89% (95% CI 85–93%), 93% (95% CI 88–97%), and 94% (95% CI 0.91–0.98%), respectively, being comparable to experts and superior to non-experts [64]. Moreover, it is worth mentioning that the use of AI will also play an important role to help in the diagnosis of blind spots and photo documentation with an accuracy of 90.4% [65,66], supporting the quality control of exams and guidelines compliance [32–34].

Although other robust prospective trials should be conducted to deeply evaluate the real performance of AI during routine EGD exams, randomized controlled trials evaluating colon AI have proven that this technology is safe to be adopted and potentially improves patient outcomes while reducing costs [67–69].

8. Other Methods

8.1. Confocal Laser Endomicroscopy

Confocal laser endomicroscopy (CLE) is an endoscopic imaging tool that provides cross-sectional real-time in vivo images at the cellular and microvascular levels at a 1000-fold magnification. Fluorescent contrast is required to obtain the imaging, and the most utilized contrast agents are intravenous fluorescein and topically applied acriflavine. Two types of CLE systems have been studied for clinical use: electronic CLE, which integrates a miniature confocal scanner into the tip of a special flexible endoscope, and probe-based CLE (pCLE), a flexible microprobe that can pass through the working channel of a standard endoscope. However, only pCLE is commercially available [70].

CLE allows histopathological diagnoses when utilized by trained endoscopists. Therefore, targeting the region with higher accuracy for a pathology assessment may have a relevant impact on treatment management, avoiding further diagnostic procedures for tissue sampling [71]. The use of pCLE optical diagnosis has been deeply studied for gastric IM, AG, and intraepithelial neoplasia [72–74], and it was shown that it contributes to distinguishing normal and neoplastic mucosa through the analysis of both cellular and vascular patterns [75]. In one study, pCLE was able to diagnose AG and IM with 98% specificity [72]. In a recent meta-analysis that included seven studies, pCLE had a sensitivity of 87.9% (95% CI 81.4–92.4%), a specificity of 96.5% (95% CI 91.5–98.6%), and an accuracy of 94.7% (95% CI 89.5–97.4%) to diagnose GC [73].

Despite these encouraging findings, pCLE is not regarded as a standard approach for diagnosis, as adoption is likely reduced for cost reasons and the lack of cost-effectiveness studies. There is also a need to define a framework for how to introduce this tool in the workflow between endoscopists and pathologists [76].

8.2. Endocytoscopy

Similarly to CLE, endocytoscopy has the capability of capturing ultra-magnified pathology images when placing the endoscope on the target mucosa during inspection [77]. It provides information about the shape of cells and individual cell nuclei of the mucosa's superficial layer. For instance, goblet cells for the evaluation of IM and the presence of structural or nuclear atypia to diagnose cancerous lesions, with a specificity of 93.3–100% [77,78].

Endocytoscopy leads to the benefits of the endoscopic optical biopsy but also requires additional training and expertise. Moreover, the need for double staining with crystal violet and methylene blue for optimal evaluation may limit the clinical application because this additional step often increases the duration of the exam, and the staining may not be easily available. However, the use of endocytoscopy with NBI (with no staining) using the vs classification has been studied with a reported accuracy higher than magnifying NBI [53]. Additionally, it has been shown that this evaluation can be simplified by assessing the microvascular pattern alone, which showed similar accuracy compared to the evaluation of microvascular and microsurface patterns [79]. Recently, Noda et al. [80] studied the application of AI to support the use of endocytoscopy. AI showed a specificity of 90.9% in diagnosing EGC, which was comparable with experts and superior to non-experts.

In summary, endoscopy plays a major role in the diagnosis of gastric cancer. It is important for the endoscopist to be aware of patients at risk, to recognize the endoscopic aspects of early neoplasms, and to understand the advantages and limitations of each endoscopic technique, as well as how to overcome them. Table 1 summarizes the advantages and limitations of the endoscopic techniques for the detection and characterization of gastric cancer.

Table 1. Comparison between endoscopic imaging methods for the diagnosis of gastric cancer.

Technique	Advantages	Limitations
White-light endoscopy	Easy to perform Readily available Low cost	Low sensitivity and specificity
Dye-based chromoendoscopy	Widely available Useful for delineating lesions margins Easy to perform	Time-consuming
Virtual chromoendoscopy	Valuable tool for evaluation of microvessels	Low accuracy to predict tumor depth
Magnifying endoscopy	High accuracy to distinguish benign and malignant lesions More specific than WLE and dyes	Low accuracy to distinguish differentiated- from undifferentiated-type adenocarcinoma Limited field of view Low accuracy to predict tumor depth
Artificial intelligence	Real-time diagnosis	High cost Lack of validation from prospective studies
Confocal laser endomicroscopy	Real-time diagnosis	High cost Steep learning curve Limited field of view Need for intravenous or topical contrast
Endocytoscopy	Real-time diagnosis Technology integrated (dedicated endoscope)	High cost Steep learning curve Limited field of view

9. Conclusions

Despite the significant improvement in endoscopic imaging for GC over the years, the adequate diagnosis and characterization of EGC is still a challenge. Consequently, GC awareness is crucial and demands a minimum proficiency among physicians, who should be encouraged to use chromoendoscopy. While novel imaging modalities are increasingly being studied, especially for optical diagnosis, specific experience, integration to current workflow, and cost-effectiveness studies should take place for major endorsement.

As modern imaging technology continues to grow, more complexity and time are added to endoscopic clinical practice. To overcome part of these challenges, AI systems will potentially play an important role in supporting physicians with more assertive diagnoses.

Author Contributions: Conceptualization, B.C.M. and R.N.M.; writing—original draft preparation, R.N.M., A.S.T.K., C.O.M. and S.B.M.; writing—review and editing, B.C.M. and A.V.S.-R.; supervision, B.C.M.; project administration, B.C.M. All authors have read and agreed to the published version of the manuscript.

Funding: This research received no external funding.

Conflicts of Interest: Carolina Ogawa Matsubayashi has full-time employment with AI Medical Service America Inc. The other authors declare no conflict of interest.

References

1. Sung, H.; Ferlay, J.; Siegel, R.L.; Laversanne, M.; Soerjomataram, I.; Jemal, A.; Bray, F. Global Cancer Statistics 2020: GLOBOCAN Estimates of Incidence and Mortality Worldwide for 36 Cancers in 185 Countries. *CA. Cancer J. Clin.* **2021**, *71*, 209–249. [CrossRef] [PubMed]
2. Crew, K.D.; Neugut, A.I. Epidemiology of Gastric Cancer. *World J. Gastroenterol.* **2006**, *12*, 354–362. [CrossRef] [PubMed]

3. Minicozzi, P.; Innos, K.; Sánchez, M.-J.; Trama, A.; Walsh, P.M.; Marcos-Gragera, R.; Dimitrova, N.; Botta, L.; Visser, O. Rossi, S.; et al. Quality Analysis of Population-Based Information on Cancer Stage at Diagnosis across Europe, with Presentation of Stage-Specific Cancer Survival Estimates: A EUROCARE-5 Study. *Eur. J. Cancer* **2017**, *84*, 335–353. [CrossRef] [PubMed]
4. Lauren, P. The two histological main types of gastric carcinoma: Diffuse and so-called intestinal-type carcinoma. an attempt at a histo-clinical classification. *Acta Pathol. Microbiol. Scand.* **1965**, *64*, 31–49. [CrossRef] [PubMed]
5. Plummer, M.; de Martel, C.; Vignat, J.; Ferlay, J.; Bray, F.; Franceschi, S. Global Burden of Cancers Attributable to Infections in 2012: A Synthetic Analysis. *Lancet Glob. Health* **2016**, *4*, e609–e616. [CrossRef]
6. Correa, P.; Houghton, J. Carcinogenesis of Helicobacter Pylori. *Gastroenterology* **2007**, *133*, 659–672. [CrossRef]
7. Mabe, K.; Inoue, K.; Kamada, T.; Kato, K.; Kato, M.; Haruma, K. Endoscopic Screening for Gastric Cancer in Japan: Current Status and Future Perspectives. *Dig. Endosc. Off. J. Jpn. Gastroenterol. Endosc. Soc.* **2022**, *34*, 412–419. [CrossRef]
8. Lee, Y.-C.; Chiang, T.-H.; Chou, C.-K.; Tu, Y.-K.; Liao, W.-C.; Wu, M.-S.; Graham, D.Y. Association Between Helicobacter Pylori Eradication and Gastric Cancer Incidence: A Systematic Review and Meta-Analysis. *Gastroenterology* **2016**, *150*, 1113–1124.e5 [CrossRef]
9. Pimenta-Melo, A.R.; Monteiro-Soares, M.; Libânio, D.; Dinis-Ribeiro, M. Missing Rate for Gastric Cancer during Upper Gastrointestinal Endoscopy: A Systematic Review and Meta-Analysis. *Eur. J. Gastroenterol. Hepatol.* **2016**, *28*, 1041–1049. [CrossRef]
10. Zhao, Z.; Yin, Z.; Wang, S.; Wang, J.; Bai, B.; Qiu, Z.; Zhao, Q. Meta-Analysis: The Diagnostic Efficacy of Chromoendoscopy for Early Gastric Cancer and Premalignant Gastric Lesions. *J. Gastroenterol. Hepatol.* **2016**, *31*, 1539–1545. [CrossRef]
11. Fiuza, F.; Maluf-Filho, F.; Ide, E.; Furuya, C.K.; Fylyk, S.N.; Ruas, J.N.; Stabach, L.; Araujo, G.A.; Matuguma, S.E.; Uemura, R.S. et al. Association between Mucosal Surface Pattern under near Focus Technology and Helicobacter Pylori Infection. *World J Gastrointest. Endosc.* **2021**, *13*, 518–528. [CrossRef]
12. Zhang, Q.; Wang, F.; Chen, Z.-Y.; Wang, Z.; Zhi, F.-C.; Liu, S.-D.; Bai, Y. Comparison of the Diagnostic Efficacy of White Light Endoscopy and Magnifying Endoscopy with Narrow Band Imaging for Early Gastric Cancer: A Meta-Analysis. *Gastric Cancer Off. J. Int. Gastric Cancer Assoc. Jpn. Gastric Cancer Assoc.* **2016**, *19*, 543–552. [CrossRef] [PubMed]
13. Oshima, A.; Hirata, N.; Ubukata, T.; Umeda, K.; Fujimoto, I. Evaluation of a Mass Screening Program for Stomach Cancer with a Case-Control Study Design. *Int. J. Cancer* **1986**, *38*, 829–833. [CrossRef] [PubMed]
14. Hamashima, C.; Shabana, M.; Okada, K.; Okamoto, M.; Osaki, Y. Mortality Reduction from Gastric Cancer by Endoscopic and Radiographic Screening. *Cancer Sci.* **2015**, *106*, 1744–1749. [CrossRef]
15. Faria, L.; Silva, J.C.; Rodríguez-Carrasco, M.; Pimentel-Nunes, P.; Dinis-Ribeiro, M.; Libânio, D. Gastric Cancer Screening: A Systematic Review and Meta-Analysis. *Scand. J. Gastroenterol.* **2022**, *57*, 1178–1188. [CrossRef]
16. Dinis-Ribeiro, M.; Areia, M.; de Vries, A.C.; Marcos-Pinto, R.; Monteiro-Soares, M.; O'Connor, A.; Pereira, C.; Pimentel-Nunes P.; Correia, R.; Ensari, A.; et al. Management of Precancerous Conditions and Lesions in the Stomach (MAPS): Guideline from the European Society of Gastrointestinal Endoscopy (ESGE), European Helicobacter Study Group (EHSG), European Society of Pathology (ESP), and the Sociedade Portuguesa de Endoscopia Digestiva (SPED). *Endoscopy* **2012**, *44*, 74–94. [CrossRef] [PubMed]
17. Pimentel-Nunes, P.; Libânio, D.; Marcos-Pinto, R.; Areia, M.; Leja, M.; Esposito, G.; Garrido, M.; Kikuste, I.; Megraud, F. Matysiak-Budnik, T.; et al. Management of Epithelial Precancerous Conditions and Lesions in the Stomach (MAPS II): European Society of Gastrointestinal Endoscopy (ESGE), European Helicobacter and Microbiota Study Group (EHMSG), European Society of Pathology (ESP), and Sociedade Portuguesa de Endoscopia Digestiva (SPED) Guideline Update 2019. *Endoscopy* **2019**, *51* 365–388. [CrossRef] [PubMed]
18. Yue, H.; Shan, L.; Bin, L. The Significance of OLGA and OLGIM Staging Systems in the Risk Assessment of Gastric Cancer: A Systematic Review and Meta-Analysis. *Gastric Cancer Off. J. Int. Gastric Cancer Assoc. Jpn. Gastric Cancer Assoc.* **2018**, *21*, 579–587 [CrossRef]
19. Gupta, S.; Li, D.; El Serag, H.B.; Davitkov, P.; Altayar, O.; Sultan, S.; Falck-Ytter, Y.; Mustafa, R.A. AGA Clinical Practice Guidelines on Management of Gastric Intestinal Metaplasia. *Gastroenterology* **2020**, *158*, 693–702. [CrossRef]
20. Shah, S.C.; Piazuelo, M.B.; Kuipers, E.J.; Li, D. AGA Clinical Practice Update on the Diagnosis and Management of Atrophic Gastritis: Expert Review. *Gastroenterology* **2021**, *161*, 1325–1332.e7. [CrossRef]
21. ASGE Standards of Practice Committee; Evans, J.A.; Chandrasekhara, V.; Chathadi, K.V.; Decker, G.A.; Early, D.S.; Fisher, D.A. Foley, K.; Hwang, J.H.; Jue, T.L.; et al. The Role of Endoscopy in the Management of Premalignant and Malignant Conditions of the Stomach. *Gastrointest. Endosc.* **2015**, *82*, 1–8. [CrossRef]
22. Banks, M.; Graham, D.; Jansen, M.; Gotoda, T.; Coda, S.; di Pietro, M.; Uedo, N.; Bhandari, P.; Pritchard, D.M.; Kuipers, E.J.; et al British Society of Gastroenterology Guidelines on the Diagnosis and Management of Patients at Risk of Gastric Adenocarcinoma *Gut* **2019**, *68*, 1545–1575. [CrossRef] [PubMed]
23. Japanese Gastric Cancer Association. Japanese Classification of Gastric Carcinoma: 3rd English Edition. *Gastric Cancer* **2011**, *14* 101–112. [CrossRef] [PubMed]
24. Kim, S.Y.; Park, J.M. Quality Indicators in Esophagogastroduodenoscopy. *Clin. Endosc.* **2022**, *55*, 319–331. [CrossRef]
25. Bisschops, R.; Areia, M.; Coron, E.; Dobru, D.; Kaskas, B.; Kuvaev, R.; Pech, O.; Ragunath, K.; Weusten, B.; Familiari, P.; et al Performance Measures for Upper Gastrointestinal Endoscopy: A European Society of Gastrointestinal Endoscopy (ESGE) Quality Improvement Initiative. *Endoscopy* **2016**, *48*, 843–864. [CrossRef]
26. Chang, W.-K.; Yeh, M.-K.; Hsu, H.-C.; Chen, H.-W.; Hu, M.-K. Efficacy of Simethicone and N-Acetylcysteine as Premedication in Improving Visibility during Upper Endoscopy. *J. Gastroenterol. Hepatol.* **2014**, *29*, 769–774. [CrossRef] [PubMed]

27. Kim, S.Y.; Park, J.M.; Cho, H.S.; Cho, Y.K.; Choi, M.-G. Assessment of Cimetropium Bromide Use for the Detection of Gastric Neoplasms During Esophagogastroduodenoscopy. *JAMA Netw. Open* **2022**, *5*, e223827. [CrossRef] [PubMed]
28. Park, J.M.; Huo, S.M.; Lee, H.H.; Lee, B.-I.; Song, H.J.; Choi, M.-G. Longer Observation Time Increases Proportion of Neoplasms Detected by Esophagogastroduodenoscopy. *Gastroenterology* **2017**, *153*, 460–469.e1. [CrossRef]
29. Kawamura, T.; Wada, H.; Sakiyama, N.; Ueda, Y.; Shirakawa, A.; Okada, Y.; Sanada, K.; Nakase, K.; Mandai, K.; Suzuki, A.; et al. Examination Time as a Quality Indicator of Screening Upper Gastrointestinal Endoscopy for Asymptomatic Examinees. *Dig. Endosc. Off. J. Jpn. Gastroenterol. Endosc. Soc.* **2017**, *29*, 569–575. [CrossRef]
30. Jang, J.-Y. The Past, Present, and Future of Image-Enhanced Endoscopy. *Clin. Endosc.* **2015**, *48*, 466–475. [CrossRef]
31. Bhat, Y.M.; Abu Dayyeh, B.K.; Chauhan, S.S.; Gottlieb, K.T.; Hwang, J.H.; Komanduri, S.; Konda, V.; Lo, S.K.; Manfredi, M.A.; Maple, J.T.; et al. High-Definition and High-Magnification Endoscopes. *Gastrointest. Endosc.* **2014**, *80*, 919–927. [CrossRef] [PubMed]
32. Rey, J.F.; Lambert, R. ESGE Quality Assurance Committee ESGE Recommendations for Quality Control in Gastrointestinal Endoscopy: Guidelines for Image Documentation in Upper and Lower GI Endoscopy. *Endoscopy* **2001**, *33*, 901–903. [CrossRef] [PubMed]
33. Yao, K. The Endoscopic Diagnosis of Early Gastric Cancer. *Ann. Gastroenterol.* **2013**, *26*, 11–22. [PubMed]
34. Emura, F.; Sharma, P.; Arantes, V.; Cerisoli, C.; Parra-Blanco, A.; Sumiyama, K.; Araya, R.; Sobrino, S.; Chiu, P.; Matsuda, K.; et al. Principles and Practice to Facilitate Complete Photodocumentation of the Upper Gastrointestinal Tract: World Endoscopy Organization Position Statement. *Dig. Endosc.* **2020**, *32*, 168–179. [CrossRef]
35. Pouw, R.E.; Barret, M.; Biermann, K.; Bisschops, R.; Czakó, L.; Gecse, K.B.; de Hertogh, G.; Hucl, T.; Iacucci, M.; Jansen, M.; et al. Endoscopic Tissue Sampling—Part 1: Upper Gastrointestinal and Hepatopancreatobiliary Tracts. European Society of Gastrointestinal Endoscopy (ESGE) Guideline. *Endoscopy* **2021**, *53*, 1174–1188. [CrossRef]
36. Chiu, P.W.Y.; Uedo, N.; Singh, R.; Gotoda, T.; Ng, E.K.W.; Yao, K.; Ang, T.L.; Ho, S.H.; Kikuchi, D.; Yao, F.; et al. An Asian Consensus on Standards of Diagnostic Upper Endoscopy for Neoplasia. *Gut* **2019**, *68*, 186–197. [CrossRef] [PubMed]
37. Dixon, M.F.; Genta, R.M.; Yardley, J.H.; Correa, P. Classification and Grading of Gastritis. The Updated Sydney System. International Workshop on the Histopathology of Gastritis, Houston 1994. *Am. J. Surg. Pathol.* **1996**, *20*, 1161–1181. [CrossRef]
38. Isajevs, S.; Liepniece-Karele, I.; Janciauskas, D.; Moisejevs, G.; Funka, K.; Kikuste, I.; Vanags, A.; Tolmanis, I.; Leja, M. The Effect of Incisura Angularis Biopsy Sampling on the Assessment of Gastritis Stage. *Eur. J. Gastroenterol. Hepatol.* **2014**, *26*, 510–513. [CrossRef]
39. Varbanova, M.; Wex, T.; Jechorek, D.; Röhl, F.W.; Langner, C.; Selgrad, M.; Malfertheiner, P. Impact of the Angulus Biopsy for the Detection of Gastric Preneoplastic Conditions and Gastric Cancer Risk Assessment. *J. Clin. Pathol.* **2016**, *69*, 19–25. [CrossRef]
40. Song, K.H.; Hwang, J.A.; Kim, S.M.; Ko, H.S.; Kang, M.K.; Ryu, K.H.; Koo, H.S.; Lee, T.H.; Huh, K.C.; Choi, Y.W.; et al. Acetic Acid Chromoendoscopy for Determining the Extent of Gastric Intestinal Metaplasia. *Gastrointest. Endosc.* **2017**, *85*, 349–356. [CrossRef]
41. Ji, R.; Liu, J.; Zhang, M.-M.; Li, Y.-Y.; Zuo, X.-L.; Wang, X.; Li, Y.-Q. Optical Enhancement Imaging versus Acetic Acid for Detecting Gastric Intestinal Metaplasia: A Randomized, Comparative Trial. *Dig. Liver Dis.* **2020**, *52*, 651–657. [CrossRef] [PubMed]
42. Taghavi, S.A.; Membari, M.E.; Eshraghian, A.; Dehghani, S.M.; Hamidpour, L.; Khademalhoseini, F. Comparison of Chromoendoscopy and Conventional Endoscopy in the Detection of Premalignant Gastric Lesions. *Can. J. Gastroenterol. J. Can. Gastroenterol.* **2009**, *23*, 105–108. [CrossRef]
43. Tada, M.; Katoh, S.; Kohli, Y.; Kawai, K. On the Dye Spraying Method in Colonofiberscopy. *Endoscopy* **1976**, *8*, 70–74. [CrossRef] [PubMed]
44. Le, H.; Wang, L.; Zhang, L.; Chen, P.; Xu, B.; Peng, D.; Yang, M.; Tan, Y.; Cai, C.; Li, H.; et al. Magnifying Endoscopy in Detecting Early Gastric Cancer: A Network Meta-Analysis of Prospective Studies. *Medicine* **2021**, *100*, e23934. [CrossRef] [PubMed]
45. Hu, Y.-Y.; Lian, Q.-W.; Lin, Z.-H.; Zhong, J.; Xue, M.; Wang, L.-J. Diagnostic Performance of Magnifying Narrow-Band Imaging for Early Gastric Cancer: A Meta-Analysis. *World J. Gastroenterol.* **2015**, *21*, 7884–7894. [CrossRef] [PubMed]
46. Hu, Y.; Chen, X.; Hendi, M.; Si, J.; Chen, S.; Deng, Y. Diagnostic Ability of Magnifying Narrow-Band Imaging for the Extent of Early Gastric Cancer: A Systematic Review and Meta-Analysis. *Gastroenterol. Res. Pract.* **2021**, *2021*, 5543556. [CrossRef]
47. Zhou, F.; Wu, L.; Huang, M.; Jin, Q.; Qin, Y.; Chen, J. The Accuracy of Magnifying Narrow Band Imaging (ME-NBI) in Distinguishing between Cancerous and Noncancerous Gastric Lesions: A Meta-Analysis. *Medicine* **2018**, *97*, e9780. [CrossRef]
48. Zhou, J.; Wu, H.; Fan, C.; Chen, S.; Liu, A. Comparison of the Diagnostic Efficacy of Blue Laser Imaging with Narrow Band Imaging for Gastric Cancer and Precancerous Lesions: A Meta-Analysis. *Rev. Esp. Enferm. Dig.* **2020**, *112*, 649–658. [CrossRef]
49. Yashima, K.; Onoyama, T.; Kurumi, H.; Takeda, Y.; Yoshida, A.; Kawaguchi, K.; Yamaguchi, N.; Isomoto, H. Current Status and Future Perspective of Linked Color Imaging for Gastric Cancer Screening: A Literature Review. *J. Gastroenterol.* **2023**, *58*, 1–13. [CrossRef]
50. Kanzaki, H.; Kawahara, Y.; Satomi, T.; Okanoue, S.; Hamada, K.; Kono, Y.; Iwamuro, M.; Kawano, S.; Okada, H. Differences in Color between Early Gastric Cancer and Cancer-Suspected Non-Cancerous Mucosa on Linked Color Imaging. *Endosc. Int. Open* **2023**, *11*, E90–E96. [CrossRef]
51. Koh, M.; Lee, J.Y.; Han, S.-H.; Jeon, S.W.; Kim, S.J.; Cho, J.Y.; Kim, S.H.; Jang, J.Y.; Baik, G.H.; Jang, J.S. Comparison Trial between I-SCAN-Optical Enhancement and Chromoendoscopy for Evaluating the Horizontal Margins of Gastric Epithelial Neoplasms. *Gut Liver* **2022**, *17*, 234–242. [CrossRef] [PubMed]

52. Uedo, N.; Ishihara, R.; Iishi, H.; Yamamoto, S.; Yamamoto, S.; Yamada, T.; Imanaka, K.; Takeuchi, Y.; Higashino, K.; Ishiguro, S.; et al. A New Method of Diagnosing Gastric Intestinal Metaplasia: Narrow-Band Imaging with Magnifying Endoscopy. *Endoscopy* **2006**, *38*, 819–824. [CrossRef] [PubMed]
53. Yao, K.; Oishi, T.; Matsui, T.; Yao, T.; Iwashita, A. Novel Magnified Endoscopic Findings of Microvascular Architecture in Intramucosal Gastric Cancer. *Gastrointest. Endosc.* **2002**, *56*, 279–284. [CrossRef]
54. Muto, M.; Yao, K.; Kaise, M.; Kato, M.; Uedo, N.; Yagi, K.; Tajiri, H. Magnifying Endoscopy Simple Diagnostic Algorithm for Early Gastric Cancer (MESDA-G). *Dig. Endosc. Off. J. Jpn. Gastroenterol. Endosc. Soc.* **2016**, *28*, 379–393. [CrossRef]
55. Yao, K.; Iwashita, A.; Tanabe, H.; Nishimata, N.; Nagahama, T.; Maki, S.; Takaki, Y.; Hirai, F.; Hisabe, T.; Nishimura, T.; et al. White Opaque Substance within Superficial Elevated Gastric Neoplasia as Visualized by Magnification Endoscopy with Narrow-Band Imaging: A New Optical Sign for Differentiating between Adenoma and Carcinoma. *Gastrointest. Endosc.* **2008**, *68*, 574–580. [CrossRef]
56. Yao, K.; Doyama, H.; Gotoda, T.; Ishikawa, H.; Nagahama, T.; Yokoi, C.; Oda, I.; Machida, H.; Uchita, K.; Tabuchi, M. Diagnostic Performance and Limitations of Magnifying Narrow-Band Imaging in Screening Endoscopy of Early Gastric Cancer: A Prospective Multicenter Feasibility Study. *Gastric Cancer Off. J. Int. Gastric Cancer Assoc. Jpn. Gastric Cancer Assoc.* **2014**, *17*, 669–679. [CrossRef] [PubMed]
57. Yao, K.; Yao, T.; Matsui, T.; Iwashita, A.; Oishi, T. Hemoglobin Content in Intramucosal Gastric Carcinoma as a Marker of Histologic Differentiation: A Clinical Application of Quantitative Electronic Endoscopy. *Gastrointest. Endosc.* **2000**, *52*, 241–245 [CrossRef]
58. Kanesaka, T.; Sekikawa, A.; Tsumura, T.; Maruo, T.; Osaki, Y.; Wakasa, T.; Shintaku, M.; Yao, K. Absent Microsurface Pattern Is Characteristic of Early Gastric Cancer of Undifferentiated Type: Magnifying Endoscopy with Narrow-Band Imaging. *Gastrointest. Endosc.* **2014**, *80*, 1194–1198.e1. [CrossRef]
59. Ochiai, K.; Ozawa, T.; Shibata, J.; Ishihara, S.; Tada, T. Current Status of Artificial Intelligence-Based Computer-Assisted Diagnosis Systems for Gastric Cancer in Endoscopy. *Diagnostics* **2022**, *12*, 3153. [CrossRef]
60. Khan, S.; Yong, S.-P. A Comparison of Deep Learning and Hand Crafted Features in Medical Image Modality Classification. In Proceedings of the 2016 3rd International Conference on Computer and Information Sciences (ICCOINS), Kuala Lumpur, Malaysia, 15–17 August 2016; pp. 633–638.
61. Sharma, P.; Hassan, C. Artificial Intelligence and Deep Learning for Upper Gastrointestinal Neoplasia. *Gastroenterology* **2022**, *162*, 1056–1066. [CrossRef]
62. Hirasawa, T.; Aoyama, K.; Tanimoto, T.; Ishihara, S.; Shichijo, S.; Ozawa, T.; Ohnishi, T.; Fujishiro, M.; Matsuo, K.; Fujisaki, J.; et al. Application of Artificial Intelligence Using a Convolutional Neural Network for Detecting Gastric Cancer in Endoscopic Images. *Gastric Cancer Off. J. Int. Gastric Cancer Assoc. Jpn. Gastric Cancer Assoc.* **2018**, *21*, 653–660. [CrossRef]
63. Wu, L.; Shang, R.; Sharma, P.; Zhou, W.; Liu, J.; Yao, L.; Dong, Z.; Yuan, J.; Zeng, Z.; Yu, Y.; et al. Effect of a Deep Learning-Based System on the Miss Rate of Gastric Neoplasms during Upper Gastrointestinal Endoscopy: A Single-Centre, Tandem, Randomised Controlled Trial. *Lancet Gastroenterol. Hepatol.* **2021**, *6*, 700–708. [CrossRef]
64. Xie, F.; Zhang, K.; Li, F.; Ma, G.; Ni, Y.; Zhang, W.; Wang, J.; Li, Y. Diagnostic Accuracy of Convolutional Neural Network-Based Endoscopic Image Analysis in Diagnosing Gastric Cancer and Predicting Its Invasion Depth: A Systematic Review and Meta-Analysis. *Gastrointest. Endosc.* **2022**, *95*, 599–609.e7. [CrossRef]
65. Wu, L.; Zhang, J.; Zhou, W.; An, P.; Shen, L.; Liu, J.; Jiang, X.; Huang, X.; Mu, G.; Wan, X.; et al. Randomised Controlled Trial of WISENSE, a Real-Time Quality Improving System for Monitoring Blind Spots during Esophagogastroduodenoscopy. *Gut* **2019**, *68*, 2161–2169. [CrossRef]
66. Choi, S.J.; Khan, M.A.; Choi, H.S.; Choo, J.; Lee, J.M.; Kwon, S.; Keum, B.; Chun, H.J. Development of Artificial Intelligence System for Quality Control of Photo Documentation in Esophagogastroduodenoscopy. *Surg. Endosc.* **2022**, *36*, 57–65. [CrossRef]
67. Hassan, C.; Spadaccini, M.; Iannone, A.; Maselli, R.; Jovani, M.; Chandrasekar, V.T.; Antonelli, G.; Yu, H.; Areia, M.; Dinis-Ribeiro, M.; et al. Performance of Artificial Intelligence in Colonoscopy for Adenoma and Polyp Detection: A Systematic Review and Meta-Analysis. *Gastrointest. Endosc.* **2021**, *93*, 77–85.e6. [CrossRef] [PubMed]
68. Shaukat, A.; Lichtenstein, D.R.; Somers, S.C.; Chung, D.C.; Perdue, D.G.; Gopal, M.; Colucci, D.R.; Phillips, S.A.; Marka, N.A.; Church, T.R.; et al. Computer-Aided Detection Improves Adenomas per Colonoscopy for Screening and Surveillance Colonoscopy: A Randomized Trial. *Gastroenterology* **2022**, *163*, 732–741. [CrossRef]
69. Areia, M.; Mori, Y.; Correale, L.; Repici, A.; Bretthauer, M.; Sharma, P.; Taveira, F.; Spadaccini, M.; Antonelli, G.; Ebigbo, A.; et al. Cost-Effectiveness of Artificial Intelligence for Screening Colonoscopy: A Modelling Study. *Lancet Digit. Health* **2022**, *4*, e436–e444. [CrossRef] [PubMed]
70. Pilonis, N.D.; Januszewicz, W.; di Pietro, M. Confocal Laser Endomicroscopy in Gastro-Intestinal Endoscopy: Technical Aspects and Clinical Applications. *Transl. Gastroenterol. Hepatol.* **2022**, *7*, 7. [CrossRef]
71. Park, C.H.; Kim, H.; Jo, J.H.; Hahn, K.Y.; Yoon, J.-H.; Kim, S.Y.; Lee, Y.C.; Noh, S.H.; Chung, H.C.; Lee, S.K. Role of Probe-Based Confocal Laser Endomicroscopy-Targeted Biopsy in the Molecular and Histopathological Study of Gastric Cancer. *J. Gastroenterol. Hepatol.* **2019**, *34*, 84–91. [CrossRef]
72. Bai, T.; Zhang, L.; Sharma, S.; Jiang, Y.D.; Xia, J.; Wang, H.; Qian, W.; Song, J.; Hou, X.H. Diagnostic Performance of Confocal Laser Endomicroscopy for Atrophy and Gastric Intestinal Metaplasia: A Meta-Analysis. *J. Dig. Dis.* **2017**, *18*, 273–282. [CrossRef] [PubMed]

73. Canakis, A.; Deliwala, S.S.; Kadiyala, J.; Bomman, S.; Canakis, J.; Bilal, M. The Diagnostic Performance of Probe-Based Confocal Laser Endomicroscopy in the Detection of Gastric Cancer: A Systematic Review and Meta-Analysis. *Ann. Gastroenterol.* **2022**, *35*, 496–502. [CrossRef] [PubMed]
74. Safatle-Ribeiro, A.V.; Ryoka Baba, E.; Corsato Scomparin, R.; Friedrich Faraj, S.; Simas de Lima, M.; Lenz, L.; Costa Martins, B.; Gusmon, C.; Shiguehissa Kawaguti, F.; Pennacchi, C.; et al. Probe-Based Confocal Endomicroscopy Is Accurate for Differentiating Gastric Lesions in Patients in a Western Center. *Chin. J. Cancer Res. Chung-Kuo Yen Cheng Yen Chiu* **2018**, *30*, 546–552. [CrossRef]
75. Li, Z.; Zuo, X.-L.; Li, C.-Q.; Liu, Z.-Y.; Ji, R.; Liu, J.; Guo, J.; Li, Y.-Q. New Classification of Gastric Pit Patterns and Vessel Architecture Using Probe-Based Confocal Laser Endomicroscopy. *J. Clin. Gastroenterol.* **2016**, *50*, 23–32. [CrossRef] [PubMed]
76. Han, W.; Kong, R.; Wang, N.; Bao, W.; Mao, X.; Lu, J. Confocal Laser Endomicroscopy for Detection of Early Upper Gastrointestinal Cancer. *Cancers* **2023**, *15*, 776. [CrossRef]
77. Misawa, M.; Kudo, S.-E.; Takashina, Y.; Akimoto, Y.; Maeda, Y.; Mori, Y.; Kudo, T.; Wakamura, K.; Miyachi, H.; Ishida, F.; et al. Clinical Efficacy of Endocytoscopy for Gastrointestinal Endoscopy. *Clin. Endosc.* **2021**, *54*, 455–463. [CrossRef]
78. Tsurudome, I.; Miyahara, R.; Funasaka, K.; Furukawa, K.; Matsushita, M.; Yamamura, T.; Ishikawa, T.; Ohno, E.; Nakamura, M.; Kawashima, H.; et al. In Vivo Histological Diagnosis for Gastric Cancer Using Endocytoscopy. *World J. Gastroenterol.* **2017**, *23*, 6894–6901. [CrossRef]
79. Horiuchi, Y.; Hirasawa, T.; Ishizuka, N.; Tokura, J.; Ishioka, M.; Tokai, Y.; Namikawa, K.; Yoshimizu, S.; Ishiyama, A.; Yoshio, T.; et al. Evaluation of Microvascular Patterns Alone Using Endocytoscopy with Narrow-Band Imaging for Diagnosing Gastric Cancer. *Digestion* **2022**, *103*, 159–168. [CrossRef]
80. Noda, H.; Kaise, M.; Higuchi, K.; Koizumi, E.; Yoshikata, K.; Habu, T.; Kirita, K.; Onda, T.; Omori, J.; Akimoto, T.; et al. Convolutional Neural Network-Based System for Endocytoscopic Diagnosis of Early Gastric Cancer. *BMC Gastroenterol.* **2022**, *22*, 237. [CrossRef]

Disclaimer/Publisher's Note: The statements, opinions and data contained in all publications are solely those of the individual author(s) and contributor(s) and not of MDPI and/or the editor(s). MDPI and/or the editor(s) disclaim responsibility for any injury to people or property resulting from any ideas, methods, instructions or products referred to in the content.

Review

Endoscopic Resection of Early Gastric Cancer and Pre-Malignant Gastric Lesions

Ana Clara Vasconcelos [1,*], Mário Dinis-Ribeiro [1,2] and Diogo Libânio [1,2]

[1] Department of Gastroenterology, Porto Comprehensive Cancer Center Raquel Seruca, and RISE@CI-IPO (Health Research Network), 4200-072 Porto, Portugal
[2] MEDCIDS (Department of Community Medicine, Health Information, and Decision), Faculty of Medicine, University of Porto, 4200-319 Porto, Portugal
* Correspondence: acvasconcelosoliveira@gmail.com

Simple Summary: Although its incidence and the mortality with which it is related seem to be decreasing, gastric cancer remains the fifth most common cause of new cancer cases and the fourth most lethal cancer worldwide. Late diagnosis occurs in a substantial portion of patients, but the increased identification of risk factors and precancerous conditions has allowed for the stratification of risk, leading to tailored patient surveillance and the early recognition of pre-malignant and malignant lesions. Since the 1990s, innovative endoscopic resection techniques have revolutionized the treatment of early gastric cancer, which would otherwise be subject to surgical resection.

Abstract: Early gastric cancer comprises gastric malignancies that are confined to the mucosa or submucosa, irrespective of lymph node metastasis. Endoscopic resection is currently pivotal for the management of such early lesions, and it is the recommended treatment for tumors presenting a very low risk of lymph node metastasis. In general, these lesions consist of two groups of differentiated mucosal adenocarcinomas: non-ulcerated lesions (regardless of their size) and small ulcerated lesions. Endoscopic submucosal dissection is the technique of choice in most cases. This procedure has high rates of complete histological resection while maintaining gastric anatomy and its functions, resulting in fewer adverse events than surgery and having a lesser impact on patient-reported quality of life. Nonetheless, approximately 20% of resected lesions do not fulfill curative criteria and demand further treatment, highlighting the importance of patient selection. Additionally, the preservation of the stomach results in a moderate risk of metachronous lesions, which underlines the need for surveillance. We review the current evidence regarding the endoscopic treatment of early gastric cancer, including the short-and long-term results and management after resection.

Keywords: gastric cancer; endoscopy; treatment; endoscopic submucosal dissection

1. Introduction

Gastric cancer (GC) remains an important cause of cancer worldwide, ranking fifth in new cancer cases and fourth in terms of mortality [1], although incidence and mortality rates have been decreasing in recent decades [2]. Nonetheless, GC was still responsible for just over 1 million new cases in 2020 [1], which is predicted to increase to 1.8 million worldwide by 2040 [3]. A recent study [4] projecting cancer incidence between 2015 and 2050 in the United States of America estimates not only an increase in the absolute number of new GC cases (explained by an aging population) but also a 7% increase in age-standardized incidence rates from 7.5 to 8.0 per 100,000. These numbers underline the importance of healthcare systems' adaptability to an increasing burden of disease, shifting focus to primary prevention and early detection.

The knowledge of gastric carcinogenesis (namely, the Correa cascade [5]) and the subsequent recognition of gastric premalignant conditions and lesions, the current widespread use of esophagogastroduodenoscopy, and the implementation of national screening programs in high-risk countries such as Japan and South Korea [1,6,7] are expected to result in an increase in the diagnosis of GC at earlier stages.

Early gastric cancer (EGC) comprises gastric malignancies that are confined to the mucosa or submucosa, irrespective of the status of lymph node metastasis (LNM) [8]. The presence of LNM constitutes one of the most relevant prognostic factors among patients with GC, including EGC, which is associated with significantly lower long-term survival [9]. While the standard curative treatment of GC had once been gastrectomy with lymphadenectomy, the development of advanced endoscopic resection techniques has surpassed surgery as a first-line curative treatment for selected early lesions presenting a minimal risk of LNM. However, up to 20% of endoscopic resections do not meet curative criteria and require further surgical treatment [10–12], highlighting the need to improve clinical staging and patient selection.

This review aims to provide a comprehensive overview of the endoscopic management of EGC, the challenges physicians still face in their daily practices, and the technical and technological advances designed to overcome these difficulties.

2. Superficial Gastric Lesions

Superficial gastric lesions are made up of premalignant neoplastic lesions and malignant lesions that do not invade beyond the submucosa [13]. The Vienna classification provides a consensus terminology of epithelial neoplasia of the gastrointestinal tract [14]. In the stomach, low-grade dysplasia, high-grade dysplasia, and carcinoma in situ (group 3 and subgroups 4-1 and 4-2 of the Vienna classification, respectively) are considered premalignant lesions in that they are confined to the epithelial layer and do not invariably progress to invasive carcinoma. Invasion into the *lamina propria* or the *muscularis mucosae* constitutes intramucosal carcinoma (subgroups 4-3 and 4-4 of the Vienna classification), which, in the stomach, is considered a malignant lesion, contrary to what is seen in the colon. Additionally, a carcinoma that invades the submucosa (group 5 of the Vienna classification) is also considered a superficial gastric lesion.

The Paris classification, a morphological classification developed in 2003 and updated in 2005, categorizes superficial neoplastic lesions of the gastrointestinal tract into three groups [13]. Type 0-I includes protruding superficial lesions, also known as polypoid lesions, and is subdivided into pedunculated (0-Ip) and sessile (0-Is) lesions. Type 0-II encompasses non-protruding non-excavating lesions, otherwise known as flat lesions, and is made up of slightly elevated (0-IIa), completely flat (0-IIb), and slightly depressed (0-IIc) lesions. It is common for mixed lesions to occur, containing concomitant depressed and elevated components, and such lesions are classified as type "0-IIa + IIc" or "0-IIc + IIa" depending on the predominant component. Finally, type 0-III lesions are excavated (or ulcerated) and can also be mixed with depressed (0-IIc) lesions. This endoscopic classification seems to correlate with histological findings and resection outcomes since a depressed morphology is associated with submucosal invasion and excavated lesions are associated with piecemeal resection. Although subject to interobserver variability, this classification's reliability is acceptable and improves both with training and the use of virtual chromoendoscopy [15].

3. Indications for Endoscopic Resection: Pre-Procedural Evaluation

The reported rate of LNM in intramucosal adenocarcinomas varies between 0% and 9% and can reach up to 25% in adenocarcinomas with submucosal invasion [16–19]. In certain circumstances, this risk is minimal or even null. The studies conducted by Gotoda et al. [16], Nakahara et al. [17], and Hirasawa et al. [18] evaluated the incidence of LNM in gastrectomy specimens, analyzing the endoscopic and histological characteristics associated with a very low risk of LNM in cases of EGC. These studies served as the cornerstone for the definition of the current criteria for endoscopic resection. More recently, the findings of Hasuike et al. [20] and Takizawa et al. [21] contributed to the expansion of indications for endoscopic resection.

The Japanese and European guidelines recommend endoscopic resection as the standard treatment for gastric lesions harboring dysplasia and for EGC when the presumed risk of lymph node metastasis is less than 1% [22,23]. The Japanese guidelines define expanded indication lesions as lesions that are presumed to have a <1% risk of LNM but for which long-term outcomes were not confirmed by a prospective confirmatory trial with 5-year survival as the primary endpoint [22]. The European guidelines state that EGC with an LNM risk presumed to be inferior to 3% can be considered for endoscopic resection as an expanded criterion, although the decision should consider the patient's characteristics and preference after the discussion of risks [23].

The absolute criteria for endoscopic resection, according to the Japanese guidelines, are gastric lesions clinically staged as (i) dysplastic regardless of size, (ii) differentiated gastric intramucosal (cT1a) adenocarcinomas of any size if not ulcerated and ≤30 mm in size if ulcerated, and (iii) poorly differentiated gastric intramucosal (cT1a) adenocarcinomas without ulcerative findings and ≤20 mm in size [22,24]. The European guidelines, on the other hand, consider the first two groups of lesions as absolute indications for endoscopic resection and the third one as an expanded indication [23]. In these cases, the decision to pursue endoscopic treatment should be individualized following the discussion of the potential risks and benefits of the different treatment options with the patient. The Japanese guidelines define lesions as expanded indications when a previously resected lesion meeting the endoscopic curability criterion eCura C-1 (see Section 6) locally recurs as a clinically staged intramucosal (cT1a) cancer [22,24] (Table 1).

Table 1. Absolute and expanded indications according to European and Japanese guidelines.

	Type of Lesion	European Guidelines	Japanese Guidelines
	Dysplasia, any size	Absolute indication	Absolute indication
Adenocarcinoma	cT1a, well-differentiated, non-ulcerated, any size	Absolute indication	Absolute indication
	cT1a, well-differentiated, ulcerated, ≤30 mm	Absolute indication	Absolute indication
	cT1a, poorly differentiated, non-ulcerated, ≤20 mm	Expanded indication	Absolute indication
	Recurrence of an eCura-C1 lesion, staged as cT1a	-	Expanded indication

cT1a: adenocarcinoma clinically staged as intramucosal.

Although endoscopic resection is considered to result in high rates of curative resection, approximately 15–20% of the resected lesions do not meet curative criteria [10–12]. Several authors have sought to establish predictive factors for non-curative resection in order to improve patient selection. A 2019 systematic review and meta-analysis identified location in the upper third of the stomach (odds ratio (OR) 1.49, 95%CI 1.24–1.79), depressed morphology (OR 1.49, 95%CI 1.04–2.12), and lesions whose identified characteristics lie outside standard criteria (OR 3.56, 95%CI 2.31–5.48) as predictors of this outcome [25]. Additional risk factors identified in individual studies include large tumor size (generally >20 mm), ulceration, undifferentiated tumors (including the presence of an undifferentiated component in differentiated-type-predominant mixed-type lesions), and old age [26–31]. Regarding lesion differentiation, a meta-analysis incorporating 5644 patients showed that undifferentiated-predominant mixed-type lesions show more aggressive biological behavior compared to pure undifferentiated-type lesions, presenting a significantly higher risk of submucosal invasion (OR 2.19, 95%CI 1.90–2.52) and LNM (OR 2.28, 95%CI 1.72–3.03) even after stratification for depth of tumor invasion [32].

Furthermore, deep submucosal invasion (>500 μm, ≥Sm2) is an independent risk factor for LNM and a major criterion of non-curability [16–18,23,33–35]. Thus, accurately estimating the depth of invasion is one of the most important components of an endoscopic preoperative assessment but also one of the most challenging. A few authors have attempted to identify macroscopic features suggestive of Sm2 invasion

Abe et al. [36] suggested that remarkable redness, an uneven surface, margin elevation, enlarged folds, a tumor size >30 mm, and ulceration were significantly associated with deeper submucosal invasion.

Magnifying endoscopy, usually applied in combination with narrow-band imaging, is an ancillary tool for the diagnosis of EGC. Several authors have evaluated whether certain vascular and surface patterns could predict the histologic type and depth of invasion of a tumor; however, there is not yet a gastric classification comparable to the ones of colonic polyps and esophageal lesions. Nakayoshi et al. [37] and Yokoyama et al. [38] found that a fine network microvascular pattern was associated with differentiated lesions, while a corkscrew pattern was associated with undifferentiated histology. What Nakayoshi et al. considered to be an unclassified pattern was designated as an intra-lobular loop pattern by Yokoyama et al., which subdivided it into type 1 (predictive of differentiated-type EGC) and type 2 (found in both differentiated and undifferentiated lesions). Tanaka et al. [39] found that a microsurface pattern of irregular arrangements and sizes was the predominant type in differentiated tubular adenocarcinomas (although depressed adenomas also presented the same pattern), while all signet-ring cell carcinomas and poorly differentiated tubular adenocarcinomas showed a destructive microsurface pattern. Ok et al. [40] concluded that the magnification patterns with narrow-band imaging could aid in predicting histopathology; specifically, a fine network or loop microvascular pattern was associated with differentiated tumors, while an absent microsurface pattern and corkscrew microvascular pattern were associated with undifferentiated tumors. Furthermore, a destructive microsurface pattern was associated with submucosal invasion. Kanesaka et al. [41] found that absent microsurface and opened-loop microvascular patterns did not improve the overall accuracy of white light endoscopy for the diagnosis of undifferentiated-type EGC in depressed or flat lesions, although it improved specificity.

Different modalities for local staging, the foremost of which is endoscopic ultrasonography, have not proven to be superior to endoscopic evaluation in assessing depth of invasion; consequently, European guidelines do not recommend such modalities' routine use [23]. Computed tomography and positron emission tomography also have no role in the pre-resection evaluation of endoscopically resectable EGC since the risk of distant metastasis is very low.

Therefore, endoscopic resection should only be proposed to a patient should after a careful evaluation of the gastric lesion by an experienced endoscopist, who should look for endoscopic features associated with non-curability and account for clinical and pathological characteristics (Figure 1).

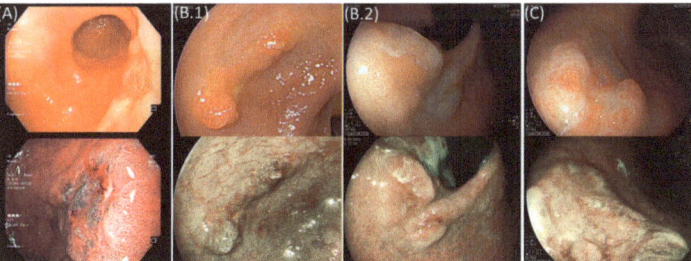

Figure 1. Superficial gastric lesions (upper image—white light; lower image—virtual chromoendoscopy). (**A**) A Paris 0-IIa+IIc lesion clinically staged as deep submucosal invasion in pre-resection endoscopic evaluation (pT1b, undifferentiated, and with lymphovascular invasion on surgical specimen). (**B**) Lesions successfully removed using ESD and meeting curative criteria. (**B.1**) A Paris IIa+IIc 40 mm lesion (pT1a, well-differentiated, and no lymphovascular invasion). (**B.2**) A Paris IIa+IIc 15 mm lesion (pT1a, well-differentiated, and no lymphovascular invasion). (**C**) A Paris 0-IIa+IIc 12 mm lesion that was endoscopically resected and did not meet curative criteria (pT1b, well-differentiated, and with lymphovascular invasion).

4. Endoscopic Resection

The endoscopic resection of gastric dysplastic lesions and EGC can be carried out by performing endoscopic mucosal resection (EMR) or endoscopic submucosal dissection (ESD).

EMR was first described in 1993 [42] for the endoscopic treatment of esophageal, gastric, and colonic lesions. Before resection, the lesion is elevated through the injection of a solution in the submucosal space to separate it from the *muscularis propria*. The lesion is then placed within a metal wire snare and resected using high-frequency diathermy. This procedure is effective and safe. However, the size of the snare generally prevents the en bloc resection of larger lesions. Piecemeal and/or incomplete resection limits proper histopathological evaluation and staging, which are crucial for post-resection management and associated with local recurrence [43]. ESD was developed in 1995 [44] to overcome the limitations of EMR, allowing for the en bloc resection of lesions of any size. In this method, the lesion is circumferentially outlined with coagulation marks and then elevated after the injection of a solution in the submucosal layer. The endoscopist makes three to four electrosurgical incisions in the coagulation marks to access the submucosa and completes a circumferential incision around the lesion. Finally, the submucosa is dissected in the submucosal plane to achieve an en bloc resection.

Several retrospective and prospective studies and meta-analyses have been carried out to compare the safety and efficacy outcomes between EMR and ESD (Table 2) [45–48]. ESD is significantly superior to EMR in achieving en bloc and complete resection for lesions of any size, resulting in significantly higher rates of curative resection and lower recurrence. Regarding safety, ESD and EMR present similar levels of post-procedural bleeding, while ESD is associated with higher perforation rates and operative time. Indeed, ESD continues to show high rates of en bloc and complete resection (over 95% and 90%, respectively) and low local recurrence (<5%) and low rates of adverse events, namely, perforation (<3%) and post-operative bleeding ($\approx$5%) [11,49,50]. The endoscopic resection of gastric superficial lesions is associated with a good long-term prognosis, with 5-year overall (OS) and disease-specific survival (DSS) rates of 89.0–95.0% and >99%, respectively [50–52].

Accordingly, ESD is the recommended first option for the endoscopic treatment of gastric superficial lesions deemed resectable [22,23]. The European guidelines state, however, that EMR should be considered for elevated lesions (Paris 0-IIa), under 10 mm in size and with a low likelihood of advanced histology.

Endoscopic resection, although safer than gastrectomy, can also present adverse events. Predicting these outcomes can assist in patient selection and help plan periprocedural measures for the prevention of such outcomes.

Post-procedural bleeding is the most frequent adverse event following ESD, occurring in 4.4–5.1% of procedures [11,53], and it is linked to prolonged hospital stays, the requirement for transfusion, endoscopic reintervention, surgery, and death. A meta-analysis identified risk factors for PPB, which were either patient-, lesion-, or procedure-related [53]. The risk factors associated with this unfavorable outcome were a male gender, cardiopathy, antithrombotic drug use, cirrhosis, chronic kidney disease, a tumor size > 20 mm, a resected specimen >30 mm size, localization in the lesser curvature, a flat or depressed morphology, carcinoma histology, ulceration, a procedure duration of >60 min, and the use of histamine-2 receptor antagonists as an acid-suppressive therapy instead of proton pump inhibitors. The latter reduce the rate of delayed bleeding [54,55], and their administration following ESD is recommended [22]; however, a meta-analysis showed that premedication with proton-pump inhibitors does not impact bleeding rates, despite significantly increasing gastric pH at the time of ESD [56]. Coagulation of visible vessels in post-ESD ulcers is also associated with reduced rates of delayed bleeding [57] and is a recommended preventive measure [22]. A network meta-analysis evaluated additional preventive measures and found that tissue shielding with polyglycolic acid significantly reduced delayed bleeding risk in high-risk patients [risk ratio (RR) 0.32; 95%CI 0.12–0.79], while hemostatic spray potentially reduced bleeding in low-risk patients, although heterogeneity was high [58].

Table 2. Summary of meta-analyses comparing short- and long-term outcomes between endoscopic mucosal resection and endoscopic submucosal dissection.

Author, Year	Type of Resection	Operation Time (in Minutes)	Perforation Rate	Local Recurrence	En Bloc Resection	Complete Resection
Tao M, 2019 [45]	-	SMD 1.12 (0.13–2.10)	OR 2.55 (1.48–4.39)	OR 0.18 (0.09–0.34)	OR 9.00 (6.66–12.17)	OR 8.43 (5.04–14.09)
Lian J, 2012 [46]	EMR	ND	17/1973 (0.9%)	126/1973 (6.4%)	1020/1973 (51.7%)	867/2053 (42.2%)
	ESD	ND	62/1438 (4.3%)	11/1438 (0.8%)	1328/1437 (92.4%)	1227/1495 (82.1%)
	-	WMD 59.4 (16.8–102.0)	OR 4.67 (2.77–7.87)	OR 0.10 (0.06–0.18)	OR 9.69 (7.74–12.13)	OR 5.66 (2.92–10.96)
Facciorusso A, 2014 [47]	EMR	ND	17/1973 (0.9%)	141/2332 (6.0%)	1020/1973 (51.7%)	867/2053 (42.2%)
	ESD	ND	62/1438 (4.3%)	12/1859 (0.6%)	1328/1437 (92.4%)	1227/1495 (82.1%)
	-	SMD 1.73 (0.52–2.95)	OR 4.67 (2.77–7.87)	OR 0.09 (0.05–0.17)	OR 9.69 (7.74–12.13)	OR 5.66 (2.92–10.96)
Zhao Y, 2018 [48]	EMR	-	26/2134 (1.2%)	116/2245 (5.2%)	1422/2551 (55.7%)	1110/1935 (57.4%)
	ESD	-	86/2676 (3.2%)	4/1932 (0.2%)	2229/2387 (93.4%)	1864/2032 (91.7%)
	-	MD −49.86 (−71.62 to −28.10)	OR 0.37 (0.24–0.57)	OR 14.94 (7.26–30.74)	OR 0.10 (0.09–0.13)	OR 0.14 (0.12–0.17)

MD: mean difference; ND: no data; OR: odds ratio; SMD: standard mean difference; WMD: weighted mean difference. The 95% confidence intervals are shown in parenthesis for MD, OR, SMD, and WMD.

Perforation is an uncommon adverse event of ESD and can be immediate (<3% of procedures) or delayed (<1%). A meta-analysis identified the following as risk factors for perforation: liver disease, location in the upper stomach, a resection size > 20 mm, submucosal invasion, operation time > 2 h, depressed or flat lesions, and piecemeal resection [59]. Another meta-analysis, this time comparing gastric ESD in elderly and non-elderly patients, found a trend for significantly increased perforation risk among patients aged >80 years [60].

5. Endoscopic Resection versus Surgery

Resection of the stomach and regional lymph nodes is the standard surgical curative treatment for GC, entailing the removal of at least two thirds of the stomach and a D2 lymph node dissection [24]. This ensures high rates of complete resection, almost negligible rates of local recurrence, a very low risk of metachronous lesions, and high disease-free and overall survival. On the other hand, surgical resection has its own adverse events; it can significantly impact the stomach's storage and digestive functions, thereby limiting nutrient absorption; and the resulting effects may impair the patient's health-related quality of life

Alternatively, ESD is a minimally invasive procedure that preserves the stomach's structure and associated functions and presents a low rate of complications and adverse outcomes. The spared mucosa constitutes, however, a sustained risk for metachronous tumors, thereby demanding long-term surveillance.

Several meta-analyses have compared the short (Table 3) and long-term (Table 4) outcomes of ESD versus surgery for the treatment of EGC [61–65]. Endoscopic treatment is associated with significantly decreased operation times, in-hospital stays, and overall postoperative complication rates, with one meta-analysis also reporting a lower risk of procedure-related death (OR 0.21, 95%CI: 0.07–0.68) [64]. On the other hand, the rates of en bloc resection, complete resection, and curative resection seem to be significantly lower for ESD compared to surgery (OR 0.07, 95%CI 0.03–0.21; OR 0.07, 95%CI 0.03–0.14 and OR 0.06, 95%CI 0.01–0.27, respectively) [64], resulting in higher rates of recurrence However, Gu et al. [62] found that the proportion of patients that were amenable to radical treatment after recurrence was higher in the ESD group incorporated in their study (OR 5.27, 95%CI 2.35–11.79). Synchronous and metachronous cancers have been found to be significantly more prevalent after ESD. Regarding long-term outcomes, the differences in 5-year disease-free survival (DFS) are not homogeneous across studies Some authors found no statistically significant differences [61,63], while others state a significantly lower DFS in their respective ESD groups [62,64,65]. This may be due to differences in defining disease-free survival. Abdelfatah et al. [61] did not incorporate the detection of metachronous lesions as a disease-defining event, Gu et al. [62] included metachronous GC occurrence in the definition of DFS, and the remaining authors [63–65] did not specify which events defined DFS. However, the ESD and surgery groups consistently showed similar 5-year overall and disease-specific survival (OS > 95% and DSS > 99% in both groups) throughout the different meta-analyses.

Gastric cancer with an undifferentiated histology presents a significantly higher risk of lymph node metastasis than differentiated tumors [16,66]. Several comparative studies have been performed to compare long-term outcomes, namely, survival, in patients undergoing ESD and surgery for undifferentiated mucosal tumors with a diameter <20 mm and without ulcerative findings. Two meta-analyses summarizing the evidence collected were recently conducted [67,68]. The results overlap with those stated above for general cohorts, with ESD showing a significantly lower 5-year DFS, no statistical difference in DSS, and similar OS (Table 4).

Table 3. Summary of meta-analyses comparing short-term outcomes between endoscopic submucosal dissection and gastrectomy patients.

Author, Year	Type of Resection	Operation Time (in Minutes)	in-Hospital Stay (in Days)	Overall Postoperative Complication	Recurrence	Synchronous Lesions	Metachronous Lesions
Abdelfatah MM, 2019 [61]	ESD				40/2943 (1.4%)	16/1082 (1.5%)	176/2943 (6%)
	Gastrectomy	ND	ND	ND	12/3116 (0.4%)	1/1485 (0.1%)	13/3116 (0.4%)
	-				OR 0.17 (0.1–4.9)	RR 5.7 (1.5–21.9)	RR 10.2 (5.9–17.1)
Gu L, 2019 [62]	-	ND	ND	ND	ND	OR 4.94 (3.04–8.03)	OR 8.64 (5.00–14.95)
Li H, 2020 [63]	-	WMD −140 (−254 to −34)	−5.41 (−5.93 to −4.89)	OR 0.39 (0.28–0.55)	OR 9.24 (5.94–14.36)	ND	ND
Liu Q, 2020 [64]	-	MD −128 (−204 to −52)	−7.13 (−7.98 to −6.28)	OR 0.47 (0.34–0.63)	OR 5.42 (2.91–10.11)	OR 6.59 (1.96–22.1)	OR 10.84 (6.43–18.26)
Xu X, 2022 [a] [65]	-	ND	ND	OR 0.49 (0.34–0.72)	ND	OR 9.09 (2.17–50)	OR 8.33 (4–20)

MD: mean difference; ND: no data; OR: odds ratio; WMD: weighted mean difference. [a] Expanded indication lesions. The 95% confidence intervals are shown in parenthesis.

A number of studies have reported on the hospital costs associated with either procedure [69–71]. ESD seems to account for significantly lower costs when compared to surgery which is mostly due to the nature of the procedure itself and differences in the length of stay. Shin et al. [70] evaluated costs related to general cases, stating that given the superior rate of adverse events following surgical resection, the difference in costs may be higher than estimated. Kim et al. [69] also compared medical costs linked to follow-up at 1-year post-discharge and did not find significant differences.

Table 4. Summary of systematic reviews with meta-analyses comparing long-term survival between endoscopic submucosal dissection and gastrectomy patients.

Author, Year	Type of Resection	Overall Survival	Disease-Specific Survival	Disease-Free Survival
Abdelfatah MM, 2019 [61]	ESD Gastrectomy -	2914/3034 (96%) 3088/3203 (96%) OR 0.96 (0.74–1.25)	2437/2451 (99.4%) 1962/1977 (99.2%) OR 0.7 (0.16–2.9)	1415/1476 (95.9%) 1816/1844 (98.5%) OR 1.86 (0.57–6.0)
Gu L, 2019 [62]	ESD Gastrectomy -	2238/2324 (96.3%) 2563/2662 (96.3%) RR 0.90 (0.68–1.19)	5/1425 (99.7%) 17/1841 (99.1%) RR 0.40 (0.15–1.03)	1241/1376 (90.2%) 1261/1298 (97.2%) RR 3.40 (2.39–4.84)
Li H, 2020 [63]	-	HR 0.51 (0.26–1.00)	ND	ND
Liu Q, 2020 [64]	-	HR 0.92 (0.71–1.19)	HR 0.73 (0.36–1.49)	HR 4.58 (2.79–7.52)
Huh CW, 2021 [a] [67]	-	OR 2.29 (0.98–5.36)	ND	ND
Xu X, 2022 [b] [65]	-	HR 1.22 (0.66–2.25)	ND	HR 3.29 (1.60–6.76)
Yang HJ, 2022 [a] [68]	ESD Gastrectomy -	383/400 (95.8%) 492/508 (96.9%) RR 1.18 (0.60–2.32)	396/400 (99.0%) 506/508 (99.6%) RR 2.49 (0.47–37.93)	362/400 (90.5%) 491/508 (96.7%) RR 2.49 (1.42–4.35)

ESD: endoscopic submucosal dissection; HR: hazard ratio; ND: no data; OR: odds ratio; RR: risk ratio [a] Undifferentiated lesions. [b] Expanded indication lesions. The 95% confidence intervals are shown in parenthesis for HR, OR, and RR.

Considering that gastrectomy with lymphadenectomy is a major surgical procedure entailing the resection of a considerable portion of the stomach and ESD is a minimally invasive and stomach-sparing procedure, a few authors have evaluated patient-reported quality of life after curative treatment. We found three comparative studies, one of which was retrospective [72] and the other two were prospective [73,74]. In all three, quality of life was assessed using the European Organization for Research and Treatment of Cancer (EORTC) Quality of Life Questionnaire Core 30 (QLQ-C30) and a GC-specific module namely, the EORTC QLQ-STO22. Song et al. [72] reported a significantly higher overall health status in the ESD group compared to the surgery group ($p < 0.05$) and a global trend in all function and symptom scales in favor of endoscopic treatment, although statistical differences were only found in relation to physical function, social function, fatigue, nausea and vomiting, appetite loss and constipation, reflux, eating restrictions, and body image. Libânio et al. [73] found, at 1-year, a net benefit in overall health favoring ESD ($p = 0.006$). ESD was not associated with worsening in any functional dimensions or symptom scales compared to baseline. This result contrasts with those regarding the surgery group, whose patients reported a significant decrease in role function and worsened fatigue, pain, appetite loss, diarrhea, dysphagia, eating restrictions, taste, and body image. ESD patients did not more frequently report fear of recurrence, new tumors, or death when compared with surgical patients. Kim et al. [74] reported significant differences between groups only with regard to physical functioning, eating restrictions, dysphagia, diarrhea, and body image.

Taking all the above into account, when a superficial gastric lesion is amenable to endoscopic resection with a high likelihood of curability, guidelines consider endoscopic resection to be a more desirable choice of curative treatment compared to surgery [23]. Especially in cases of expanded indications, this should be a shared decision between a patient and their physician that is finalized after a discussion of the advantages and downsides of both treatment modalities with respect to both short- and long-term outcomes [75].

6. Management after Resection

After endoscopic resection, a pathological examination is essential in order to properly characterize the resected lesion and classify the resection as curative or non-curative, thereby guiding posterior management. The criteria for curability regarding resections have been defined according to the risk of LNM based on the histological findings of surgical specimens. Several studies throughout the years have consistently identified lymphovascular invasion, deep submucosal invasion (>500 µm), undifferentiated histology, and a size ≥30 mm as independent risk factors for LNM [16–18,23,33–35], and this evidence is the cornerstone for the definition of current curative criteria.

European guidelines [23] consider two groups of curative resections (Figure 2):

- Very-low-risk resections (LNM risk < 0.5–1%), i.e., when a differentiated mucosal (pT1a) lesion, without lymphovascular invasion, and independent of size if there are no ulceration findings or ≤30 mm in size if ulcerated, is resected en bloc and with negative margins;
- Low-risk resections (LNM risk <3%), i.e., when a poorly differentiated pT1a lesion ≤ 20 mm in size or a differentiated pT1b lesion (submucosal invasion ≤ 500 µm) ≤30 mm in size, that present neither ulceration nor lymphovascular invasion, is resected en bloc with negative margins.

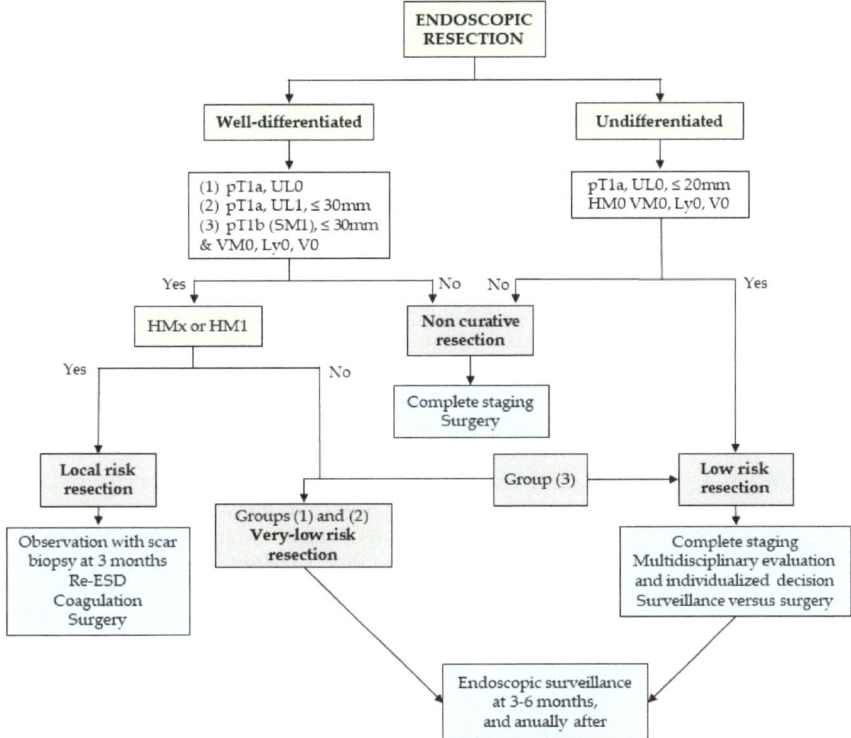

Figure 2. Post-resection management according to the European Society of Gastrointestinal Endoscopy guidelines on ESD. pT1a: intramucosal adenocarcinoma. pT1b (SM1): adenocarcinoma with superficial submucosal invasion (≤500 µm). UL0: non-ulcerated. UL1: ulcerated. VM0: negative vertical margin. Ly0, V0: no lymphovascular invasion. HM0: negative horizontal margin. HMx: piecemeal resection. HM1: positive horizontal margin.

A very-low-risk resection does not require any further radiological staging or treatment, whereas for lesions meeting low-risk criteria, further treatment is generally not recommended, but the patient should undergo complete staging, and the decision to pursue additional surgical therapy should be individualized after discussion with a multidisciplinary team.

A third group of lesions is classified as local-risk resections—due to a very low risk of LNM but an increased risk of local recurrence—when a piecemeal resection or tumor-positive horizontal margin occurs in (i) lesions otherwise meeting very low risk criteria and (ii) differentiated pT1b lesions with submucosal invasion ≤ 500 µm, a size ≤ 30 mm, and negative vertical margins, provided that there is no evidence of submucosal invasion at the resection margin. Management in such situations should be tailored, for which patient preferences should be considered, with guidelines preferring either close observation with scar biopsy or re-ESD/scar ablation over surgery given its poorer safety profile. However, surgery is an adequate alternative, especially for cases of recurrence that are not amenable to endoscopic re-intervention.

Finally, endoscopic resections are classified as noncurative for any lesion with: positive vertical margins; lymphovascular invasion; deep submucosal invasion (>500 µm from the *muscularis mucosae*); ulceration or a size > 20 mm in poorly differentiated lesions; a size > 30 mm in pT1b differentiated lesions with submucosal invasion $\leq$500 µm and in intramucosal ulcerated lesions. In these cases, complete staging is recommended, and a further curative resection should generally be pursued, namely, gastrectomy and lymphadenectomy, since the presence of LNM is linked to a poor prognosis. For patients who refuse salvage surgery or are unfit for a major surgical procedure, surveillance may be an acceptable alternative.

The Japanese guidelines [22,24], on the other hand, use the eCura grading system to categorize the curability of resected lesions. Lesions are classified as endoscopic curability A (eCuraA) when the effect of endoscopic resection is equal to or superior to surgery with respect to long-term outcomes. These include the same resections classified as very low risk in European guidelines as well as the en bloc resection of intramucosal (pT1a) predominantly undifferentiated-type lesions that are $\leq$20 mm and non-ulcerated, possess negative horizontal and vertical margins, and do not present lymphovascular invasion. However, predominantly differentiated lesions with an undifferentiated component > 20 mm are considered non-curative resections (endoscopic curability C-2). When curability can be expected, although there is not yet sufficient evidence of long-term results, lesions are graded as endoscopic curability B (eCuraB), and are constituted by en bloc resection of predominantly differentiated-type lesions with a minute degree of submucosal invasion ($\leq$500 µm from the *muscularis mucosae*, pT1b1), negative horizontal and vertical margins, and no lymphovascular invasion. If an undifferentiated component is present in the submucosal portion of the lesion, the resection is considered non-curative. Every other lesion not fulfilling eCuraA or eCuraB criteria is a non-curative resection and classified as endoscopic curability C lesions (eCuraC). This group subdivides into eCuraC-1, which encompasses differentiated eCuraA or eCuraB lesions that were either not resected en bloc or had positive horizontal margins, and eCuraC-2, which is made up of all other non-curative resections.

Regarding non-curative resections, Libânio et al. [76] reported that 75% of the gastrectomy specimens of such cases did not show residual lesions, and the 5-year DSS did not seem to differ between patients in the surgical and non-surgical groups [76,77]. Thus, indiscriminately recommending surgical treatment to all non-curative resections may be excessive. Accordingly Hatta et al. [78] created a scoring system for non-curative resections, attributing the following points to five different risk factors for LNM: three points for lymphatic invasion, and one point each for tumors > 30 mm, presenting positive vertical margins, presenting venous invasion, and whose level of submucosal invasion is >500 µm. Patients were then stratified into three groups corresponding to LNM risk: low (zero points to one point: 2.5% risk), intermediate (two to four points: 6.7% risk), and high (five to seven points: 22.7% risk). A validation arm verified that this categorization is associated with significantly different DSS between risk groups (99.6, 96.0, and 90.1% at 5 years, respectively; $p < 0.001$) and that the low-risk group presents very high DSS, which is comparable to that of EGC patients who fulfill curative criteria after endoscopic resection. This tool may be helpful in attempting to predict which patients will receive the most benefit from salvage surgery after non-curative ESD and for whom surgical treatment may represent a riskier option than surveillance.

As stated before, endoscopic resection preserves the stomach at the expense of maintaining gastric mucosa at risk for metachronous lesions and recurrence. The rate of metachronous lesions after curative endoscopic resection described in the literature varies between 3% and 20%. In a recent meta-analysis, Ortigão et al. [79] determined a value of metachronous gastric lesion cumulative incidence at 5 years of 9.5% after endoscopic resection, which was significantly higher than that of 0.7% for surgery, with the meta-regression model predicting an increase in the metachronous rate with time, namely, up to 14.9% at 10 years for endoscopic resection versus 2.3% for surgery. This highlights the need for endoscopic surveillance post-resection.

European guidelines [23] recommend a follow-up endoscopy 3–6 months after a curative resection or local-risk resection without local recurrence and annually thereafter, while Japanese guidelines [24] recommend annual endoscopy for an eCuraA resection and annual or biannual endoscopic surveillance for an eCuraB resection. There are no studies comparing annual and biannual surveillance, but an endoscopy interval less than 12 months does not seem to increase the proportion of metachronous lesions amenable to endoscopic resection [79]. On the other hand, one study found that a surveillance interval greater than 12 months was significantly linked to the recurrence of adenocarcinoma, larger lesions, and a higher proportion of patients undergoing surgical treatment [80].

Still regarding the surveillance interval, multiple studies have tried to find risk factors for metachronous GC to enable the tailoring of surveillance according to individual risk. The aforementioned meta-analysis found the following to be significantly associated with metachronous: older age (mean difference 1.08 years, 95%CI 0.21–1.96), male sex (OR 1.43, 95%CI 1.22–1.66), a family history of GC (OR 1.88, 95%CI 1.03–3.41), synchronous lesions (OR 1.72, 95%CI 1.30–2.28), severe gastric mucosal atrophy (OR 2.77, 95%CI 1.22–6.29), intestinal metaplasia in corpus (OR 3.15, 95%CI 1.67–5.96), a persistent Helicobacter pylori infection (OR 2.08, 95%CI 1.60–2.72), and a lower pepsinogen I/II ratio (mean difference–0.54, 95%CI -0.86 to -0.22) [79].

Several meta-analyses have evaluated the impact of *H. pylori* eradication on the risk of metachronous lesions following an endoscopic resection of EGC and generally concluded that eradication is associated with reduced rates of metachronous GC [RR 0.46, 95%CI 0.37–0.57 [81]; RR 0.467, 95%CI 0.362–0.602 [82]; RR 0.50, 95%CI 0.41–0.61 [83]; OR 0.42, 95%CI 0.32–0.56 [84]; OR 0.47, 95%CI 0.33–0.67 [85]; hazard ratio (HR) 0.43, 95%CI 0.26–0.70 [86]. One meta-analysis [85] incorporating 6967 patients from nine randomized controlled trials found that there was no difference in metachronous incidence when patients had already-established atrophic gastritis and intestinal metaplasia at baseline. International guidelines [22,24,87] recommend that a patient's *H. pylori* status be determined after the endoscopic resection of EGC, with reflex eradication.

Finally, the required duration of a follow-up after resection has not not clearly defined, and neither is the level of expertise of the endoscopists assigned to this task. The risk of metachronous lesions is higher for older patients but also seems to increase with time for up to 10 years after resection (even among younger patients). In one study, a survival analysis showed a stable cumulative incidence of metachronous cancer 10 years post-resection [88].

7. Future Perspectives

Predicting the depth of invasion of EGC is one of the most challenging aspects of the endoscopic assessment of superficial gastric lesions. Artificial intelligence (AI) systems have been used in several medical fields. A few studies have undertaken the evaluation of the accuracy of AI systems in predicting the depth of invasion of EGC. Zhu et al. [89] and Tang et al. [90] report an accuracy of around 88–89% for predicting tumor depth, while Yoon et al. [91] report a sensitivity and specificity of 79.2% and 77.8%, respectively. Nagao et al. [92] evaluated an AI system's ability to predict depth of invasion using conventional white-light imaging, non-magnifying narrow-band imaging, and indigo carmine dye contrast imaging and found no differences, with accuracies varying between 94.5% and 95.5%. Wu et al. [93] report a lower accuracy of 78.57% for predicting EGC invasion depth, which is still comparable to endoscopists' results, and Hamada et al. [94] present similar accuracy values (78.9–82.4%, depending on whether evaluations were image-based or lesion-based). Two systematic reviews with meta-analyses have assessed the performance of AI systems with respect to estimating depth invasion [95,96]. The pooled sensitivity and specificity for predicting deep submucosal invasion were 72–82% and 79–90%, respectively. Jiang et al. concluded that AI-assisted depth diagnosis is more accurate than that of experts, while Xie et al. did not find differences on this matter. Kim et al. [97] compared two AI models, one developed from static images and the second from video clips, and concluded that models developed from videos could predict EGC depth invasion more precisely than image-trained models. A recent study [98] suggests that human–machine cooperation improves performance when compared to the individual results of either one. Although promising, AI systems have yet to prove themselves more accurate than experts at predicting depth of invasion. Therefore, they have not been implemented in clinical practice; however, the technology is expected to improve quickly.

There also seems to be room for improvement in cases of non-curative resection, as the search for the less invasive management of GC continues. As mentioned previously, a great portion of lesions that do not meet curative criteria fail to show residual disease or LNM after rescue surgery. Given the post-surgery morbidity and impact on quality of life of gastrectomy, it would be desirable to avoid surgery among patients who have not yet developed LNM. In this regard, Abe et al. [99] first described in 2005 a minimally invasive strategy combining ESD followed by laparoscopic lymph node dissection (LLND). Theoretically, in a patient with a lesion that has been completely resected via ESD but with a clinically significant risk of LNM, LLND would offer the potential to confirm the absence of LNM, hence obviating the need for gastric resection. The same group evaluated the long-term outcomes of combining ESD and LLND in a group of 21 patients whose lesions were completely removed but presented at least one risk factor for LNM [100]. Fourteen patients had undifferentiated-type lesions, eight had deep submucosal invasion and two had lymphatic invasion. After a median follow-up of 61 months, none showed evidence of metastatic disease, including two patients with positive lymph node metastasis as determined via LLND who refused salvage surgery and were followed for 78–85 months. The authors also evaluated adverse events resulting from the procedure. Gastric lymph node dissection usually implies the division of major feeding arteries and the resection of vagal trunks, which may result in early or delayed gastric ischemia on the one hand and gastritis, perforation or ulcers, and impaired gastric motility on the other. In this study, one patient suffered gastric perforation from early ischemic gastritis, three patients presented a moderate amount of gastric residue following gastroscopy, and two patients complained of postprandial static symptoms such as abdominal distention and belching.

The consequences related to an extended lymph node dissection may be partially curbed by further limiting the number of patients submitted to radical lymphadenectomy. As already conducted for other cancer types, a strategy of lymph node mapping in GC patients has been under study. A lymph node metastasis diagnosis based on the sentinel lymph node biopsy (SLNB) of patients with a significant risk of LNM after ESD could theoretically avoid unnecessary gastrectomy and/or radical lymphadenectomy. Several meta-analyses have evaluated the diagnostic accuracy of SLNB [101–105]. The identification rate of sentinel nodes varied between 93.7–99.0%, and sensitivity varied between 76.9–92.0%. However, the studies were highly heterogenous, with stark differences in the clinical staging of GC patients, tracers used, methods of injection, comparison groups, and the extent of lymphadenectomy. False negative rates of up to nearly 25% seem unreasonable considering the prognosis of GC patients with LNM. Sensitivity seems to be higher in earlier T stages, with a meta-analysis of cT1N0M0 gastric cancer reporting a sensitivity of 92% [104], and a cohort of two randomized controlled trials reporting a pooled sensitivity of 97.7% for pT1 tumors after subgroup analysis [106]. A single-arm study of the long-term oncologic outcomes of SLNB in cT1 gastric cancer cases, incorporating 100 patients and employing a median follow-up period of 47.5 months, showed a 3-year recurrence-free survival rate of 96.0% (95%CI 92.2–100.0%) and an OS of 98.0% (95%CI 95.2–100.0%) [107].

New minimally invasive strategies such as SLNB or LLND after ESD could eventually lead to the expansion of the indications for the endoscopic resection of EGC. However, there are few studies evaluating the combination of LLND with ESD, and SLNB has not yet shown consistent and satisfactory results, with a high heterogeneity of methods among studies.

8. Conclusions

ESD is now established as the preferential endoscopic resection technique for gastric superficial lesions (when compared to EMR) and is also preferable to surgery, offering advantages in terms of morbidity and quality of life. ESD is being successfully implemented in western countries, and in the stomach, the corresponding efficacy and safety outcomes are comparable to eastern studies. As ESD is now recommended as a first-line treatment for lesions with a low risk of LNM, three aspects should drive future research:

1. Prediction of and decrease in adverse events: The identification of patients at higher risk of adverse outcomes is important in order to provide patients with more comprehensive information and implement preventive strategies such as defect closure or defect shielding.
2. Better patient selection: Up to 20% of endoscopically resected lesions still do not meet curative criteria, and it is desirable to improve pre-resection endoscopic assessments to avoid unnecessary procedures conducted on patients who would not benefit from them and to better allocate scarce resources. In this regard, AI will probably have a clear role in assisting endoscopists in treatment allocation.
3. The optimization of the management of patients with non-curative resection: The stratification of the risk of LNM, with individualized predictions, should be pursued; this can be achieved through the refinement of existing scoring systems (eCura) and possibly by incorporating additional variables (and possibly molecular features that can help predict this undesirable outcome of LNM). Less invasive alternatives to gastrectomy with lymphadenectomy among patients with non-curative resections should also be pursued, but more studies are needed to clarify the potential role of LLND and SLNB.

The efficacy of a follow-up after resection is also a matter of debate, with sparse evidence backing such intensive and longstanding protocols. We hope that trials evaluating different surveillance protocols according to a patient's individual risk of developing metachronous lesions will soon be found.

Author Contributions: Conceptualization, D.L. and M.D.-R.; writing—original draft preparation, A.C.V.; writing—review and editing, D.L., M.D.-R and A.C.V.; supervision, D.L. and M.D.-R. All authors have read and agreed to the published version of the manuscript.

Funding: This research received no external funding.

Conflicts of Interest: The authors declare no conflict of interest.

References

1. Sung, H.; Ferlay, J.; Siegel, R.L.; Laversanne, M.; Soerjomataram, I.; Jemal, A.; Bray, F. Global Cancer Statistics 2020: GLOBOCAN Estimates of Incidence and Mortality Worldwide for 36 Cancers in 185 Countries. *CA Cancer J. Clin.* **2021**, *71*, 209–249. [CrossRef] [PubMed]
2. Stewart, B.W.; Wild, C.P. *World Cancer Report 2014*; IARC Publications; International Agency for Research on Cancer: Lyon, France, 2014.
3. World Health Organization. *Cancer Tomorrow—International Agency for Research on Cancer*; World Health Organization: Lyon, France, 2020.
4. Weir, H.K.; Thompson, T.D.; Stewart, S.L.; White, M.C. Cancer Incidence Projections in the United States Between 2015 and 2050. *Prev. Chronic Dis.* **2021**, *18*, E59. [CrossRef] [PubMed]
5. Correa, P.; Piazuelo, M.B. The Gastric Precancerous Cascade. *J. Dig. Dis.* **2012**, *13*, 2–9. [CrossRef] [PubMed]

6. Hamashima, C.; Kato, K.; Miyashiro, I.; Nishida, H.; Takaku, R.; Terasawa, T.; Yoshikawa, T.; Honjo, S.; Inoue, K.; Nakayama, T.; et al. Update Version of the Japanese Guidelines for Gastric Cancer Screening. *Jpn. J. Clin. Oncol.* **2018**, *48*, 673–683 [CrossRef]
7. Park, H.A.; Nam, S.Y.; Lee, S.K.; Kim, S.G.; Shim, K.N.; Park, S.M.; Lee, S.Y.; Han, H.S.; Shin, Y.M.; Kim, K.M.; et al. The Korean Guideline for Gastric Cancer Screening. *J. Korean Med. Assoc.* **2015**, *58*, 373–384. [CrossRef]
8. Japanese Gastric Cancer Association. Japanese Classification of Gastric Carcinoma—2nd English Edition. *Gastric Cancer* **1998**, *1*, 10–24. [CrossRef]
9. Yanzhang, W.; Guanghua, L.; Zhihao, Z.; Zhixiong, W.; Zhao, W. The Risk of Lymph Node Metastasis in Gastric Cancer Conforming to Indications of Endoscopic Resection and Pylorus-Preserving Gastrectomy: A Single-Center Retrospective Study. *BMC Cancer* **2021**, *21*, 1280. [CrossRef]
10. Park, Y.M.; Cho, E.; Kang, H.Y.; Kim, J.M. The Effectiveness and Safety of Endoscopic Submucosal Dissection Compared with Endoscopic Mucosal Resection for Early Gastric Cancer: A Systematic Review and Metaanalysis. *Surg. Endosc.* **2011**, *25*, 2666–2677 [CrossRef]
11. Suzuki, H.; Takizawa, K.; Hirasawa, T.; Takeuchi, Y.; Ishido, K.; Hoteya, S.; Yano, T.; Tanaka, S.; Endo, M.; Nakagawa, M.; et al. Short-Term Outcomes of Multicenter Prospective Cohort Study of Gastric Endoscopic Resection: 'Real-World Evidence' in Japan. *Dig. Endosc.* **2019**, *31*, 30–39. [CrossRef]
12. Kim, S.G.; Park, C.M.; Lee, N.R.; Kim, J.; Lyu, D.H.; Park, S.H.; Choi, I.J.; Lee, W.S.; Park, S.J.; Kim, J.J.; et al. Long-Term Clinical Outcomes of Endoscopic Submucosal Dissection in Patients with Early Gastric Cancer: A Prospective Multicenter Cohort Study. *Gut Liver* **2018**, *12*, 402–410. [CrossRef]
13. Endoscopic Classification Review Group. Update on the Paris Classification of Superficial Neoplastic Lesions in the Digestive Tract. *Endoscopy* **2005**, *37*, 570–578. [CrossRef] [PubMed]
14. Dixon, M.F. Gastrointestinal Epithelial Neoplasia: Vienna Revisited. *Gut* **2002**, *51*, 130–131. [CrossRef] [PubMed]
15. Ribeiro, H.; Libânio, D.; Castro, R.; Ferreira, A.; Barreiro, P.; Boal Carvalho, P.; Capela, T.; Pimentel-Nunes, P.; Santos, C.; Dinis-Ribeiro, M. Reliability of Paris Classification for Superficial Neoplastic Gastric Lesions Improves with Training and Narrow Band Imaging. *Endosc. Int. Open* **2019**, *07*, E633–E640. [CrossRef] [PubMed]
16. Gotoda, T.; Yanagisawa, A.; Sasako, M.; Ono, H.; Nakanishi, Y.; Shimoda, T.; Kato, Y. Incidence of Lymph Node Metastasis from Early Gastric Cancer: Estimation with a Large Number of Cases at Two Large Centers. *Gastric Cancer* **2000**, *3*, 219–225. [CrossRef] [PubMed]
17. Nakahara, K.; Tsuruta, O.; Tateishi, H.; Arima, N.; Takeda, J.; Toyonaga, A.; Sata, M. Extended Indication Criteria for Endoscopic Mucosal Resection of Early Gastric Cancer with Special Reference to Lymph Node Metastasis Examination by Multivariate Analysis. *Kurume Med. J.* **2004**, *51*, 9–14. [CrossRef]
18. Hirasawa, T.; Gotoda, T.; Miyata, S.; Kato, Y.; Shimoda, T.; Taniguchi, H.; Fujisaki, J.; Sano, T.; Yamaguchi, T. Incidence of Lymph Node Metastasis and the Feasibility of Endoscopic Resection for Undifferentiated-Type Early Gastric Cancer. *Gastric Cancer* **2009**, *12*, 148–152. [CrossRef]
19. Chen, J.; Zhao, G.; Wang, Y. Analysis of Lymph Node Metastasis in Early Gastric Cancer: A Single Institutional Experience from China. *World J. Surg. Oncol.* **2020**, *18*, 57. [CrossRef]
20. Hasuike, N.; Ono, H.; Boku, N.; Mizusawa, J.; Takizawa, K.; Fukuda, H.; Oda, I.; Doyama, H.; Kaneko, K.; Hori, S.; et al. A Non-Randomized Confirmatory Trial of an Expanded Indication for Endoscopic Submucosal Dissection for Intestinal-Type Gastric Cancer (CT1a): The Japan Clinical Oncology Group Study (JCOG0607). *Gastric Cancer* **2017**, *21*, 114–123. [CrossRef]
21. Takizawa, K.; Ono, H.; Hasuike, N.; Takashima, A.; Minashi, K.; Boku, N.; Kushima, R.; Katayama, H.; Ogawa, G.; Fukuda, H.; et al. A Nonrandomized, Single-Arm Confirmatory Trial of Expanded Endoscopic Submucosal Dissection Indication for Undifferentiated Early Gastric Cancer: Japan Clinical Oncology Group Study (JCOG1009/1010). *Gastric Cancer* **2021**, *24*, 479–491. [CrossRef]
22. Ono, H.; Yao, K.; Fujishiro, M.; Oda, I.; Uedo, N.; Nimura, S.; Yahagi, N.; Iishi, H.; Oka, M.; Ajioka, Y.; et al. Guidelines for Endoscopic Submucosal Dissection and Endoscopic Mucosal Resection for Early Gastric Cancer (Second Edition). *Dig. Endosc.* **2021**, *33*, 4–20. [CrossRef]
23. Pimentel-Nunes, P.; Libânio, D.; Bastiaansen, B.A.J.; Bhandari, P.; Bisschops, R.; Bourke, M.J.; Esposito, G.; Lemmers, A.; Maselli, R.; Messmann, H.; et al. Endoscopic Submucosal Dissection for Superficial Gastrointestinal Lesions: European Society of Gastrointestinal Endoscopy (ESGE) Guideline—Update 2022. *Endoscopy* **2022**, *54*, 591–622. [CrossRef] [PubMed]
24. Japanese Gastric Cancer Association. Japanese Gastric Cancer Treatment Guidelines 2021 (6th Edition). *Gastric Cancer* **2023**, *26*, 1–25. [CrossRef] [PubMed]
25. Figueirôa, G.; Pimentel-Nunes, P.; Dinis-Ribeiro, M.; Libânio, D. Gastric Endoscopic Submucosal Dissection: A Systematic Review and Meta-Analysis on Risk Factors for Poor Short-Term Outcomes. *Eur. J. Gastroenterol. Hepatol.* **2019**, *31*, 1234–1246. [CrossRef] [PubMed]
26. Kim, E.H.; Park, J.C.; Song, I.J.; Kim, Y.J.; Joh, D.H.; Hahn, K.Y.; Lee, Y.K.; Kim, H.Y.; Chung, H.; Shin, S.K.; et al. Prediction Model for Non-Curative Resection of Endoscopic Submucosal Dissection in Patients with Early Gastric Cancer. *Gastrointest. Endosc.* **2016**, *85*, 976–983. [CrossRef]
27. Nam, H.S.; Choi, C.W.; Kim, S.J.; Kang, D.H.; Kim, H.W.; Park, S.B.; Ryu, D.G.; Choi, J.S. Preprocedural Prediction of Non-Curative Endoscopic Submucosal Dissection for Early Gastric Cancer. *PLoS ONE* **2018**, *13*, e0206179. [CrossRef]

28. Horiuchi, Y.; Fujisaki, J.; Yamamoto, N.; Ishizuka, N.; Omae, M.; Ishiyama, A.; Yoshio, T.; Hirasawa, T.; Yamamoto, Y.; Nagahama, M.; et al. Undifferentiated-Type Component Mixed with Differentiated-Type Early Gastric Cancer Is a Significant Risk Factor for Endoscopic Non-Curative Resection. *Dig. Endosc.* **2018**, *30*, 624–632. [CrossRef]
29. Xu, P.; Wang, Y.; Dang, Y.; Huang, Q.; Wang, J.; Zhang, W.; Zhang, Y.; Zhang, G. Predictive Factors and Long-Term Outcomes of Early Gastric Carcinomas in Patients with Non-Curative Resection by Endoscopic Submucosal Dissection. *Cancer Manag. Res.* **2020**, *12*, 8037–8046. [CrossRef]
30. Ma, X.; Zhang, Q.; Zhu, S.; Zhang, S.; Sun, X. Risk Factors and Prediction Model for Non-Curative Resection of Early Gastric Cancer With Endoscopic Resection and the Evaluation. *Front. Med.* **2021**, *8*, 637875. [CrossRef]
31. Kadota, T.; Hasuike, N.; Ono, H.; Boku, N.; Mizusawa, J.; Oda, I.; Oyama, T.; Horiuchi, Y.; Hirasawa, K.; Yoshio, T.; et al. Clinical Factors Associated with Noncurative Endoscopic Submucosal Dissection for the Expanded Indication of Intestinal-type Early Gastric Cancer: Post Hoc Analysis of a Multi-institutional, Single-arm, Confirmatory Trial (JCOG0607). *Dig. Endosc.* **2022**, *35*, 494–502. [CrossRef]
32. Yang, P.; Zheng, X.-D.; Wang, J.-M.; Geng, W.-B.; Wang, X. Undifferentiated-Predominant Mixed-Type Early Gastric Cancer Is More Aggressive than Pure Undifferentiated Type: A Systematic Review and Meta-Analysis. *BMJ Open* **2022**, *12*, e054473. [CrossRef]
33. Du, M.Z.; Gan, W.J.; Yu, J.; Liu, W.; Zhan, S.H.; Huang, S.; Huang, R.P.; Chuan Guo, L.; Huang, Q. Risk Factors of Lymph Node Metastasis in 734 Early Gastric Carcinoma Radical Resections in a Chinese Population: Nodes Metastasis in Early Gastric Cancer. *J. Dig. Dis.* **2018**, *19*, 586–595. [CrossRef] [PubMed]
34. Milhomem, L.M.; Milhomem-Cardoso, D.M.; da Mota, O.M.; Mota, E.D.; Kagan, A.; Filho, J.B.S. Risk of Lymph Node Metastasis in Early Gastric Cancer and Indications for Endoscopic Resection: Is It Worth Applying the East Rules to the West? *Surg. Endosc.* **2021**, *35*, 4380–4388. [CrossRef] [PubMed]
35. Oh, Y.J.; Kim, D.H.; Han, W.H.; Eom, B.W.; Kim, Y.I.; Yoon, H.M.; Lee, J.Y.; Kim, C.G.; Kook, M.-C.; Choi, I.J.; et al. Risk Factors for Lymph Node Metastasis in Early Gastric Cancer without Lymphatic Invasion after Endoscopic Submucosal Dissection. *Eur. J. Surg. Oncol.* **2021**, *47*, 3059–3063. [CrossRef] [PubMed]
36. Abe, S.; Oda, I.; Shimazu, T.; Kinjo, T.; Tada, K.; Sakamoto, T.; Kusano, C.; Gotoda, T. Depth-Predicting Score for Differentiated Early Gastric Cancer. *Gastric Cancer* **2011**, *14*, 35–40. [CrossRef] [PubMed]
37. Nakayoshi, T.; Tajiri, H.; Matsuda, K.; Kaise, M.; Ikegami, M.; Sasaki, H. Magnifying Endoscopy Combined with Narrow Band Imaging System for Early Gastric Cancer: Correlation of Vascular Pattern with Histo-pathology (Including Video). *Endoscopy* **2004**, *36*, 1080–1084. [CrossRef]
38. Yokoyama, A.; Inoue, H.; Minami, H.; Wada, Y.; Sato, Y.; Satodate, H.; Hamatani, S.; Kudo, S. Novel Nar-row-Band Imaging Magnifying Endoscopic Classification for Early Gastric Cancer. *Dig. Liver Dis.* **2010**, *42*, 704–708. [CrossRef]
39. Tanaka, K.; Toyoda, H.; Kadowaki, S.; Kosaka, R.; Shiraishi, T.; Imoto, I.; Shiku, H.; Adachi, Y. Features of Early Gastric Cancer and Gastric Adenoma by Enhanced-Magnification Endoscopy. *J. Gastroenterol.* **2006**, *41*, 332–338. [CrossRef]
40. Ok, K.-S.; Kim, G.H.; Park, D.Y.; Lee, H.J.; Jeon, H.K.; Baek, D.H.; Lee, B.E.; Song, G.A. Magnifying En-doscopy with Narrow Band Imaging of Early Gastric Cancer: Correlation with Histopathology and Mucin Phenotype. *Gut Liver* **2016**, *10*, 532–541. [CrossRef]
41. Kanesaka, T.; Uedo, N.; Doyama, H.; Yoshida, N.; Nagahama, T.; Ohtsu, K.; Uchita, K.; Kojima, K.; Ueo, T.; Takahashi, H.; et al. Diagnosis of Histological Type of Early Gastric Cancer by Magnifying Narrow-band Imaging: A Multicenter Prospective Study. *DEN Open* **2021**, *2*, e61. [CrossRef]
42. Inoue, H.; Takeshita, K.; Hori, H.; Muraoka, Y.; Yoneshima, H.; Endo, M. Endoscopic Mucosal Resection with a Cap-Fitted Panendoscope for Esophagus, Stomach, and Colon Mucosal Lesions. *Gastrointest. Endosc.* **1993**, *39*, 58–62. [CrossRef]
43. Ono, H.; Kondo, H.; Gotoda, T.; Shirao, K.; Yamaguchi, H.; Saito, D.; Hosokawa, K.; Shimoda, T.; Yoshida, S. Endoscopic Mucosal Resection for Treatment of Early Gastric Cancer. *Gut* **2001**, *48*, 225–229. [CrossRef] [PubMed]
44. Gotoda, T.; Kondo, H.; Ono, H.; Saito, Y.; Yamaguchi, H.; Saito, D.; Yokota, T. A New Endoscopic Mucosal Resection Procedure Using an Insulation-Tipped Electrosurgical Knife for Rectal Flat Lesions: Report of Two Cases. *Gastrointest. Endosc.* **1999**, *50*, 560–563. [CrossRef] [PubMed]
45. Tao, M.; Zhou, X.; Hu, M.; Pan, J. Endoscopic Submucosal Dissection versus Endoscopic Mucosal Resection for Patients with Early Gastric Cancer: A Meta-Analysis. *BMJ Open* **2019**, *9*, e025803. [CrossRef] [PubMed]
46. Lian, J.; Chen, S.; Zhang, Y.; Qiu, F. A Meta-Analysis of Endoscopic Submucosal Dissection and EMR for Early Gastric Cancer. *Gastrointest. Endosc.* **2012**, *76*, 763–770. [CrossRef]
47. Facciorusso, A.; Antonino, M.; Di Maso, M.; Muscatiello, N. Endoscopic Submucosal Dissection vs Endoscopic Mucosal Resection for Early Gastric Cancer: A Meta-Analysis. *World J. Gastrointest. Endosc.* **2014**, *6*, 555. [CrossRef]
48. Zhao, Y.; Wang, C. Long-Term Clinical Efficacy and Perioperative Safety of Endoscopic Submucosal Dissection versus Endoscopic Mucosal Resection for Early Gastric Cancer: An Updated Meta-Analysis. *BioMed Res. Int.* **2018**, *2018*, 3152346. [CrossRef]
49. Tanabe, S.; Ishido, K.; Matsumoto, T.; Kosaka, T.; Oda, I.; Suzuki, H.; Fujisaki, J.; Ono, H.; Kawata, N.; Oyama, T.; et al. Long-Term Outcomes of Endoscopic Submucosal Dissection for Early Gastric Cancer: A Multicenter Collaborative Study. *Gastric Cancer* **2017**, *20*, 45–52. [CrossRef]
50. Peng, L.J.; Tian, S.N.; Lu, L.; Chen, H.; Ouyang, Y.Y.; Wu, Y.J. Outcome of Endoscopic Submucosal Dissection for Early Gastric Cancer of Conventional and Expanded Indications: Systematic Review and Meta-Analysis. *J. Dig. Dis.* **2015**, *16*, 67–74. [CrossRef]

51. Suzuki, H.; Ono, H.; Hirasawa, T.; Takeuchi, Y.; Ishido, K.; Hoteya, S.; Yano, T.; Tanaka, S.; Toya, Y.; Nakagawa, M.; et al. Long-Term Survival After Endoscopic Resection For Gastric Cancer: Real-World Evidence From a Multicenter Prospective Cohort. *Clin. Gastroenterol. Hepatol.* **2022**, *21*, 307–318.e2. [CrossRef]
52. Shichijo, S.; Uedo, N.; Kanesaka, T.; Ohta, T.; Nakagawa, K.; Shimamoto, Y.; Ohmori, M.; Arao, M.; Iwatsubo, T.; Suzuki, S.; et al. Long-term Outcomes after Endoscopic Submucosal Dissection for Differentiated-type Early Gastric Cancer That Fulfilled Expanded Indication Criteria: A Prospective Cohort Study. *J. Gastroenterol. Hepatol.* **2020**, *36*, 664–670. [CrossRef]
53. Libânio, D.; Costa, M.N.; Pimentel-Nunes, P.; Dinis-Ribeiro, M. Risk Factors for Bleeding after Gastric Endoscopic Submucosal Dissection: A Systematic Review and Meta-Analysis. *Gastrointest. Endosc.* **2016**, *84*, 572–586. [CrossRef] [PubMed]
54. Uedo, N.; Takeuchi, Y.; Yamada, T.; Ishihara, R.; Ogiyama, H.; Yamamoto, S.; Kato, M.; Tatsumi, K.; Masuda, E.; Tamai, C.; et al. Effect of a Proton Pump Inhibitor or an H2-Receptor Antagonist on Prevention of Bleeding From Ulcer After Endoscopic Submucosal Dissection of Early Gastric Cancer: A Prospective Randomized Controlled Trial. *Am. J. Gastroenterol.* **2007**, *102*, 1610–1616. [CrossRef] [PubMed]
55. Yang, Z.; Wu, Q.; Liu, Z.; Wu, K.; Fan, D. Proton Pump Inhibitors versus Histamine-2-Receptor Antagonists for the Management of Iatrogenic Gastric Ulcer after Endoscopic Mucosal Resection or Endoscopic Submucosal Dissection: A Meta-Analysis of Randomized Trials. *Digestion* **2011**, *84*, 315–320. [CrossRef]
56. Nishizawa, T.; Suzuki, H.; Akimoto, T.; Maehata, T.; Morizane, T.; Kanai, T.; Yahagi, N. Effects of Preoperative Proton Pump Inhibitor Administration on Bleeding after Gastric Endoscopic Submucosal Dissection: A Systematic Review and Meta-analysis. *United Eur. Gastroenterol. J.* **2016**, *4*, 5–10. [CrossRef]
57. Takizawa, K.; Oda, I.; Gotoda, T.; Yokoi, C.; Matsuda, T.; Saito, Y.; Saito, D.; Ono, H. Routine Coagulation of Visible Vessels May Prevent Delayed Bleeding after Endoscopic Submucosal Dissection—An Analysis of Risk Factors. *Endoscopy* **2008**, *40*, 179–183. [CrossRef] [PubMed]
58. Chen, Y.; Zhao, X.; Wang, D.; Liu, X.; Chen, J.; Song, J.; Bai, T.; Hou, X. Endoscopic Delivery of Polymers Reduces Delayed Bleeding after Gastric Endoscopic Submucosal Dissection: A Systematic Review and Meta-Analysis. *Polymers* **2022**, *14*, 2387. [CrossRef] [PubMed]
59. Ding, X.; Luo, H.; Duan, H. Risk Factors for Perforation of Gastric Endoscopic Submucosal Dissection: A Systematic Review and Meta-Analysis. *Eur. J. Gastroenterol. Hepatol.* **2019**, *31*, 1481–1488. [CrossRef]
60. Zhao, J.; Sun, Z.; Liang, J.; Guo, S.; Huang, D. Endoscopic Submucosal Dissection for Early Gastric Cancer in Elderly vs Non-Elderly Patients: A Systematic Review and Meta-Analysis. *Front. Oncol.* **2022**, *11*, 5767. [CrossRef]
61. Abdelfatah, M.M.; Barakat, M.; Ahmad, D.; Ibrahim, M.; Ahmed, Y.; Kurdi, Y.; Grimm, I.S.; Othman, M.O. Long-term Outcomes of Endoscopic Submucosal Dissection versus Surgery in Early Gastric Cancer: A Systematic Review and Meta-Analysis. *Eur. J. Gastroenterol. Hepatol.* **2019**, *31*, 418–424. [CrossRef]
62. Gu, L.; Khadaroo, P.A.; Chen, L.; Li, X.; Zhu, H.; Zhong, X.; Pan, J.; Chen, M. Comparison of Long-Term Outcomes of Endoscopic Submucosal Dissection and Surgery for Early Gastric Cancer: A Systematic Review and Meta-Analysis. *J. Gastrointest. Surg.* **2019**, *23*, 1493–1501. [CrossRef]
63. Li, H.; Feng, L.-Q.; Bian, Y.-Y.; Yang, L.-L.; Liu, D.-X.; Huo, Z.-B.; Zeng, L. Comparison of Endoscopic Submucosal Dissection with Surgical Gastrectomy for Early Gastric Cancer: An Updated Meta-Analysis. *World J. Gastrointest. Oncol.* **2019**, *11*, 161–171. [CrossRef] [PubMed]
64. Liu, Q.; Ding, L.; Qiu, X.; Meng, F. Updated Evaluation of Endoscopic Submucosal Dissection versus Surgery for Early Gastric Cancer: A Systematic Review and Meta-Analysis. *Int. J. Surg.* **2019**, *73*, 28–41. [CrossRef] [PubMed]
65. Xu, X.; Zheng, G.; Gao, N.; Zheng, Z. Long-Term Outcomes and Clinical Safety of Expanded Indication Early Gastric Cancer Treated with Endoscopic Submucosal Dissection versus Surgical Resection: A Meta-Analysis. *BMJ Open* **2022**, *12*, e055406. [CrossRef]
66. Nakamura, R.; Omori, T.; Mayanagi, S.; Irino, T.; Wada, N.; Kawakubo, H.; Kameyama, K.; Kitagawa, Y. Risk of Lymph Node Metastasis in Undifferentiated-Type Mucosal Gastric Carcinoma. *World J. Surg. Oncol.* **2019**, *17*, 32. [CrossRef]
67. Huh, C.-W.; Ma, D.W.; Kim, B.-W.; Kim, J.S.; Lee, S.J. Endoscopic Submucosal Dissection versus Surgery for Undifferentiated-Type Early Gastric Cancer: A Systematic Review and Meta-Analysis. *Clin. Endosc.* **2021**, *54*, 202–210. [CrossRef] [PubMed]
68. Yang, H.-J.; Kim, J.-H.; Kim, N.W.; Choi, I.J. Comparison of Long-Term Outcomes of Endoscopic Submucosal Dissection and Surgery for Undifferentiated-Type Early Gastric Cancer Meeting the Expanded Criteria: A Systematic Review and Meta-Analysis. *Surg. Endosc.* **2022**, *36*, 3686–3697. [CrossRef] [PubMed]
69. Kim, Y.; Kim, Y.-W.; Choi, I.J.; Cho, J.Y.; Kim, J.H.; Kwon, J.-W.; Lee, J.Y.; Lee, N.R.; Seol, S.-Y. Cost Comparison between Surgical Treatments and Endoscopic Submucosal Dissection in Patients with Early Gastric Cancer in Korea. *Gut Liver* **2015**, *9*, 174–180. [CrossRef]
70. Shin, D.W.; Hwang, H.Y.; Jeon, S.W. Comparison of Endoscopic Submucosal Dissection and Surgery for Differentiated Type Early Gastric Cancer within the Expanded Criteria. *Clin. Endosc.* **2017**, *50*, 170–178. [CrossRef]
71. Qian, M.; Sheng, Y.; Wu, M.; Wang, S.; Zhang, K. Comparison between Endoscopic Submucosal Dissection and Surgery in Patients with Early Gastric Cancer. *Cancers* **2022**, *14*, 3603. [CrossRef]
72. Song, W.; Qiao, X.; Gao, X. A Comparison of Endoscopic Submucosal Dissection (ESD) and Radical Surgery for Early Gastric Cancer: A Retrospective Study. *World J. Surg. Oncol.* **2015**, *13*, 309. [CrossRef]

73. Libânio, D.; Braga, V.; Ferraz, S.; Castro, R.; Lage, J.; Pita, I.; Ribeiro, C.; Abreu De Sousa, J.; Dinis-Ribeiro, M.; Pimentel-Nunes, P. Prospective Comparative Study of Endoscopic Submucosal Dissection and Gastrectomy for Early Neoplastic Lesions Including Patients' Perspectives. *Endoscopy* **2019**, *51*, 30–39. [CrossRef] [PubMed]
74. Kim, Y.-I.; Kim, Y.A.; Kim, C.G.; Ryu, K.W.; Kim, Y.-W.; Sim, J.A.; Yun, Y.H.; Choi, I.J. Serial Intermediate-Term Quality of Life Comparison after Endoscopic Submucosal Dissection versus Surgery in Early Gastric Cancer Patients. *Surg. Endosc.* **2018**, *32*, 2114–2122. [CrossRef] [PubMed]
75. Libânio, D.; Ortigão, R.; Pimentel-Nunes, P.; Dinis-Ribeiro, M. Improving the Diagnosis and Treatment of Early Gastric Cancer in the West. *GE Port J. Gastroenterol.* **2022**, *29*, 299–310. [CrossRef] [PubMed]
76. Libânio, D.; Pimentel-Nunes, P.; Afonso, L.P.; Henrique, R.; Dinis-Ribeiro, M. Long-Term Outcomes of Gastric Endoscopic Submucosal Dissection: Focus on Metachronous and Non-Curative Resection Management. *GE Port. J. Gastroenterol.* **2016**, *24*, 31–39. [CrossRef] [PubMed]
77. Kawata, N.; Kakushima, N.; Takizawa, K.; Tanaka, M.; Makuuchi, R.; Tokunaga, M.; Tanizawa, Y.; Bando, E.; Kawamura, T.; Sugino, T.; et al. Risk Factors for Lymph Node Metastasis and Long-Term Outcomes of Patients with Early Gastric Cancer after Non-Curative Endoscopic Submucosal Dissection. *Surg. Endosc.* **2016**, *31*, 1607–1616. [CrossRef]
78. Hatta, W.; Gotoda, T.; Oyama, T.; Kawata, N.; Takahashi, A.; Yoshifuku, Y.; Hoteya, S.; Nakagawa, M.; Hirano, M.; Esaki, M.; et al. A Scoring System to Stratify Curability after Endoscopic Submucosal Dissection for Early Gastric Cancer: "ECura System". *Am. J. Gastroenterol.* **2017**, *112*, 874–881. [CrossRef] [PubMed]
79. Ortigão, R.; Figueirôa, G.; Frazzoni, L.; Pimentel-Nunes, P.; Hassan, C.; Dinis-Ribeiro, M.; Fuccio, L.; Libânio, D. Risk Factors for Gastric Metachronous Lesions after Endoscopic or Surgical Resection: A Systematic Review and Meta-Analysis. *Endoscopy* **2022**, *54*, 892–901. [CrossRef]
80. Hahn, K.Y.; Park, J.C.; Kim, E.H.; Shin, S.; Park, C.H.; Chung, H.; Shin, S.K.; Lee, S.K.; Lee, Y.C. Incidence and Impact of Scheduled Endoscopic Surveillance on Recurrence after Curative Endoscopic Resection for Early Gastric Cancer. *Gastrointest. Endosc.* **2016**, *84*, 628–638.e1. [CrossRef]
81. Fan, F.; Wang, Z.; Li, B.; Zhang, H. Effects of Eradicating Helicobacter Pylori on Metachronous Gastric Cancer Prevention: A Systematic Review and Meta-analysis. *J. Eval. Clin. Pract.* **2020**, *26*, 308–315. [CrossRef]
82. Bang, C.S.; Baik, G.H.; Shin, I.S.; Kim, J.B.; Suk, K.T.; Yoon, J.H.; Kim, Y.S.; Kim, D.J. Helicobacter Pylori Eradication for Prevention of Metachronous Recurrence after Endoscopic Resection of Early Gastric Cancer. *J. Korean Med. Sci.* **2015**, *30*, 749–756. [CrossRef]
83. Xiao, S.; Li, S.; Zhou, L.; Jiang, W.; Liu, J. Helicobacter Pylori Status and Risks of Metachronous Recurrence after Endoscopic Resection of Early Gastric Cancer: A Systematic Review and Meta-Analysis. *J. Gastroenterol.* **2019**, *54*, 226–237. [CrossRef] [PubMed]
84. Yoon, S.B.; Park, J.M.; Lim, C.-H.; Cho, Y.K.; Choi, M.-G. Effect of Helicobacter Pylori Eradication on Metachronous Gastric Cancer after Endoscopic Resection of Gastric Tumors: A Meta-Analysis. *Helicobacter* **2014**, *19*, 243–248. [CrossRef] [PubMed]
85. Khan, M.Y.; Aslam, A.; Mihali, A.B.; Shabbir Rawala, M.; Dirweesh, A.; Khan, S.; Adler, D.G.; Siddiqui, A. Effectiveness of Helicobacter Pylori Eradication in Preventing Metachronous Gastric Cancer and Preneoplastic Lesions. A Systematic Review and Meta-Analysis. *Eur. J. Gastroenterol. Hepatol.* **2020**, *32*, 686–694. [CrossRef] [PubMed]
86. Zhao, B.; Zhang, J.; Mei, D.; Luo, R.; Lu, H.; Xu, H.; Huang, B. Does Helicobacter Pylori Eradication Reduce the Incidence of Metachronous Gastric Cancer After Curative Endoscopic Resection of Early Gastric Cancer. *J. Clin. Gastroenterol.* **2020**, *54*, 235–241. [CrossRef]
87. Pimentel-Nunes, P.; Libânio, D.; Marcos-Pinto, R.; Areia, M.; Leja, M.; Esposito, G.; Garrido, M.; Kikuste, I.; Megraud, F.; Matysiak-Budnik, T.; et al. Management of Epithelial Precancerous Conditions and Lesions in the Stomach (MAPS II): European Society of Gastrointestinal Endoscopy (ESGE), European Helicobacter and Microbiota Study Group (EHMSG), European Society of Pathology (ESP), and Sociedade Portuguesa de Endoscopia Digestiva (SPED) Guideline Update 2019. *Endoscopy* **2019**, *51*, 365–388. [PubMed]
88. Kobayashi, M.; Narisawa, R.; Sato, Y.; Takeuchi, M.; Aoyagi, Y. Self-Limiting Risk of Metachronous Gastric Cancers after Endoscopic Resection. *Dig. Endosc.* **2010**, *22*, 169–173. [CrossRef]
89. Zhu, Y.; Wang, Q.-C.; Xu, M.-D.; Zhang, Z.; Cheng, J.; Zhong, Y.-S.; Zhang, Y.-Q.; Chen, W.-F.; Yao, L.-Q.; Zhou, P.-H.; et al. Application of Convolutional Neural Network in the Diagnosis of the Invasion Depth of Gastric Cancer Based on Conventional Endoscopy. *Gastrointest. Endosc.* **2019**, *89*, 806–815.e1. [CrossRef]
90. Tang, D.; Zhou, J.; Wang, L.; Ni, M.; Chen, M.; Hassan, S.; Luo, R.; Chen, X.; He, X.; Zhang, L.; et al. A Novel Model Based on Deep Convolutional Neural Network Improves Diagnostic Accuracy of Intramucosal Gastric Cancer (With Video). *Front. Oncol.* **2021**, *11*, 622827. [CrossRef]
91. Yoon, H.J.; Kim, S.; Kim, J.-H.; Keum, J.-S.; Oh, S.-I.; Jo, J.; Chun, J.; Youn, Y.H.; Park, H.; Kwon, I.G.; et al. A Lesion-Based Convolutional Neural Network Improves Endoscopic Detection and Depth Prediction of Early Gastric Cancer. *J. Clin. Med.* **2019**, *8*, 1310. [CrossRef]
92. Nagao, S.; Tsuji, Y.; Sakaguchi, Y.; Takahashi, Y.; Minatsuki, C.; Niimi, K.; Yamashita, H.; Yamamichi, N.; Seto, Y.; Tada, T.; et al. Highly Accurate Artificial Intelligence Systems to Predict the Invasion Depth of Gastric Cancer: Efficacy of Conventional White-Light Imaging, Nonmagnifying Narrow-Band Imaging, and Indigo-Carmine Dye Contrast Imaging. *Gastrointest. Endosc.* **2020**, *92*, 866–873.e1. [CrossRef]

93. Wu, L.; Wang, J.; He, X.; Zhu, Y.; Jiang, X.; Chen, Y.; Wang, Y.; Huang, L.; Shang, R.; Dong, Z.; et al. Deep Learning System Compared with Expert Endoscopists in Predicting Early Gastric Cancer and Its Invasion Depth and Differentiation Status (with Videos). *Gastrointest. Endosc.* **2022**, *95*, 92–104.e3. [CrossRef] [PubMed]
94. Hamada, K.; Kawahara, Y.; Tanimoto, T.; Ohto, A.; Toda, A.; Aida, T.; Yamasaki, Y.; Gotoda, T.; Ogawa, T.; Abe, M.; et al. Application of Convolutional Neural Networks for Evaluating the Depth of Invasion of Early Gastric Cancer Based on Endoscopic Images. *J. Gastroenterol. Hepatol.* **2021**, *37*, 352–357. [CrossRef] [PubMed]
95. Jiang, K.; Jiang, X.; Pan, J.; Wen, Y.; Huang, Y.; Weng, S.; Lan, S.; Nie, K.; Zheng, Z.; Ji, S.; et al. Current Evidence and Future Perspective of Accuracy of Artificial Intelligence Application for Early Gastric Cancer Diagnosis With Endoscopy: A Systematic and Meta-Analysis. *Front. Med. (Lausanne)* **2021**, *8*, 629080. [CrossRef] [PubMed]
96. Xie, F.; Zhang, K.; Li, F.; Ma, G.; Ni, Y.; Zhang, W.; Wang, J.; Li, Y. Diagnostic Accuracy of Convolutional Neural Network-Based Endoscopic Image Analysis in Diagnosing Gastric Cancer and Predicting Its Invasion Depth: A Systematic Review and Meta-Analysis. *Gastrointest. Endosc.* **2021**, *95*, 599–609.e7. [CrossRef]
97. Kim, J.-H.; Oh, S.-I.; Han, S.-Y.; Keum, J.-S.; Kim, K.-N.; Chun, J.-Y.; Youn, Y.-H.; Park, H. An Optimal Artificial Intelligence System for Real-Time Endoscopic Prediction of Invasion Depth in Early Gastric Cancer. *Cancers* **2022**, *14*, 6000. [CrossRef] [PubMed]
98. Goto, A.; Kubota, N.; Nishikawa, J.; Ogawa, R.; Hamabe, K.; Hashimoto, S.; Ogihara, H.; Hamamoto, Y.; Yanai, H.; Miura, O.; et al. Cooperation between Artificial Intelligence and Endoscopists for Diagnosing Invasion Depth of Early Gastric Cancer. *Gastric Cancer* **2023**, *26*, 116–122. [CrossRef] [PubMed]
99. Abe, N.; Mori, T.; Takeuchi, H.; Yoshida, T.; Ohki, A.; Ueki, H.; Yanagida, O.; Masaki, T.; Sugiyama, M.; Atomi, Y. Laparoscopic Lymph Node Dissection after Endoscopic Submucosal Dissection: A Novel and Minimally Invasive Approach to Treating Early-Stage Gastric Cancer. *Am. J. Surg.* **2005**, *190*, 496–503. [CrossRef]
100. Abe, N.; Takeuchi, H.; Ohki, A.; Yanagida, O.; Masaki, T.; Mori, T.; Sugiyama, M. Long-Term Outcomes of Combination of Endoscopic Submucosal Dissection and Laparoscopic Lymph Node Dissection without Gastrectomy for Early Gastric Cancer Patients Who Have a Potential Risk of Lymph Node Metastasis. *Gastrointest. Endosc.* **2011**, *74*, 792–797. [CrossRef]
101. Wang, Z.; Dong, Z.-Y.; Chen, J.-Q.; Liu, J.-L. Diagnostic Value of Sentinel Lymph Node Biopsy in Gastric Cancer: A Meta-Analysis. *Ann. Surg. Oncol.* **2011**, *19*, 1541–1550. [CrossRef]
102. Huang, L.; Wei, T.; Chen, J.; Zhou, D. Feasibility and Diagnostic Performance of Dual-Tracer-Guided Sentinel Lymph Node Biopsy in CT1-2N0M0 Gastric Cancer: A Systematic Review and Meta-Analysis of Diagnostic Studies. *World J. Surg. Oncol.* **2017**, *15*, 103. [CrossRef]
103. Skubleny, D.; Dang, J.T.; Skulsky, S.; Switzer, N.; Tian, C.; Shi, X.; de Gara, C.; Birch, D.W.; Karmali, S. Diagnostic Evaluation of Sentinel Lymph Node Biopsy Using Indocyanine Green and Infrared or Fluorescent Imaging in Gastric Cancer: A Systematic Review and Meta-Analysis. *Surg. Endosc.* **2018**, *32*, 2620–2631. [CrossRef] [PubMed]
104. Huang, Y.; Pan, M.; Deng, Z.; Ji, Y.; Chen, B. How Useful Is Sentinel Lymph Node Biopsy for the Status of Lymph Node Metastasis in CT1N0M0 Gastric Cancer? A Systematic Review and Meta-Analysis. *Updates Surg.* **2021**, *73*, 1275–1284. [CrossRef] [PubMed]
105. Huang, Y.; Pan, M.; Chen, B. A Systematic Review and Meta-Analysis of Sentinel Lymph Node Biopsy in Gastric Cancer, an Optimization of Imaging Protocol for Tracer Mapping. *World J. Surg.* **2021**, *45*, 1126–1134. [CrossRef] [PubMed]
106. Zhong, Q.; Chen, Q.-Y.; Huang, X.-B.; Lin, G.-T.; Liu, Z.-Y.; Chen, J.-Y.; Wang, H.-G.; Weng, K.; Li, P.; Xie, J.-W.; et al. Clinical Implications of Indocyanine Green Fluorescence Imaging-Guided Laparoscopic Lymphadenectomy for Patients with Gastric Cancer: A Cohort Study from Two Randomized, Controlled Trials Using Individual Patient Data. *Int. J. Surg.* **2021**, *94*, 106120 [CrossRef]
107. Park, D.J.; Park, Y.S.; Son, S.Y.; Lee, J.-H.; Lee, H.S.; Park, Y.S.; Lee, K.H.; Kim, Y.H.; Park, K.U.; Lee, W.W.; et al. Long-Term Oncologic Outcomes of Laparoscopic Sentinel Node Navigation Surgery in Early Gastric Cancer: A Single-Center, Single-Arm, Phase II Trial. *Ann. Surg. Oncol.* **2018**, *25*, 2357–2365. [CrossRef]

Disclaimer/Publisher's Note: The statements, opinions and data contained in all publications are solely those of the individual author(s) and contributor(s) and not of MDPI and/or the editor(s). MDPI and/or the editor(s) disclaim responsibility for any injury to people or property resulting from any ideas, methods, instructions or products referred to in the content.

Review

Using Endoscopy in the Diagnosis of Pancreato-Biliary Cancers

Julia Chaves, Michael Fernandez Y Viesca and Marianna Arvanitakis *

Department of Gastroenterology, Hepatopancreatology, and Digestive Oncology, CUB Hôpital Erasme, Université Libre de Bruxelles, 1070 Brussels, Belgium; j.chavesrodriguez@hubruxelles.be (J.C.); michael.fernandez.y.viesca@erasme.ulb.ac.be (M.F.Y.V.)
* Correspondence: marianna.arvanitaki@hubruxelles.be

Simple Summary: Endoscopic modalities have a central role in the diagnosis of pancreato-biliary cancers. Endoscopic ultrasound (EUS) is crucial in the diagnosis of both solid and cystic pancreatic lesions through tissue acquisition and fluid sampling. Intraductal brushings and biopsies performed during endoscopic retrograde cholangiopancreatography (ERCP) can provide diagnosis for biliary strictures, and additionally, cholangioscopy can allow direct visualization and image-directed biopsies. Moreover, advances in molecular markers can increase diagnostic accuracy and assist in risk stratification for premalignant lesions, such as pancreatic cystic lesions. The present review focuses on recent developments in the field of endoscopic modalities for the exploration of pancreato-biliary malignant and premalignant lesions.

Abstract: Pancreatic cancer and cholangiocarcinoma are life threatening oncological conditions with poor prognosis and outcome. Pancreatic cystic lesions are considered precursors of pancreatic cancer as some of them have the potential to progress to malignancy. Therefore, accurate identification and classification of these lesions is important to prevent the development of invasive cancer. In the biliary tract, the accurate characterization of biliary strictures is essential for providing appropriate management and avoiding unnecessary surgery. Techniques have been developed to improve the diagnosis, risk stratification, and management of pancreato-biliary lesions. Endoscopic ultrasound (EUS) and associated techniques, such as elastography, contrasted-enhanced EUS, and EUS-guided needle confocal laser endomicroscopy, may improve diagnostic accuracy. In addition, intraductal techniques applied during endoscopic retrograde cholangiopancreatography (ERCP), such as new generation cholangioscopy and in vivo cellular evaluation through probe-based confocal laser endomicroscopy, can increase the diagnostic yield in characterizing indeterminate biliary strictures. Both EUS-guided and intraductal approaches can provide the possibility for tissue sampling with new tools, such as needles, biopsies forceps, and brushes. At the molecular level, novel biomarkers have been explored that provide new insights into diagnosis, risk stratification, and management of these lesions.

Keywords: endoscopic ultrasound; tissue sampling; intraductal biopsies; pancreatic cystic lesions; cyst fluid analysis

1. Introduction

Pancreatic adenocarcinoma (PADC) and cholangiocarcinoma (CC) are life threatening oncological conditions with poor prognosis and outcome. These cancers are frequently diagnosed at a later, inoperable, stage and have low 5-year survival rates of 11.5% and 20.8%, respectively [1–3]. PADC is the tenth most common cancer but has the third highest mortality in the United States [3]. Survival depends on the stage of cancer, illustrated by 5-year survival rates for pancreatic cancer which span from 43.9% in cases of resectable disease, to 14.7% in cases of locally advanced disease, and 3.1% in cases of metastatic disease [3]. Unfortunately, only 20% of patients are eligible for surgical resection at the time of diagnosis, which underlines the importance of early diagnosis [4].

Obstacles to timely diagnosis include a potentially indolent clinical presentation, inaccurate serum biomarkers, and low specificity and sensitivity of cross-sectional imaging techniques to detect these lesions at an early stage [4,5]. In addition, CC not only has a poor prognosis but can also be very difficult to palliate with optimal biliary drainage, therefore impeding proper oncological management [6].

Pancreatic cystic lesions (PCLs) are mostly benign entities, but some types have a potential for malignant transformation, making characterization and stratification of these lesions crucial to offer appropriate management and surveillance [7,8]. As PCLs are being increasingly detected via cross-sectional imaging techniques performed even in patients without symptoms, evidence-based recommendations are more important than ever for the clinician [9].

Both endoscopic ultrasound (EUS)- and endoscopic retrograde cholangiopancreatography (ERCP)-based techniques can offer substantial information that can be used for determining the diagnosis of both solid lesions and PCLs, as well as undetermined biliary and/or pancreatic strictures. Additionally, a multidisciplinary approach is required to offer appropriate management and avoid misdiagnosis and unnecessary surgical resections that are potentially related to morbidity and mortality [10–12].

The scope of this review is to focus on recent advances in the endoscopic diagnosis of malignant and premalignant pancreato-biliary lesions. This includes potential application of EUS-related modalities, as well as ERCP with intraductal visualization and assessment. Advances in tissue acquisition, both EUS-guided and that obtained during ERCP, are also explored. Finally, developments in the molecular field with new biomarkers and next generation sequencing (NGS) are also discussed.

2. EUS Techniques

EUS provides pancreato-biliary imaging that is complementary to cross-sectional imaging for both solid lesions and PCLs. It has been proven that computed tomography (CT) and magnetic resonance imaging (MRI) have less sensitivity in detecting smaller pancreatic solid lesions measuring less than 2 cm compared to EUS [13–15]. Novel techniques have been developed in the EUS field to improve diagnostic accuracy, such as elastography and contrast-enhanced EUS (CE-EUS). Furthermore, EUS offers the possibility of acquiring tissue or fluid, which is the cornerstone for decision-making regarding management. New needle designs and tissue-acquisition modalities have been developed that have improved tissue specimen quality and diagnostic yield.

2.1. Contrast-Enhanced EUS

CE-EUS is a complementary technique to the traditional B-mode EUS imaging that involves a contrast agent which creates microbubbles in the target tissue area once injected intravenously in order to assess local micro-vascularization [16]. The main parameters evaluated are type of enhancement (hyper-, hypo- or non-enhanced), contrast distribution (heterogeneous or homogenous), and speed of wash-out. CE-EUS can allow for better evaluation of a solid lesion or a cystic lesion with a suspected solid component [17].

Regarding solid lesions, CE-EUS can play an important role in differentiating PDAC from other types of lesions, such as neuroendocrine tumors (pNETs) or inflammatory lesions, as seen in patients with chronic pancreatitis and autoimmune pancreatitis [17]. PDAC appears as a hypo-enhanced, homogenous, or non-homogenous lesion, with a fast wash-out, while pNETs appear hyper-enhanced with a slow wash-out, and inflammatory masses present as hyper- or iso-enhanced [17–19]. In a recent meta-analysis, the pooled sensitivity, specificity, and diagnostic odds ratio of CE-EUS for the differential diagnosis of PDAC were 0.91 (95% confidence interval (CI), 0.89–0.93), 0.86 (95% CI, 0.83–0.89), and 69.50 (95% CI, 48.89–98.80), respectively [18]. Although CE-EUS does not seem to have a better diagnostic yield than tissue acquisition, it may help if cytology is inconclusive [20]. Furthermore, CE-EUS can help guide tissue acquisition by targeting the most suspicious component of the lesion and avoiding necrotic areas, biopsies of which may yield inconclusive results [16,17,20].

Regarding PCLs, the cystic wall, septae, and mural nodules are assessed for vascularization with CE-EUS. Cystic pNETs, mucinous cystic neoplasms (MCNs), and intraductal papillary mucinous neoplasms (IPMNs) present with a hyper-enhanced wall, whereas pseudocysts have an avascular wall [16,17]. Furthermore, the mural nodules encountered in MCNs or IPMNs at risk of becoming malignant appear as hyper/isoechoic without a hyperechoic rim, whereas mucus or debris are not enhanced [21,22]. Based on recent recommendations, the presence of an enhancing mural nodule over 5 mm is an indication for surgical resection if the patient is deemed fit for surgery. Therefore, CE-EUS can offer crucial input in this setting [9] (Figure 1).

2.2. EUS Elastography

Elastography evaluates tissue stiffness by measuring its elasticity [23]. The compression of a target tissue via the EUS probe produces a displacement of the tissue called "strain", which correlates with the hardness of the structure and may differentiate between benign lesions (soft tissue) and malignant lesions (hard tissue) [16,23]. Additionally, it can be used to guide the biopsy to the optimal area of the lesion to increase diagnostic accuracy [23].

Qualitative and quantitative methods of measurement have been described [24]. Qualitative differentiation is based on a color distinction in which green, blue and red represent normal, hard and soft pancreatic tissue stiffness, respectively. Nevertheless, this measurement is highly operator-dependent and subjective. A quantitative measure, called the strain ratio, is an objective method of stiffness comparison between the target area and a reference area in a grayscale image [24]. Finally, the strain histogram is a computer-enhanced method for dynamic analysis, where color images are transformed into a grayscale of 256 tones. These two aforementioned quantitative measurements allow a more objective assessment. Interestingly, a meta-analysis did not show any difference in diagnostic accuracy between qualitative and quantitative evaluations, with a pooled sensitivity/specificity of 98%/63% and 95%/61%, respectively [24].

EUS techniques can be combined, and it has been reported that EUS-elastography and contrast-enhanced EUS together can improve the accuracy of the diagnosis [19].

2.3. EUS-Guided Tissue Acquisition

Despite all the aforementioned advances, the final diagnosis is still based on histopathological sampling. EUS fine-needle aspiration (EUS-FNA) was initially developed to provide tissue for cytological analysis [25–27]. On the other hand, fine-needle biopsy (FNB) provides a larger segment of tissue allowing assessment of the architecture and subsequent histological analysis. This is due to an adapted needle tip design that allows more tissue to be sampled and preserves the architectural structure [27] (Table 1).

Rapid onsite cytopathological evaluation (ROSE) consists of the preparation of cytology slides, staining, and assessment of sample adequacy by a pathologist onsite and directly in the procedure room [25]. Macroscopic on-site evaluation (MOSE) consists of the direct macroscopic evaluation of the core tissue obtained from EUS-FNB by the operator [26].

Overall, the diagnostic yield does not differ between FNA and FNB needles, but it seems that FNB needles provide higher sample adequacy [28–31]. A recent meta-analysis suggested the non-superiority of 22G FNB needles over 22G FNA, with the only advantage being a similar diagnostic yield as FNA but with fewer passes [28]. Regarding FNB needle tip design, a recent review and network meta-analysis showed that Franseen and Fork-tip needles (new generation FNB needles) significantly outperformed the older reverse-bevel FNB needles [30]. Moreover, a multicenter randomized controlled trial confirmed the noninferiority of EUS-FNB without ROSE compared to FNB with ROSE in solid pancreatic lesions when new-generation FNB needles are used, thus highlighting the benefit of the use of FNB needles when the pathologist is not available [31]. Finally, MOSE has an overall diagnostic yield of 90%, sensitivity of 86.5%, and a specificity of 100% for solid pancreatic lesions and may represent a valid alternative when ROSE is not feasible [26].

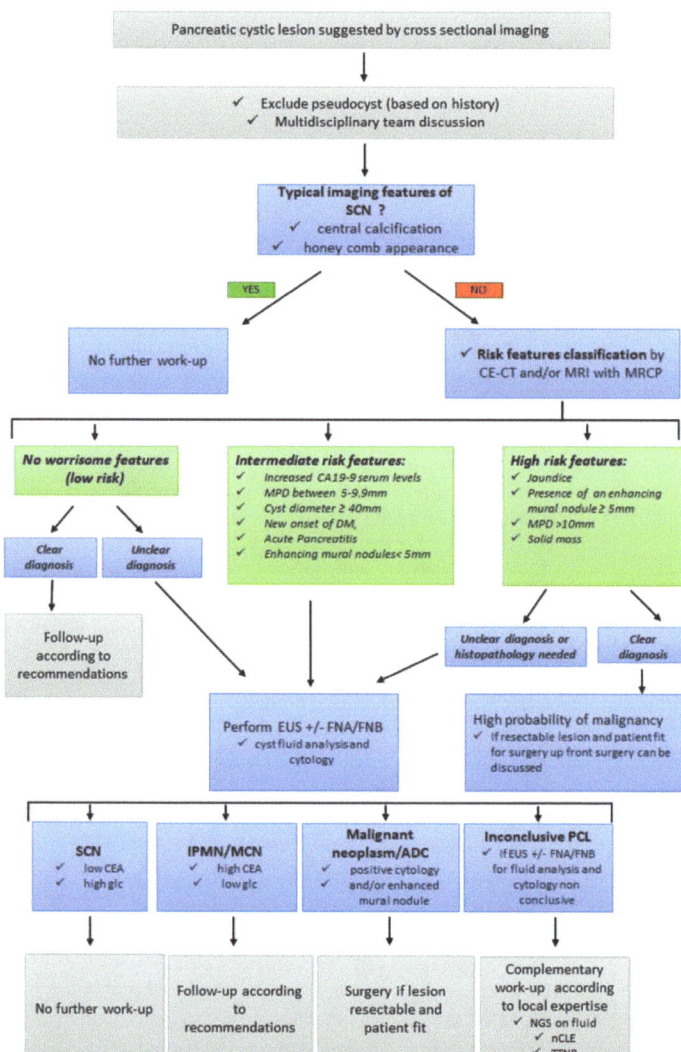

Figure 1. Algorithm for the assessment and management of PCLs. Abbreviations: ADC: adenocarcinoma; CA19-9: carbohydrate antigen 19-9; CE-EUS: contrast-enhanced endoscopic ultrasound; DM: diabetes mellitus; EUS-FNA: EUS fine-needle aspiration; EUS-FNB: EUS fine-needle biopsy; Glc: glucose; IPMN: intraductal papillary mucinous *neoplasm*; MCN: mucinous cystic neoplasm; MPD: main pancreatic duct; PCLs: pancreatic cystic lesions; SCN: serous cystic neoplasm; nCLE: needle-based confocal laser endomicroscopy; TTNB: through-the-needle microforceps biopsy.

Different sampling techniques applied during EUS-guided tissue acquisition have been described [32,33]. The classic technique to obtain tissue sampling is the fanning technique (during a single-needle pass, the endoscopist targets different areas to biopsy). A randomized controlled trial compared EUS-FNB using the fanning technique to CE-EUS-guided FNB and revealed similar rates of diagnostic accuracy for solid pancreatic lesions [32]. A recent network meta-analysis showed that the application of suction (specif ically wet suction involving saline infusion through the needle) seemed to provide high rates of adequate samples, although with high blood contamination, compared to "no suction" [33]. Adverse events related to EUS-guided tissue acquisition are rare, but may include acute pancreatitis, infection, perforation, and bleeding, with rates estimated to be 0.5–3% of cases [34].

Table 1. Characteristics of EUS-FNA and EUS-FNB needles.

	Needle Type	Characteristics
FNA	Conventional needles (19G, 22G, 25G)	End-cutting needle. Sharply pointed tip to facilitate puncture.
	Menghini-tip needle	End-cutting needle. Tapered bevel edge that facilitates the tissue being withdrawn into the lumen.
FNB	Franseen needle (22G, 25G) *	End-cutting needle. Crown tip with three-plane symmetric cutting edges. No side-slot.
	Reverse-bevel needle (19G, 22G, 25G)	Modified Menghini-type needle with a beveled side-slot near the needle tip. Tissue collected during retrograde movement of the needle.
	Fork-type needle (19G, 22G, 25G) *	End-cutting needle. Fork-shaped distal tip including six cutting edges and an opposing bevel. No side-slot.
	Antegrade core trap (20G)	Modified Menghini-type needle with a beveled side-slot near the needle tip. Tissue collected during antegrade movement of the needle.

Abbreviations: EUS: endoscopic ultrasound; FNA: fine-needle aspiration; FNB: fine-needle biopsy. * new generation FNB needles.

2.4. Fluid Analysis for PCLs

EUS-FNA for fluid aspiration and analysis plays an essential role in determining the type of PCL, and, in particular, whether it has a mucinous component, and therefore malignant potential, in cases of non-contributive cross-sectional imaging [35]. High levels of carcinoembryonic antigen (CEA) and mucin staining are consistent with a mucinous PCL, such as MCN or IPMN [35]. Intracystic glucose measurement, which is easily available and inexpensive, has been studied as an additional diagnostic tool. A recent multicenter study in 93 patients showed that a glucose concentration of ≤ 25 mg/dL had a sensitivity and specificity of 88.1% and 91.2%, respectively, for differentiating mucinous PCLs, whereas a CEA concentration of ≥ 192 ng/mL had a sensitivity of 62.7% and a specificity of 88.2% [36]. Furthermore, cyst wall sampling using EUS-FNA may also increase the diagnostic yield [35]. Cuboidal epithelial cells, clear cytoplasm, and excess glycogen can diagnose serous PCLs. The presence of mucin, ovarian-like stroma with a degree of cell atypia, is mainly found with mucinous PCLs, such as MCN (Figure 1, Table 2).

Finally, there is an immediate on-site method to improve the diagnostic accuracy of PCL fluid analysis called the string test. A drop of cystic fluid is placed between the examiner's fingers and then it is stretched, and the maximal length of mucus is measured. It is considered positive if ≥ 1 cm string is formed and lasts for ≥ 1 s [37]. The string test has been shown to have a high positive predictive value for correctly diagnosing mucinous PCLs [37].

The analysis of mutations from fluid-containing DNA is increasingly applied in clinical practice. KRAS and GNAS mutations have a good accuracy for the diagnosis of IPMNs and MCNs, based on a recent meta-analysis [38]. Finally, a recent multi-center prospective study showed that NGS of PCL fluid has a high sensitivity and specificity for differentiating between cystic lesions and advanced neoplasia or pNETS [39]. Combining different markers, such as MAPK/GNAS and P53/SMAD4/CTNNB1/mTOR, increased the sensitivity to 89% and specificity to 98% for the diagnosis of advanced neoplasia [39].

Table 2. Different predictors in cyst fluid analysis of PCLs.

	Cyst Fluid Analysis	Mucinous PCLs	Serous/Non Mucinous PCLs	Advanced Neoplasia (Predictors of Degenerate PCLs)
Biomarkers	Intracystic glucose	↘	↗	↘
	CEA	↗	↘	↗
Cytology	Mucin staining	↗	/	+
	Cuboidal epithelial cells	/	+	/
	Clear cytoplasm	/	+	/
	Excess glycogen	/	+	/
NGS	KRAS mutation	+	/	+
	GNAS mutation	+	/	+
	MAPK/GNAS	/	/	+
	P53/SMAD4/CTNNB1/mTOR	/	/	+

Abbreviations: CEA: carcino embryonic antigen; PCLs: pancreatic cystic lesions; NGS: next generation sequencing. Arrow up: increase, Arrow down; decrease, +: presence, /: absence

2.5. EUS-Guided Needle Confocal Laser Endomicroscopy

Needle-based confocal laser endomicroscopy (nCLE) is a novel technique that uses EUS to guide a thin CLE probe through a 19-gauge EUS needle, allowing evaluation of the inner walls of PCLs [8], real-time imaging of intracystic epithelium within a single plane and in vivo pathological analysis [22].

In a recent prospective observational study, it was shown that the addition of EUS nCLE to EUS-FNA improved the specificity, sensitivity, and diagnostic accuracy for PCLs [40] as well as an increased diagnostic yield compared to EUS-FNA alone [40].

Typical nCLE features include papillary projections in IPMNs and the superficial vascular network in serous cystic neoplasm (SCN) [40]. Moreover, a recent study identified two criteria related to dysplasia and malignant degeneration: papillary epithelial thickness and darkness [41]. These worrisome features can help in risk stratification for IPMNs [41].

2.6. Through-the-Needle Microforceps Biopsy (TTNB)

Although the main diagnostic tool for characterization of PCLs is cyst fluid analysis cytology of the liquid and the wall can be also be obtained with EUS-FNA but, unfortunately, this is not sufficient for molecular testing and diagnostic yields remain low [38]. Through the-needle microforceps biopsy (TTNB) allows biopsies to be obtained from the cyst wall with the aid of a miniforceps that is passed through a 19-Gauge EUS needle under EUS guidance [42]. A meta-analysis and systematic review of TTNB including 11 studies [43] demonstrated that TTNB is a superior diagnostic technique compared to FNA for EUS guided sampling of PCL walls. The most common adverse events (AE) included post procedural acute pancreatitis (AP) and mild intracystic bleeding. In a prospective study [44] the feasibility of molecular analysis by NGS via TTNB was assessed in 101 patients. The authors demonstrated that TTNB was superior to cyst fluid analysis for differentiating between mucinous and non-mucinous PCLs with a higher sensitivity and specificity, albeit with a 10% AE rate. In addition to the beneficial diagnostic yield of this technique, the rate and severity of AE are not negligible. Indeed, in another prospective open-label controlled study on 101 patients, Kovacevic et al. [45] reported an AE rate of 9.9%, the majority of which was AP. Among these complications, four were considered severe, and one was fatal. More recently, Facciorusso et al. [46] attempted to identify the risk factors for AE in a retrospective study of 506 patients. The AE rate was 11%, including three patients with AP requiring ICU hospitalization and one patient undergoing surgical necrosectomy.

Four independent risk factors were highlighted: the type of cysts (IPMN), the number of passages, the complete aspiration of the cyst and the age (>64 years).

Therefore, this technique must only be selected when the benefit of accurate diagnosis outweighs the potential AE, especially since there has been recent development on the identification of molecular markers in the cyst fluid [39,47].

3. Intraductal Biliopancreatic Techniques

Biliary strictures are classified as distal when the common bile duct is involved, and proximal when located at the level of the hepatic hilum and intrahepatic ducts [48,49]. Common causes of malignant distal biliary stricture are PDAC followed by CC and ampullary cancer [48,49]. Proximal malignant strictures are mostly related to CC, hepatocarcinoma, gallbladder cancer, and compression due to metastatic lymph nodes [49]. Distal strictures related to PDAC with a mass can be explored with EUS-guided tissue acquisition; however, proximal strictures with no clear mass, as frequently encountered in patients with CC, are more challenging [49].

Biliary strictures are considered as indeterminate when the diagnosis is unclear after cross-sectional imaging and ERCP with biliary sampling. Determining diagnosis is crucial to avoid unnecessary high-risk surgeries as well as a progression to an advanced stage cancer [50,51] (Figure 2).

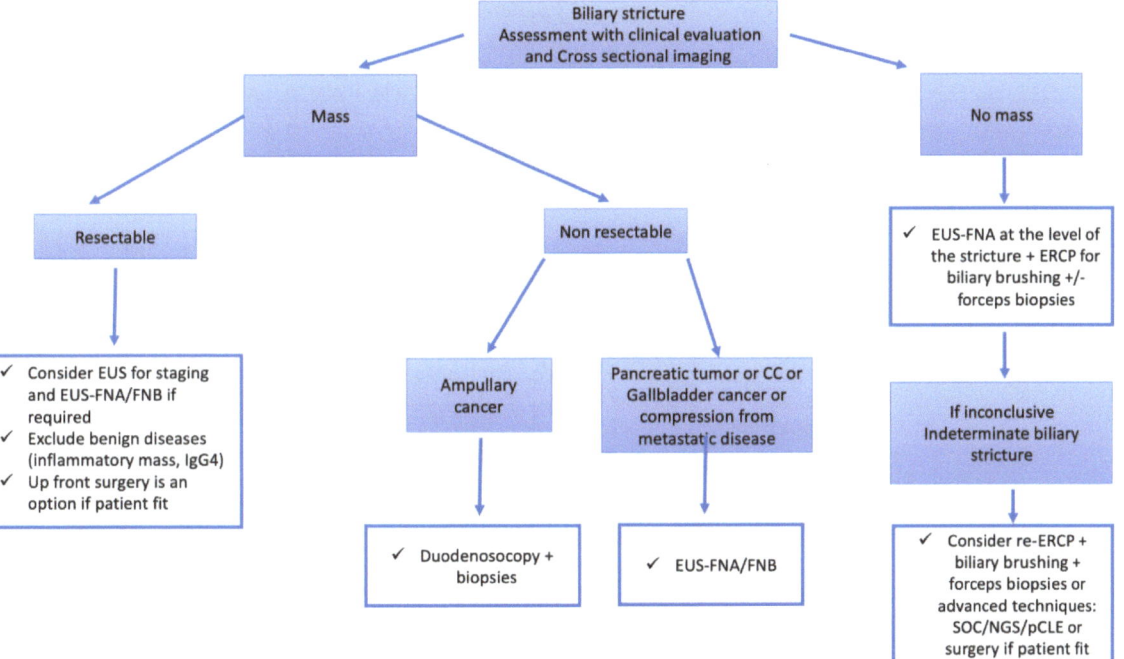

Figure 2. Algorithm for the assessment of biliary strictures. Abbreviations: CC: Cholangiocarcinoma; pCLE: probe-based confocal laser endomicroscopy; ERCP: endoscopic retrograde cholangiopancreatography; EUS: endoscopic ultrasound; FNA: fine-needle aspiration; FNB: fine-needle biopsy; NGS: next generation sequencing; SOC: single-operator cholangioscopy.

3.1. Intraductal Tissue Acquisition

During ERCP, it is possible to obtain tissue from strictures, under fluoroscopy guidance, via brush cytology and forceps biopsy [50,51]. The yield for brushings varies from 40% to 80%, and can be increased when combined with forceps biopsy [50]. A recent randomized trial confirmed that EUS-FNA had superior accuracy compared to combined brush cytology and forceps biopsy (94% vs. 62%, $p = 0.003$) in cases of extraductal lesions larger than 1.5 cm, but accuracy was similar when considering intraductal lesions less than 1.5 cm [52]. Immunohistochemical staining is a widely available method for investigating specific tumorigenesis-related protein expression patterns in brush and biopsy samples [53,54].

An endoscopic scraper has been developed with a wire-guided system and three scraping loops to obtain tissue and cell samples for histology and cytology. A recently published study including 435 patients with biliary strictures showed that the diagnostic performance of the endoscopic scraper combined with the cell block is better than brush cytology alone or brush with cell block [55]. Nevertheless, sensitivity does not exceed 53%, highlighting the need for complementary investigations.

Malignant biliary strictures lead to chromosomal alterations which can be detected using specific techniques, such as fluorescence in situ hybridization (FISH) and NGS [54]. FISH uses fluorescently labeled complementary DNA probes that allow detection of aneuploidy of chromosomes in biliary brushings or biopsies in order to distinguish between CC and benign bile duct strictures [54,56,57]. Addition of FISH and mutational analysis can increase the level of sensitivity from 32% to 73% for the detection of malignancy, and reach 100% specificity [56]. Nevertheless, overall sensitivities obtained via this technique vary from 31% to 88% according to studies, with a better yield for primary sclerosing cholangitis (PSC)-related strictures and CC [54,56,57].

NGS can detect chromosomal mutations, even in small amounts of tissue or fluid. In a recent study including 252 patients and 346 biliary specimens, the authors identified mutations in a considerable number of biliary brushings and biopsies by performing targeted NGS with a large gene panel [57]. The most prevalent genomic alterations consisted of mutations in KRAS, TP53, CDKN2A, SMAD4, PIK3CA, and GNAS. NGS increased the sensitivity to 83% and maintained a specificity of 99%. The increase in the diagnostic yield was particularly observed for patients with PSC [57].

3.2. Cholangioscopy

The development of the single-use, single-operator cholangioscopy device (SOC) that allows direct visualization of the bile tract and targeted biopsies has replaced the previous "mother-baby" peroral cholangioscopy (POCS) system, a device which required two operators and had significant fragility [58].

The first-generation SOC was a fiberoptic device that was replaced by the digital version with improved high-resolution imaging, dedicated aspiration and irrigation channels, and an operating channel that allows the passage of a microforceps to perform biopsies [59,60].

A recent meta-analysis including 13 studies and 876 patients reported an overall sensitivity and specificity of 88% and 95%, respectively [60]. Subgroup analysis showed that SOC image impression provided higher sensitivity but lower specificity than SOC-guided tissue diagnosis with the forceps biopsy.

Direct cholangioscopy can also be applied with ultra-thin endoscopes through direct insertion in the bile ducts. Although this is a challenging procedure, this system may offer the potential of digital chromoendoscopy, like narrow-band imaging, which may increase visualization quality and differentiation of surface structures and architecture [61].

Concerning adverse events, SOC has higher rates of cholangitis related to the need for intraductal perfusion, therefore, the use of prophylactic antibiotics during the procedure is required [58].

3.3. Probe-Based Confocal Laser Endomicroscopy

Probe-based confocal laser endomicroscopy (pCLE), also known as optical biopsy, is an endoscopic technique that provides real-time magnification of $1000\times$ microscopic tissue information to diagnose indeterminate biliary strictures [62].

A recent meta-analysis including 18 studies showed that pCLE had a higher sensitivity but lower specificity than tissue sampling during ERCP for the diagnosis of indeterminate biliary strictures [62]. Nevertheless, correct interpretation of the real-time microscopic images can be challenging. Consequently, classification systems have been developed to differentiate between the patterns [63]. A recent prospective study focused on patients with primary sclerosing cholangitis reported high sensitivity for diagnosis of CC, especially at the level of the bifurcation; nevertheless, technical aspects of the probe may limit evaluation of the common bile duct [64]. Major limitations of generalizing the use of pCLE include availability, cost, and lack of expertise.

3.4. Pancreatoscopy

Compared to cholangioscopy, peroral pancreatoscopy (POPS) allows direct visualization of the main pancreatic duct (MPD) and tissue acquisition under visual control [65]. Access to the MPD occurs through the major papilla, with or without sphincterotomy, depending on the diameter of the pancreatic orifice and the indication (example, fish mouth encountered in patients with main duct IPMN) [66]. The pancreatoscope is advanced in the duct on a guidewire under regular irrigation and fluoroscopy [65,66]. Therefore, prophylactic antibiotics are recommended in patients undergoing pancreatoscopy because there is a risk of bacterial translocation by irrigating the pancreatic duct with saline solution [65,66]. The two recognized indications are the diagnostic assessment of IPMN (diagnosis, localization, and extension of the disease before surgery) and secondarily for the evaluation of indeterminate pancreatic strictures to discriminate between benign or malignant etiology [65]. Hara et al. first classified pancreatoscopy findings according to the pit-pattern to differentiate benign and malignant aspect with a good accuracy (88% in main duct IPMN and 66% for branch duct IPMN) [67]. A recent meta-analysis (25 studies) showed an excellent diagnostic yield in the diagnostic work-up of IPMN (88–100%). The disease extent of IPMN changed the surgery in 13–62% of the patients. The reported AE event rate was 12%, majority of which was acute pancreatitis (most mild and moderate) [68]. Future studies are needed to better define the role of POP in the diagnostic work-up of IPMN. There are few data regarding the assessment of indeterminate pancreatic strictures when conventional imaging techniques are sometimes insufficient to distinguish between benign and malignant strictures, particularly in patients with chronic pancreatitis [66]. The classification between benign and malignant can be challenging, especially when there is no associated lesion. El Hajj et al. [69] highlighted the role of pancreatoscopy to evaluate pancreatic duct strictures and ductal dilation with different lesions as adenocarcinoma, main and branch duct IPMN and inflammatory strictures. The overall accuracy of visual assessment via POP was 87%, which increased to 94% when pancreatoscopy-guided tissue acquisition was performed [69]. Therefore, current data suggest that in selected patients, pancreatoscopy may play an essential role in characterizing indeterminate pancreatic duct strictures and mapping IPMN before surgery. Nevertheless, pancreatoscopy should be reserved for specific groups of patients due to a narrow range of advantages that this technique allows and also because they can be performed only at expert centers.

4. Conclusions and Future Directions

The detection of pancreato-biliary lesions is rising due to an increased use of cross-sectional imaging, even in patients without symptoms. Management of these lesions is crucial due to the potential for malignant transformation. Misdiagnosis can lead to development of advanced neoplasia or unnecessary surgery. Advances have been made in the field of EUS, ERCP, cholangioscopy, as well as in biochemical and molecular detection, to improve diagnosis, risk stratification, and management of these lesions. However, there is a need for prospective, multicenter studies to provide evidence and establish standard guidelines for diagnosis and overall management.

Funding: This research received no external funding.

Data Availability Statement: Not applicable.

Conflicts of Interest: The authors have no conflict of interest in relation to the present work.

Abbreviations

AP	Acute pancreatitis
AE	Adverse event
CEA	Carcinoembryonic antigen
CC	Cholangiocarcinoma
CT	Computed tomography
CI	Confidence interval
CE-EUS	Contrast-enhanced EUS
ERCP	Endoscopic retrograde cholangiopancreatography
EUS	Endoscopic ultrasound
EUS-FNA	EUS fine-needle aspiration
EUS-FNB	EUS fine-needle biopsy
FISH	Fluorescence in situ hybridization
IPMN	Intraductal papillary mucinous neoplasms
MCN	Mucinous cystic neoplasms
MOSE	Macroscopic on-site evaluation
MPD	Main pancreatic duct
MRI	Magnetic resonance imaging
nCLE	Needle-based confocal laser endomicroscopy
NGS	Next generation sequencing
PADC	Pancreatic adenocarcinoma
PCLs	Pancreatic cystic lesions
pNET	Pancreatic neuroendocrine tumors
POCS	Peroral cholangioscopy
POPS	Peroral pancreatoscopy
PSC	Primary sclerosing cholangitis
pCLE	Probe-based confocal laser endomicroscopy
ROSE	Rapid onsite cytopathological evaluation
SOC	Single-operator cholangioscopy device
TTNB	Through-the-needle microforceps biopsy

References

1. Wood, L.D.; Canto, M.I.; Jaffee, E.M.; Simeone, D.M. Pancreatic Cancer: Pathogenesis, Screening, Diagnosis, and Treatment *Gastroenterology* **2022**, *163*, 386–402.e1. [CrossRef]
2. Izquierdo-Sanchez, L.; Lamarca, A.; La Casta, A.; Buettner, S.; Utpatel, K.; Klümpen, H.-J.; Adeva, J.; Vogel, A.; Lleo, A.; Fabris, L.; et al. Cholangiocarcinoma landscape in Europe: Diagnostic, prognostic and therapeutic insights from the ENSCCA Registry. *J Hepatol.* **2022**, *76*, 1109–1121. [CrossRef] [PubMed]
3. Cancer of the Pancreas—Cancer Stat Facts. Available online: https://seer.cancer.gov/statfacts/html/pancreas.html (accessed on 29 March 2023).
4. Salom, F.; Prat, F. Current role of endoscopic ultrasound in the diagnosis and management of pancreatic cancer. *World J Gastrointest. Endosc.* **2022**, *14*, 35–48. [CrossRef] [PubMed]
5. Goggins, M.; Overbeek, K.A.; Brand, R.; Syngal, S.; Del Chiaro, M.; Bartsch, D.K.; Bassi, C.; Carrato, A.; Farrell, J.; Fishman, E.K.; et al. Management of patients with increased risk for familial pancreatic cancer: Updated recommendations from the International Cancer of the Pancreas Screening (CAPS) Consortium. *Gut* **2020**, *69*, 7–17. [CrossRef]
6. Kato, H.; Matsumoto, K.; Okada, H. Recent advances regarding endoscopic biliary drainage for unresectable malignant hilar biliary obstruction. *DEN Open* **2022**, *2*, e33. [CrossRef]
7. Nista, E.C.; Schepis, T.; Candelli, M.; Giuli, L.; Pignataro, G.; Franceschi, F.; Gasbarrini, A.; Ojetti, V. Humoral Predictors of Malignancy in IPMN: A Review of the Literature. *Int. J. Mol. Sci.* **2021**, *22*, 12839. [CrossRef]
8. Elta, G.H.; Enestvedt, B.K.; Sauer, B.G.; Lennon, A.M. ACG Clinical Guideline: Diagnosis and Management of Pancreatic Cysts *Am. J. Gastroenterol.* **2018**, *113*, 464–479. [CrossRef] [PubMed]
9. The European Study Group on Cystic Tumours of the Pancreas. European evidence-based guidelines on pancreatic cystic neoplasms. *Gut* **2018**, *67*, 789–804. [CrossRef]
10. Siddiqui, A.A.; Kowalski, T.E.; Shahid, H.; O'Donnell, S.; Tolin, J.; Loren, D.E.; Infantolino, A.; Hong, S.-K.; Eloubeidi, M.A. False-positive EUS-guided FNA cytology for solid pancreatic lesions. *Gastrointest. Endosc.* **2011**, *74*, 535–540. [CrossRef]
11. Ohno, E.; Ishikawa, T.; Mizutani, Y.; Iida, T.; Uetsuki, K.; Yashika, J.; Yamada, K.; Gibo, N.; Aoki, T.; Kawashima, H. Factors associated with misdiagnosis of preoperative endoscopic ultrasound in patients with pancreatic cystic neoplasms undergoing surgical resection. *J. Med. Ultrason.* **2022**, *49*, 433–441. [CrossRef]

12. Okasha, H.H.; Abdellatef, A.; Elkholy, S.; Mogawer, M.-S.; Yosry, A.; Elserafy, M.; Medhat, E.; Khalaf, H.; Fouad, M.; Elbaz, T.; et al. Role of endoscopic ultrasound and cyst fluid tumor markers in diagnosis of pancreatic cystic lesions. *World J. Gastrointest. Endosc.* **2022**, *14*, 402–415. [CrossRef] [PubMed]
13. Matthaei, H.; Schulick, R.D.; Hruban, R.H.; Maitra, A. Cystic precursors to invasive pancreatic cancer. *Nat. Rev. Gastroenterol. Hepatol.* **2011**, *8*, 141–150. [CrossRef] [PubMed]
14. Yamashita, Y.; Shimokawa, T.; Napoléon, B.; Fusaroli, P.; Gincul, R.; Kudo, M.; Kitano, M. Value of contrast-enhanced harmonic endoscopic ultrasonography with enhancement pattern for diagnosis of pancreatic cancer: A meta-analysis. *Dig. Endosc.* **2019**, *31*, 125–133. [CrossRef]
15. Du, C.; Chai, N.-L.; Linghu, E.-Q.; Li, H.-K.; Sun, L.-H.; Jiang, L.; Wang, X.-D.; Tang, P.; Yang, J. Comparison of endoscopic ultrasound, computed tomography and magnetic resonance imaging in assessment of detailed structures of pancreatic cystic neoplasms. *World J. Gastroenterol.* **2017**, *23*, 3184. [CrossRef] [PubMed]
16. Seicean, A.; Mosteanu, O.; Seicean, R. Maximizing the endosonography: The role of contrast harmonics, elastography and confocal endomicroscopy. *World J. Gastroenterol.* **2017**, *23*, 25. [CrossRef]
17. Fusaroli, P.; Spada, A.; Mancino, M.G.; Caletti, G. Contrast Harmonic Echo–Endoscopic Ultrasound Improves Accuracy in Diagnosis of Solid Pancreatic Masses. *Clin. Gastroenterol. Hepatol.* **2010**, *8*, 629–634.e2. [CrossRef]
18. Mei, S.; Wang, M.; Sun, L. Contrast-Enhanced EUS for Differential Diagnosis of Pancreatic Masses: A Meta-Analysis. *Gastroenterol. Res. Pract.* **2019**, *2019*, 1670183. [CrossRef]
19. Iordache, S.; Costache, M.I.; Popescu, C.F.; Streba, C.T.; Cazacu, S.; Săftoiu, A. Clinical impact of EUS elastography followed by contrast-enhanced EUS in patients with focal pancreatic masses and negative EUS-guided FNA. *Med. Ultrason.* **2016**, *18*, 18. [CrossRef]
20. Lai, J.-H.; Lin, C.-C.; Lin, H.-H.; Chen, M.-J. Is contrast-enhanced endoscopic ultrasound-guided fine needle biopsy better than conventional fine needle biopsy? A retrospective study in a medical center. *Surg. Endosc.* **2022**, *36*, 6138–6143. [CrossRef]
21. Lisotti, A.; Napoleon, B.; Facciorusso, A.; Cominardi, A.; Crinò, S.F.; Brighi, N.; Gincul, R.; Kitano, M.; Yamashita, Y.; Marchegiani, G.; et al. Contrast-enhanced EUS for the characterization of mural nodules within pancreatic cystic neoplasms: Systematic review and meta-analysis. *Gastrointest. Endosc.* **2021**, *94*, 881–889.e5. [CrossRef] [PubMed]
22. Ardeshna, D.R.; Cao, T.; Rodgers, B.; Onongaya, C.; Jones, D.; Chen, W.; Koay, E.J.; Krishna, S.G. Recent advances in the diagnostic evaluation of pancreatic cystic lesions. *World J. Gastroenterol.* **2022**, *28*, 624–634. [CrossRef] [PubMed]
23. Dhar, J.; Samanta, J. The expanding role of endoscopic ultrasound elastography. *Clin. J. Gastroenterol.* **2022**, *15*, 841–858. [CrossRef] [PubMed]
24. Zhang, B.; Zhu, F.; Li, P.; Yu, S.; Zhao, Y.; Li, M. Endoscopic ultrasound elastography in the diagnosis of pancreatic masses: A meta-analysis. *Pancreatology* **2018**, *18*, 833–840. [CrossRef] [PubMed]
25. Polkowski, M.; Jenssen, C.; Kaye, P.; Carrara, S.; Deprez, P.; Gines, A.; Fernández-Esparrach, G.; Eisendrath, P.; Aithal, G.; Arcidiacono, P.; et al. Technical aspects of endoscopic ultrasound (EUS)-guided sampling in gastroenterology: European Society of Gastrointestinal Endoscopy (ESGE) Technical Guideline–March 2017. *Endoscopy* **2017**, *49*, 989–1006. [CrossRef]
26. Mangiavillano, B.; Frazzoni, L.; Togliani, T.; Fabbri, C.; Tarantino, I.; De Luca, L.; Staiano, T.; Binda, C.; Signoretti, M.; Eusebi, L.H.; et al. Macroscopic on-site evaluation (MOSE) of specimens from solid lesions acquired during EUS-FNB: Multicenter study and comparison between needle gauges. *Endosc. Int. Open* **2021**, *9*, E901–E906. [CrossRef]
27. Bang, J.Y.; Hebert-Magee, S.; Navaneethan, U.; Hasan, M.K.; Hawes, R.; Varadarajulu, S. EUS-guided fine needle biopsy of pancreatic masses can yield true histology. *Gut* **2018**, *67*, 2081–2084. [CrossRef]
28. Facciorusso, A.; Bajwa, H.; Menon, K.; Buccino, V.; Muscatiello, N. Comparison between 22G aspiration and 22G biopsy needles for EUS-guided sampling of pancreatic lesions: A meta-analysis. *Endosc. Ultrasound* **2019**, *9*, 167–174. [CrossRef]
29. Yang, M.J.; Kim, J.; Park, S.W.; Cho, J.H.; Kim, E.J.; Lee, Y.N.; Lee, D.W.; Park, C.H.; Lee, S.S. Comparison between three types of needles for endoscopic ultrasound-guided tissue acquisition of pancreatic solid masses: A multicenter observational study. *Sci. Rep.* **2023**, *13*, 3677. [CrossRef]
30. Gkolfakis, P.; Crinò, S.F.; Tziatzios, G.; Ramai, D.; Papaefthymiou, A.; Papanikolaou, I.S.; Triantafyllou, K.; Arvanitakis, M.; Lisotti, A.; Fusaroli, P.; et al. Comparative diagnostic performance of end-cutting fine-needle biopsy needles for EUS tissue sampling of solid pancreatic masses: A network meta-analysis. *Gastrointest. Endosc.* **2022**, *95*, 1067–1077.e15. [CrossRef]
31. Crinò, S.F.; Di Mitri, R.; Nguyen, N.Q.; Tarantino, I.; de Nucci, G.; Deprez, P.H.; Carrara, S.; Kitano, M.; Shami, V.M.; Fernández-Esparrach, G.; et al. Endoscopic Ultrasound–guided Fine-needle Biopsy With or Without Rapid On-site Evaluation for Diagnosis of Solid Pancreatic Lesions: A Randomized Controlled Non-Inferiority Trial. *Gastroenterology* **2021**, *161*, 899–909.e5. [CrossRef]
32. Kuo, Y.T.; Chu, Y.L.; Wong, W.F.; Han, M.L.; Chen, C.C.; Jan, I.S.; Cheng, W.C.; Shun, C.T.; Tsai, M.C.; Cheng, T.Y.; et al. Randomized trial of contrast-enhanced harmonic guidance versus fanning technique for EUS-guided fine-needle biopsy sampling of solid pancreatic lesions. *Gastrointest. Endosc.* **2023**, *97*, 732–740. [CrossRef]
33. Facciorusso, A.; Crinò, S.F.; Ramai, D.; Madhu, D.; Fugazza, A.; Carrara, S.; Spadaccini, M.; Mangiavillano, B.; Gkolfakis, P.; Mohan, B.P.; et al. Comparative Diagnostic Performance of Different Techniques for Endoscopic Ultrasound-Guided Fine-Needle Biopsy of Solid Pancreatic Masses: A Network Meta-analysis. *Gastrointest. Endosc.* **2023**, *97*, 839–848.e5. [CrossRef]
34. Katanuma, A.; Maguchi, H.; Yane, K.; Hashigo, S.; Kin, T.; Kaneko, M.; Kato, S.; Kato, R.; Harada, R.; Osanai, M.; et al. Factors Predictive of Adverse Events Associated with Endoscopic Ultrasound-Guided Fine Needle Aspiration of Pancreatic Solid Lesions. *Dig. Dis. Sci.* **2013**, *58*, 2093–2099. [CrossRef]

35. Okasha, H.H.; Awad, A.; El-meligui, A.; Ezzat, R.; Aboubakr, A.; AbouElenin, S.; El-Husseiny, R.; Alzamzamy, A. Cystic pancreatic lesions, the endless dilemma. *World J. Gastroenterol.* **2021**, *27*, 2664–2680. [CrossRef] [PubMed]
36. Smith, Z.L.; Satyavada, S.; Simons-Linares, R.; Mok, S.R.S.; Martinez Moreno, B.; Aparicio, J.R.; Chahal, P. Intracystic Glucose and Carcinoembryonic Antigen in Differentiating Histologically Confirmed Pancreatic Mucinous Neoplastic Cysts. *Am. J. Gastroenterol.* **2022**, *117*, 478–485. [CrossRef] [PubMed]
37. Bick, B.; Enders, F.; Levy, M.; Zhang, L.; Henry, M.; Dayyeh, B.; Chari, S.; Clain, J.; Farnell, M.; Gleeson, F.; et al. The string sign for diagnosis of mucinous pancreatic cysts. *Endoscopy* **2015**, *47*, 626–631. [CrossRef]
38. McCarty, T.R.; Paleti, S.; Rustagi, T. Molecular analysis of EUS-acquired pancreatic cyst fluid for KRAS and GNAS mutations for diagnosis of intraductal papillary mucinous neoplasia and mucinous cystic lesions: A systematic review and meta-analysis. *Gastrointest. Endosc.* **2021**, *93*, 1019–1033.e5. [CrossRef]
39. Paniccia, A.; Polanco, P.M.; Boone, B.A.; Wald, A.I.; McGrath, K.; Brand, R.E.; Khalid, A.; Kubiliun, N.; O'Broin-Lennon, A.M.; Park, W.G.; et al. Prospective, Multi-Institutional, Real-Time Next-Generation Sequencing of Pancreatic Cyst Fluid Reveals Diverse Genomic Alterations That Improve the Clinical Management of Pancreatic Cysts. *Gastroenterology* **2023**, *164*, 117–133.e7. [CrossRef] [PubMed]
40. Bertani, H.; Pezzilli, R.; Pigò, F.; Bruno, M.; De Angelis, C.; Manfredi, G.; Delconte, G.; Conigliaro, R.; Buscarini, E. Needle-based confocal endomicroscopy in the discrimination of mucinous from non-mucinous pancreatic cystic lesions. *World J. Gastrointest. Endosc.* **2021**, *13*, 555–564. [CrossRef] [PubMed]
41. Machicado, J.D.; Chao, W.-L.; Carlyn, D.E.; Pan, T.-Y.; Poland, S.; Alexander, V.L.; Maloof, T.G.; Dubay, K.; Ueltschi, O.; Middendorf, D.M.; et al. High performance in risk stratification of intraductal papillary mucinous neoplasms by confocal laser endomicroscopy image analysis with convolutional neural networks (with video). *Gastrointest. Endosc.* **2021**, *94*, 78–87.e2. [CrossRef]
42. Robles-Medranda, C.; Olmos, J.I.; Puga-Tejada, M.; Oleas, R.; Baquerizo-Burgos, J.; Arevalo-Mora, M.; Del Valle Zavala, R.; Nebel, J.A.; Calle Loffredo, D.; Pitanga-Lukashok, H. Endoscopic ultrasound-guided through-the-needle microforceps biopsy and needle-based confocal laser-endomicroscopy increase detection of potentially malignant pancreatic cystic lesions: A single-center study. *World J. Gastrointest. Endosc.* **2022**, *14*, 129–141. [CrossRef] [PubMed]
43. McCarty, T.; Rustagi, T. Endoscopic ultrasound-guided through-the-needle microforceps biopsy improves diagnostic yield for pancreatic cystic lesions: A systematic review and meta-analysis. *Endosc. Int. Open* **2020**, *8*, E1280–E1290. [CrossRef] [PubMed]
44. Rift, C.V.; Melchior, L.C.; Kovacevic, B.; Klausen, P.; Toxværd, A.; Grossjohann, H.; Karstensen, J.G.; Brink, L.; Hassan, H.; Kalaitzakis, E.; et al. Targeted next-generation sequencing of EUS-guided through-the-needle-biopsy sampling from pancreatic cystic lesions. *Gastrointest. Endosc.* **2023**, *97*, 50–58.e4. [CrossRef]
45. Kovacevic, B.; Klausen, P.; Rift, C.V.; Toxvaerd, A.; Grossjohann, H.; Karstensen, J.G.; Brink, L.; Hassan, H.; Kalaitzakis, E.; Storkolm, J.; et al. Clinical Impact of endoscopic ultrasounf-guided through-thhe-needle microbiopsy in patients with papancreatic cysts. *Endoscopy* **2021**, *53*, 44–52. [CrossRef]
46. Facciorusso, A.; Kovacevic, B.; Yang, D.; Vilas-Boas, F.; Martinez-Moreno, B.; Stigliano, S.; Rizzatti, G.; Sacco, M.; Arevalo-Mora, M.; Villareal-Sanchez, L.; et al. Predictors of adverse events after endoscopic ultrasound-guided through-the-needle biopsy of pancreaticysts: A recursive partitioning analysis. *Endoscopy* **2022**, *54*, 1158–1168. [CrossRef]
47. Turner, R.C.; Melnychuk, J.T.; Chen, W.; Jones, D.; Krishna, S.G. Molecular Analysis of Pancreatic Cyst Fluid for the Management of Intraductal Papillary Mucinous Neoplasms. *Diagnostics* **2022**, *12*, 2573. [CrossRef]
48. Dumonceau, J.-M.; Tringali, A.; Papanikolaou, I.; Blero, D.; Mangiavillano, B.; Schmidt, A.; Vanbiervliet, G.; Costamagna, G.; Devière, J.; García-Cano, J.; et al. Endoscopic biliary stenting: Indications, choice of stents, and results: European Society of Gastrointestinal Endoscopy (ESGE) Clinical Guideline–Updated October 2017. *Endoscopy* **2018**, *50*, 910–930. [CrossRef]
49. Fernandez, Y.; Viesca, M.; Arvanitakis, M. Early Diagnosis And Management of Malignant Distal Biliary Obstruction: A Review On Current Recommendations And Guidelines. *Clin. Exp. Gastroenterol.* **2019**, *12*, 415–432. [CrossRef] [PubMed]
50. Inchingolo, R.; Acquafredda, F.; Posa, A.; Nunes, T.F.; Spiliopoulos, S.; Panzera, F.; Praticò, C.A. Endobiliary biopsy. *World J. Gastrointest. Endosc.* **2022**, *14*, 291–301. [CrossRef]
51. Korc, P.; Sherman, S. ERCP tissue sampling. *Gastrointest. Endosc.* **2016**, *84*, 557–571. [CrossRef]
52. Moura, D.; de Moura, E.; Matuguma, S.; dos Santos, M.; Moura, E.; Baracat, F.; Artifon, E.; Cheng, S.; Bernardo, W.; Chacon, D.; et al. EUS-FNA versus ERCP for tissue diagnosis of suspect malignant biliary strictures: A prospective comparative study. *Endosc. Int. Open* **2018**, *6*, E769–E777. [CrossRef]
53. Kamp, E.J.C.A.; Dinjens, W.N.M.; Doukas, M.; Bruno, M.J.; de Jonge, P.J.F.; Peppelenbosch, M.P.; de Vries, A.C. Optimal tissue sampling during ERCP and emerging molecular techniques for the differentiation of benign and malignant biliary strictures. *Ther. Adv. Gastroenterol.* **2021**, *14*, 175628482110020. [CrossRef] [PubMed]
54. Brooks, C.; Gausman, V.; Kokoy-Mondragon, C.; Munot, K.; Amin, S.P.; Desai, A.; Kipp, C.; Poneros, J.; Sethi, A.; Gress, F.G.; et al. Role of Fluorescent In Situ Hybridization, Cholangioscopic Biopsies, and EUS-FNA in the Evaluation of Biliary Strictures. *Dig. Dis. Sci.* **2018**, *63*, 636–644. [CrossRef]
55. Kato, A.; Kato, H.; Naitoh, I.; Hayashi, K.; Yoshida, M.; Hori, Y.; Kachi, K.; Asano, G.; Sahashi, H.; Toyohara, T.; et al. Use of Endoscopic Scraper and Cell Block Technique as a Replacement for Conventional Brush for Diagnosing Malignant Biliary Strictures. *Cancers* **2022**, *14*, 4147. [CrossRef]

56. Gonda, T.A.; Viterbo, D.; Gausman, V.; Kipp, C.; Sethi, A.; Poneros, J.M.; Gress, F.; Park, T.; Khan, A.; Jackson, S.A.; et al. Mutation Profile and Fluorescence In Situ Hybridization Analyses Increase Detection of Malignancies in Biliary Strictures. *Clin. Gastroenterol. Hepatol.* **2017**, *15*, 913–919.e1. [CrossRef]
57. Singhi, A.D.; Nikiforova, M.N.; Chennat, J.; Papachristou, G.I.; Khalid, A.; Rabinovitz, M.; Das, R.; Sarkaria, S.; Ayasso, M.S.; Wald, A.I.; et al. Integrating next-generation sequencing to endoscopic retrograde cholangiopancreatography (ERCP)-obtained biliary specimens improves the detection and management of patients with malignant bile duct strictures. *Gut* **2020**, *69*, 52–61. [CrossRef]
58. Subhash, A.; Buxbaum, J.L.; Tabibian, J.H. Peroral cholangioscopy: Update on the state-of-the-art. *World J. Gastrointest. Endosc.* **2022**, *14*, 63–76. [CrossRef]
59. Oleas, R.; Alcívar-Vasquez, J.; Robles-Medranda, C. New technologies for indeterminate biliary strictures. *Transl. Gastroenterol. Hepatol.* **2022**, *7*, 22. [CrossRef]
60. Kulpatcharapong, S.; Pittayanon, R.; Kerr, S.J.; Rerknimitr, R. Diagnostic performance of digital and video cholangioscopes in patients with suspected malignant biliary strictures: A systematic review and meta-analysis. *Surg. Endosc.* **2022**, *36*, 2827–2841. [CrossRef] [PubMed]
61. Shin, I.S.; Moon, J.H.; Lee, Y.N.; Kim, H.K.; Lee, T.H.; Yang, J.K.; Cha, S.-W.; Cho, Y.D.; Park, S.-H. Efficacy of narrow-band imaging during peroral cholangioscopy for predicting malignancy of indeterminate biliary strictures (with videos). *Gastrointest. Endosc.* **2022**, *96*, 512–521. [CrossRef] [PubMed]
62. Mi, J.; Han, X.; Wang, R.; Ma, R.; Zhao, D. Diagnostic accuracy of probe-based confocal laser endomicroscopy and tissue sampling by endoscopic retrograde cholangiopancreatography in indeterminate biliary strictures: A meta-analysis. *Sci. Rep.* **2022**, *12*, 7257. [CrossRef] [PubMed]
63. Caillol, F.; Filoche, B.; Gaidhane, M.; Kahaleh, M. Refined Probe-Based Confocal Laser Endomicroscopy Classification for Biliary Strictures: The Paris Classification. *Dig. Dis. Sci.* **2013**, *58*, 1784–1789. [CrossRef]
64. Han, S.; Kahaleh, M.; Sharaiha, R.Z.; Tarnasky, P.R.; Kedia, P.; Slivka, A.; Chennat, J.S.; Joshi, V.; Sejpal, D.V.; Sethi, A.; et al. Probe-based confocal laser endomicroscopy in the evaluation of dominant strictures in patients with primary sclerosing cholangitis: Results of a U.S. multicenter prospective trial. *Gastrointest. Endosc.* **2021**, *94*, 569–576.e1. [CrossRef]
65. Zhou, S.; Buxbaum, J. Advanced Imaging of the Biliary System and Pancreas. *Dig. Dis. Sci.* **2022**, *67*, 1599–1612. [CrossRef]
66. Pérez-Cuadrado Robles, E.; Deprez, P.H. Indications for Single-Operator Cholangioscopy and Pancreatoscopy: An expert review. *Curr. Treat. Options Gastro.* **2019**, *17*, 408–419. [CrossRef] [PubMed]
67. Hara, T.; Yamaguchi, T.; Ishihara, T.; Tsuyuguchi, T.; Kondo, F.; Kato, K.; Asano, T.; Saisho, T. Diagnosis and patient management of 55. intraductal papillary-mucinous tumor of the pancreas by using peroral pancreatoscopy and intraductal ultra-sonography. *Gastroenterology* **2002**, *122*, 34–43. [CrossRef]
68. de Jong, D.M.; Stassen, P.M.C.; Groot Koerkamp, B.; Ellrichmann, M.; Karagyozov, P.I.; Anderloni, A.; Kylänpää, L.; Webster, G.J.M.; Van Driel, L.M.J.W.; Bruno, M.J.; et al. The role of pancreatoscopy in the diagnostic work-up of intraductal papillary mucinous neoplasms: A systematic review and meta-analysis. *Endoscopy* **2023**, *55*, 25–35. [CrossRef] [PubMed]
69. El Hajj, I.I.; Brauer, B.C.; Wani, S.; Fukami, N.; Attwell, A.R.; Shah, R.J. Role of per-oral pancreatoscopy in the evaluation of suspected pancreatic duct neoplasia: A 13-year U.S. single-center experience. *Gastrointest. Endosc.* **2017**, *85*, 737–774. [CrossRef] [PubMed]

Disclaimer/Publisher's Note: The statements, opinions and data contained in all publications are solely those of the individual author(s) and contributor(s) and not of MDPI and/or the editor(s). MDPI and/or the editor(s) disclaim responsibility for any injury to people or property resulting from any ideas, methods, instructions or products referred to in the content.

Review

Endoscopic Management of Large Non-Pedunculated Colorectal Polyps

Oliver Cronin [1,2] and Michael J. Bourke [1,2,*]

[1] Department of Gastroenterology and Hepatology, Westmead Hospital, Sydney, NSW 2145, Australia
[2] Westmead Clinical School, University of Sydney, Sydney, NSW 2145, Australia
* Correspondence: michael@citywestgastro.com.au

Simple Summary: Endoscopic resection (ER) of large non-pedunculated colorectal polyps ≥ 20 mm (LNPCPs) is safe, effective and the preferred treatment compared to surgery. Predicted histopathology of an LNPCP based on size, morphology, granularity, pit pattern and location in the colo-rectum is essential when deciding upon resection technique. Post resection defect inspection and adjuvant techniques, such as thermal ablation of the margin, have been demonstrated to reduce recurrence rates. Follow-up surveillance colonoscopy can accurately identify recurrence. Endoscopic treatment of recurrence is effective.

Abstract: Large non-pedunculated colorectal polyps ≥20 mm (LNPCPs) comprise approximately 1% of all colorectal polyps. LNPCPs more commonly contain high-grade dysplasia, covert and overt cancer. These lesions can be resected using several means, including conventional endoscopic mucosal resection (EMR), cold-snare EMR (C-EMR) and endoscopic submucosal dissection (ESD). This review aimed to provide a comprehensive, critical and objective analysis of ER techniques. Evidence-based, selective resection algorithms should be used when choosing the most appropriate technique to ensure the safe and effective removal of LNPCPs. Due to its enhanced safety and comparable efficacy, there has been a paradigm shift towards cold-snare polypectomy (CSP) for the removal of small polyps (<10 mm). This technique is now being applied to the management of LNPCPs; however, further research is required to define the optimal LNPCP subtypes to target and the viable upper size limit. Adjuvant techniques, such as thermal ablation of the resection margin significantly reduce recurrence risk. Bleeding risk can be mitigated using through-the-scope clips to close defects in the right colon. Endoscopic surveillance is important to detect recurrence and synchronous lesions. Recurrence can be readily managed using an endoscopic approach.

Keywords: colonoscopy; polyp; polypectomy; colorectal cancer; endoscopic mucosal resection; endoscopic submucosal dissection

Citation: Cronin, O.; Bourke, M.J. Endoscopic Management of Large Non-Pedunculated Colorectal Polyps. *Cancers* 2023, *15*, 3805. https://doi.org/10.3390/cancers15153805

Academic Editor: Hajime Isomoto

Received: 13 June 2023
Revised: 14 July 2023
Accepted: 20 July 2023
Published: 27 July 2023

Copyright: © 2023 by the authors. Licensee MDPI, Basel, Switzerland. This article is an open access article distributed under the terms and conditions of the Creative Commons Attribution (CC BY) license (https://creativecommons.org/licenses/by/4.0/).

1. Introduction

Colorectal cancer (CRC) is the third most commonly diagnosed malignancy and the second most frequent cause of cancer-related death [1,2]. The majority of CRCs arise via the stepwise acquisition of molecular abnormalities in the adenoma–carcinoma and serrated pathways [3–5]. This creates the opportunity for intervention to remove premalignant polyps. Endoscopic resection (ER) of pre-malignant polyps has been shown to reduce the incidence of CRC [6–8]. Moreover, screening colonoscopy and polypectomy have been shown to reduce the risk of death from CRC at 10 years (risk ratio 0.82, 95% confidence interval (CI) 0.70–0.93) [9]. In a large study (*n* = 2602) with follow-up over 23 years, a 53% reduction (relative risk (RR) 0.47; 95% CI 0.26–0.80) in mortality was demonstrated in those who had undergone polypectomy [7].

The majority (90%) of colorectal polyps are <10 mm in size, do not contain advanced pathology and can be removed either en bloc or piecemeal using cold-snare polypectomy (CSP) [10–12]. Large non-pedunculated colorectal polyps ≥ 20 mm (LNPCPs) comprise

~1% of all colorectal polyps. These lesions have varied risk of overt and covert submucosal invasive cancer (SMIC), and therefore require a detailed, methodical optical assessment before deciding on the most suitable resection technique [13–16]. This algorithm needs to account for LNPCP size, morphology, location and pit pattern in addition to any patient-specific factors, such as co-morbidities and anticoagulation or anti-platelet medications [17].

Consensus recommendations favour an endoscopic approach as first line for the resection of LNPCPs (based on high-quality evidence) [18,19]. Compared to surgical resection, EMR has been demonstrated to have reduced morbidity and mortality and lower healthcare costs [20,21].

ER can be divided into three discrete phases: pre-resection, resection and post-resection. The technical success of ER requires a methodical, collaborative approach, ideally at a centre with access to the complete range of ER techniques, including conventional endoscopic mucosal resection (EMR), cold-snare EMR (C-EMR) and endoscopic submucosal dissection (ESD) (Figure 1). This review aimed to provide a comprehensive, critical and objective analysis of ER techniques. Herein, we outline an evidence-based approach to the ER of colorectal polyps.

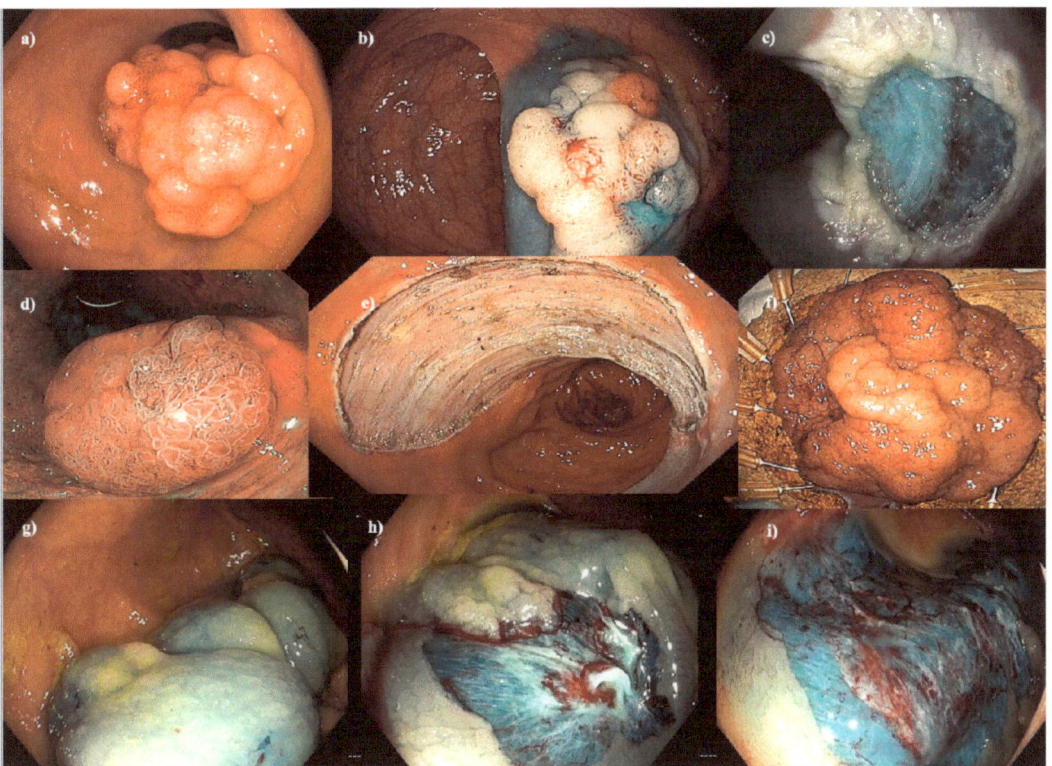

Figure 1. Endoscopic mucosal resection (EMR), endoscopic submucosal dissection (ESD) and cold-snare EMR (C-EMR). (**a–c**) EMR of a 40 mm Paris 0-IIa+Is granular hepatic flexure lesion. (**d–f**) ESD of a hemi-circumferential 45 mm Paris 0-IIa+Is granular rectal lesion. (**g–i**) C-EMR of a 50 mm serrated lesion without dysplasia in a patient with serrated polyposis syndrome.

2. Pre-Resection

Planning is essential to ensure technical success. The planning phase can be subdivided into pre-procedure and intra-procedure.

Pre-procedural planning starts with patient assessment, accounting for frailty, functional status, co-morbidities and medications. Consent must include the risks and benefits of ER and a discussion around alternative modalities, such as surgery. Predicted lesion histopathology, including the risk of SMIC, should influence ER modality, and any related imaging should be reviewed. Pre-procedural planning also includes an in-room discussion with the endoscopy team to ensure that nursing and anaesthetic staff are aware of the various stages of the procedure, including any site-specific challenges, such as those seen with ileocaecal valve (ICV) lesions [17,22]. The pre-procedure discussion with the endoscopy team should also include the expected procedure time, any required medications, such as surgical antibiotic prophylaxis or local anaesthetic for anorectal junction (ARJ) lesions, and a check to ensure appropriate snares and ESD knives are available [23,24]. ER should only be performed using carbon dioxide insufflation [25]. Required ancillary devices should be in the room pre-procedure, including closure devices, such as through-the-scope clips, and those used to treat intra-procedural bleeding, such as haemostatic forceps.

Intra-procedural planning starts with patient positioning. The optimal patient position is to have the fluid pool opposite the lesion to maximise the effect of gravity on lesion elevation and achieve a clear working field during tissue resection or for management of any complications. Therefore, a supine or right lateral position may be required. Position of the colonoscope to align the lesion at a 6 o'clock position is essential. Dependent on location, a retroflexed position may improve access and optical assessment.

Thorough optical assessment is key. The risk of overt (optical features of SMIC present) or covert (optical features of SMIC absent) cancer can be predicted based on LNPCP size, location, morphology, granularity, and microvascular and surface pit patterns [13,26]. Several classification systems exist, including the Kudo pit pattern (KPP) and the Japan Narrow-Band Imaging Expert Team (JNET) classification [27,28]. An understanding of these systems is useful. A simple innovation to assist with familiarly and use is to place large posters of these classification systems in endoscopy rooms and reporting areas. Benign lesions have surface homogeneity with a regular pit and microvascular pattern (Figure 2). High grade dysplasia or cancer within a benign lesion appears as a demarcated area of disruption within this regular pattern (Figure 3). Such areas need to be very carefully examined to ensure the correct optical diagnosis and treatment strategy.

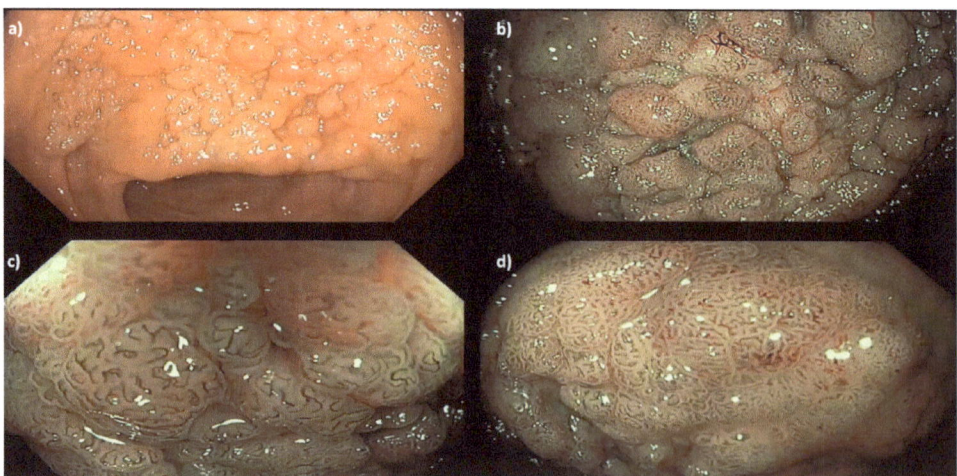

Figure 2. A 35 mm granular Paris 0-IIa LNPCP in the mid-ascending colon, assessed using (**a**) high definition white light, (**b**) narrow band imaging (NBI) and (**c**,**d**) near-focus with NBI, demonstrating a homogenous pit pattern (Kudo pit pattern IV).

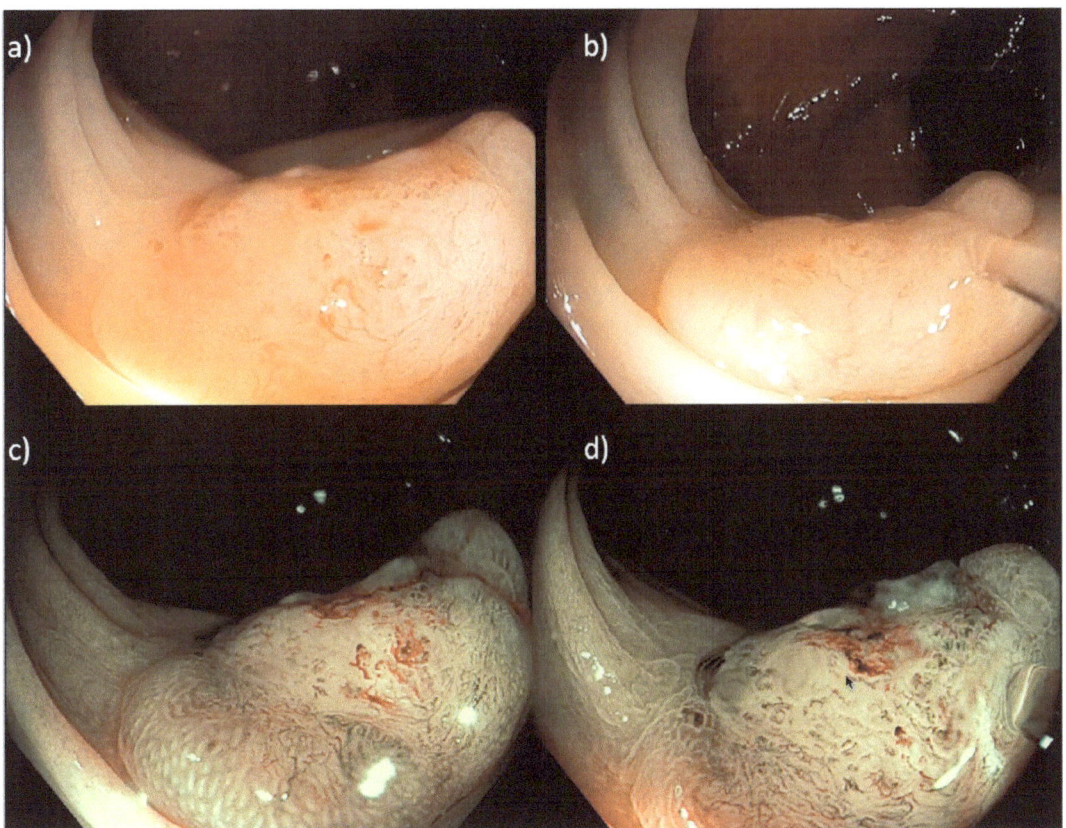

Figure 3. A 20 mm sessile serrated Paris 0-IIa LNPCP in the proximal ascending colon, assessed using (**a,b**) near focus and (**c,d**) near focus with narrow band imaging (NBI). There is a central well-demarcated area with loss of homogeneity, neovascularization, dilated vessels and a non-structural pit pattern (Kudo pit pattern V_N), suggestive of a deeply invasive cancer.

Traditionally, the accuracy of optical diagnosis for SMIC was evaluated across the entire LNPCP spectrum and was found to have suboptimal utility. Recently, optical assessment of flat (Paris 0-IIa) LNPCPs has been proven to be highly accurate [15]. In a large, prospective, single-centre cohort study (n = 1583), the sensitivity and specificity for predicting cancer in Paris 0-IIa LNPCPs was 91% and 96%, respectively. The likelihood that cancer would be missed in this study was 6 in 1000 cases. Optical diagnosis for SMIC in nodular lesions is less accurate (sensitivity 53%, specificity 94%, missed SMIC 6%) [15]. Excluding those lesions with overt SMIC, a large multicentre, prospective study (n = 2277) found that covert SMIC was associated with Paris 0-Is and Paris 0-IIa+Is morphology, non-granularity, size and distal location [13]. Supporting this, a large prospective cohort study (n = 3405) demonstrated that nodular rectal LNPCPs are more likely to contain SMIC than non-rectal colonic LNPCPs (15% vs. 6%, $p < 0.001$) [26].

3. Resection

In 2023, a selective resection algorithm should be employed when considering a therapeutic strategy for any colorectal polyp or neoplasm. This is based on optical diagnosis for predicted histology, lesion size, morphology, surface granularity and location in the colon.

3.1. Diminutive (<5 mm) and Small (5–9 mm) Colorectal Polyps

The overwhelming majority of colonic polyps are diminutive (<5 mm) or small (5–9 mm). CSP is safer and equi-efficacious compared to hot-snare polypectomy (HSP) for the removal of these colorectal polyps. The absence of electrocautery all but eliminates the risks of perforation and post-polypectomy bleeding [29,30]. Based on high quality data, en bloc or oligo-piecemeal CSP should be used to resect these polyps [18,19].

3.2. Medium (10–19 mm) Colorectal Polyps

There is a paradigm shift toward C-EMR given its superior safety profile. A large prospective, multicentre cohort study (n = 286 lesions) comparing conventional EMR to C-EMR for 6–15 mm polyps favoured the use of C-EMR over EMR [31]. At present, US consensus guidelines recommend either EMR or C-EMR for resection of lesions 10–19 mm [18]

3.3. Large (>20 mm) Non-Pedunculated Colorectal Polyps

Conventional EMR is the mainstay for ER of LNPCPs due to its superior safety efficacy and cost effectiveness compared to surgery and ESD [17–19,22]. High-quality studies over the past 10–15 years have lead to improvements in the safety and efficacy of EMR. These include the use of CO_2 for insufflation; addition of chromo-injectate into the submucosal space; use of a systematic inject and sequential snare resection technique, removing a 2–3 mm margin of normal mucosa; water expansion of the defect to identify any residual adenoma; and recognition and management of significant DMI [25,32,33]. When all visible adenoma has been excised, thermal ablation of the margin should be completed by gently applying snare-tip soft coagulation (Effect 4, 80 W: ERBE Electromedizin, Tubingen, Germany), aiming for a 3–5 mm rim of ablated mucosa [34]. In a large, prospective cohort (n = 390) comparing conventional EMR without and with thermal ablation, recurrence rates reduced from 21.0% (37/176) to 5.2% (10/192), p < 0.001. No adverse events were attributed to margin thermal ablation. Since its inception, application of this adjuvant technique has improved. In a recent, larger multicentre cohort (n = 1049), recurrence rates at 6-month follow-up colonoscopy (SC1) were 1.4% (10/707) [35].

At present, given the paucity of data, conventional EMR is recommended over C-EMR for LNPCP resection. The safety profile of C-EMR is appealing for the piecemeal resection of Paris 0-IIa (flat, sessile) LNPCPs; however, the upper size limit that can be effectively removed using C-EMR without excessive burden of recurrence is unknown. Several ongoing large randomised controlled trials comparing EMR and C-EMR for non serrated LNPCPs (clinicaltrials.gov identifier: NCT04138030; NCT04418843) aim to provide clarity on this issue. The next important RCT will compare C-EMR to C-EMR with thermal ablation of the margin (clinicaltrials.gov identifier NCT05041478).

In contrast to adenomatous LNPCPs, C-EMR is always the primary modality for ER of serrated LNPCPs, irrespective of size [36]. A large study (n = 562) of serrated lesions found no difference in technical success and recurrence rates between EMR and C-EMR groups; however, bleeding (0% vs. 5.1%) and significant deep mural injury (DMI) (0% vs 2.8%) were more common in the EMR group.

3.4. Special Considerations

Site specific considerations and technique modifications may be needed for LNPCPs located at the ICV, appendiceal orifice, surgical anastomosis, or an ARJ or those which are circumferential [16,37–39].

The rectum should be regarded as a complex high-risk site, with distinct challenges compared to the colon. This is not due to its technical limitations, but due to its increased risk of covert SMIC [26,40]. Furthermore, the consequences of failed endoscopic cure include consideration of the most hazardous and complicated forms of colorectal surgery, including permanent ostomy formation [41]. Patients with rectal lesions removed using a low or ultra-low anterior resection have an increased risk of incontinence (12%) [42] and sexual dysfunction (20–46%) [43], and a 10–20% risk of permanent stoma [44,45]. Low anterior resection has a 30-day morbidity and mortality of 25% and 6%, respectively [46]. Postoperative complications have been associated with negative economic impact, increased morbidity, extended postoperative hospital stay, readmission, sepsis and death. ER is organ-sparing and minimally invasive, which enables avoiding wound infections as well as other postoperative complications after open surgery, which cause pain and suffering to patients [47].

In a large, multicentre observational study ($n = 618$), rectal LNPCPs were more likely to have nodular morphology (53% vs. 17%, $p < 0.001$) and contain cancer (15% vs. 6%, $p < 0.001$) compared to LNPCPs in the remainder of the colon [26]. Endoscopic en bloc resection for any LNPCP with a nodular component is critical with the aim of achieving an R0 (curative) resection. This requires meticulous planning.

ESD was developed as an ER technique for the curative treatment of early gastric cancer. ESD is now an established technique in the colo-rectum. It is typically performed with a generous submucosal injection, in a retroflexed position for improved scope stability, and using an improved more parallel angle of the cutting plane. Dissection is performed using an electrosurgical knife. Technique has improved over the past 10 years, aided by internal and external traction devices as well as techniques such as pocket creation ESD.

EMR and ESD are complementary techniques for resection of rectal LNPCPs. A selective resection algorithm (SRA) has demonstrated superior outcomes compared to a universal EMR algorithm (UEA). In a large study ($n = 480$) comparing an SRA to a UEA, LNPCPs underwent ESD if they had features suggestive of superficial overt SMIC (1000 μm, KPP V_1) or covert SMIC (Paris 0-Is or a dominant nodule). All ($n = 7$, 100%) LNPCPs with SMIC amenable to R0 resection that underwent ESD were cured [16]. A rectum-specific SRA avoids the piecemeal resection of cancer.

Until recently, the management of covert SMIC discovered after piecemeal ER has been challenging. A recent observational study ($n = 3372$) identified 143 (4.2%) cases with covert SMIC post piecemeal resection [48]; 109 cases underwent surgical resection, and 62 (63%) cases had no residual cancer. All cases with residual intramucosal cancer ($n = 24$) could be identified by a R1 histological deep margin. Cases with poor differentiation and/or lymphovascular invasion had a high risk of lymph node metastases (12/33); there was a very low risk without these features (<1%, 0/35). The majority of patients with covert SMIC resected piecemeal had no residual malignancy. The risk of malignancy can be predicted by poor differentiation, lymphovascular invasion and an R1 deep margin.

Prevention of bleeding by prophylactic treatment of medium and large vessels with coagulating forceps is key. Bleeding stains the mucosa, impeding views, and leading to a higher risk of incomplete resection. Treatment of bleeding can char the mucosa, also obscuring views. Given its resource intensive, time consuming nature, this technique is best reserved for lesions with superficial overt SMIC or a high risk of covert SMIC. In clinical practice, this limits its use predominantly to the rectum [24].

Previously attempted LNPCPs are common and present a unique set of challenges. Due to the dense submucosal fibrosis, submucosal lift if often unsuccessful. A large observational study ($n = 1292$) demonstrated that with the use of auxiliary these lesions can be effectively resected by EMR. CAST was used in 73 (46.2%) cases. No recurrence ($n = 0$, 0%) was identified in any previously attempted LNPCPs that underwent margin thermal ablation, demonstrating that EMR is effective for resection of these lesions [49].

3.5. Complications

3.5.1. Deep Mural Injury

Significant DMI (Deep Mural Injury Types III–V) was previously a feared intra-procedural complication, with a frequency of approximately 3% [40]. However, due to an improved understanding of risk factors, earlier recognition and advances in closure devices, such as through-the-scope clips, significant DMIs can now be successfully managed [40,50]. In a large, prospective cohort ($n = 911$), significant DMI was associated with attempted en bloc resection, advanced histopathology and transverse colon location [50]. In a large, prospective cohort ($n = 3717$), significant DMI occurred in 2.7% (101/3717) of EMR resections (median lesion size 35 mm, interquartile range 25–45 mm). Successful defect closure occurred in 97.0% (98/101) of cases. There were no differences found between DMI and non-DMI cases in terms of technical success or recurrence [40].

3.5.2. Post-Procedural Bleeding

Prophylactic treatment of visible vessels within a defect post EMR has been previously investigated. In a multicentre RCT ($n = 347$, 55.3% proximal colonic lesions), prophylactic endoscopic coagulation of all visible vessels within the post-EMR defect did not reduce clinically significant post-EMR bleeding compared to no treatment (5.2% vs. 8.0%; $p = 0.30$) [51]. Post-resection defect closure for right-sided lesions using through-the-scope clips has been shown to reduce clinically significant post-EMR bleeding from 10.6% (12/113) to 3.4% (4/118), $p = 0.031$ [52].

Post-ER bleeding has a frequency of 6–7%, dependent on defect location and the selected ER modality. Bleeding typically does not require intervention, and these cases are managed conservatively in >50% of cases [53].

4. Post-Resection

4.1. Post-Operative Care

Post-resection instructions and communication with nursing staff, patients and their next-of-kin are important to ensure early recognition and management of any adverse events or complications. Recovery staff should receive a verbal handover and a written endoscopy report from the proceduralist, including any complexities or nuances of the case. Dependent on the procedure type, patients should remain fasting for at least 2 h or until they have been examined by the proceduralist. After clinical assessment, if the patient is well, they can commence a clear fluid diet.

The patient should receive a copy of their report. Dietary instructions should be highlighted and details of the best hospital contact should be clear, should the patient have any issues or questions overnight. Most patients can be discharged home the same day, but an endoscopy team member should contact the patient the following day for a telehealth assessment.

4.2. Surveillance

Guidelines recommend a follow-up surveillance colonoscopy 6 months post-ER [18,19,54]. Surveillance post-ER is essential to evaluate the previous resection site and to exclude synchronous lesions [55]. Co-existent advanced pathology (polyps > 10 mm or with a villous component or high-grade dysplasia) is reported to occur at surveillance in 10–20% of cases [55,56].

The previous ER site can be identified by a bland pale area, sometimes with anatomic distortion of the mucosal folds [57]. A standardised imaging protocol for optical assessment of the scar should include high definition white light and narrow band imaging (NBI, Olympus, Inc, Tokyo, Japan) [57]. Optical scar assessment is accurate. A recent multicentre single-blind cross-over trial ($n = 203$) to compare NBI and high definition white light for the assessment of recurrence or residual adenoma at a post-EMR scar reported a negative predictive value (NPV) > 90% (NPV 96% using NBI, NPV 93% using high definition white light) [58]. Use of NBI was not superior to high definition white light ($p = 0.06$) [58]. Expert consensus is that a biopsy is not needed for a bland scar with a uniform pit pattern [57]. Common mimics of recurrence include clip artefact and inflammatory nodules. If an abnormality is suspected, this area should be excised and ablated, as described in a proposed Westmead algorithm for evaluating recurrence [59]. Techniques include cold-snare resection or cold-forceps avulsion with adjuvant snare-tip soft coagulation (CAST), margin ablation and clip closure if any DMI $\geq$ Type 2 [34,50,60].

5. Conclusions

ER is organ-sparing and minimally invasive. It is the recommended primary management strategy for the excision of LNPCPs, supported by high-quality studies. Referral to an expert endoscopist, rather than for surgery, is the standard of care for all patients with an LNPCP. Predicted histopathology underpins the selective resection algorithm and accounts for lesion size, site, granularity, pit pattern and morphology. These resection decision strategies have revolutionised management of LNPCPs. Compared to surgery, they have a lower morbidity and mortality, and are more cost-effective. Unnecessary surgery remains an important issue, and can be overcome by greater awareness of the efficacy and superior risk profiles of ER.

Author Contributions: Conceptualization, M.J.B.; methodology, O.C. and M.J.B.; software, O.C.; validation, O.C. and M.J.B.; formal analysis, O.C. and M.J.B.; data curation, O.C.; writing—original draft preparation, O.C.; writing—review and editing, O.C. and M.J.B.; supervision, M.J.B.; project administration, M.J.B.; funding acquisition, M.J.B. All authors have read and agreed to the published version of the manuscript.

Funding: This research received no external funding.

Conflicts of Interest: The authors declare no conflict of interest.

References

1. Keum, N.; Giovannucci, E. Global burden of colorectal cancer: Emerging trends, risk factors and prevention strategies. *Nat. Rev. Gastroenterol. Hepatol.* **2019**, *16*, 713–732. [CrossRef]
2. Patel, S.G.; Karlitz, J.J.; Yen, T.; Lieu, C.H.; Boland, C.R. The rising tide of early-onset colorectal cancer: A comprehensive review of epidemiology, clinical features, biology, risk factors, prevention, and early detection. *Lancet Gastroenterol. Hepatol.* **2022**, *7*, 262–274. [CrossRef]
3. Levin, B.; Lieberman, D.A.; McFarland, B.; Andrews, K.S.; Brooks, D.; Bond, J.; Dash, C.; Giardiello, F.M.; Glick, S.; Johnson, D.; et al. Screening and surveillance for the early detection of colorectal cancer and adenomatous polyps, 2008: A joint guideline from the American Cancer Society, the US Multi-Society Task Force on Colorectal Cancer, and the American College of Radiology. *Gastroenterology* **2008**, *134*, 1570–1595. [CrossRef] [PubMed]
4. Kedrin, D.; Gala, M.K. Genetics of the serrated pathway to colorectal cancer. *Clin. Transl. Gastroenterol.* **2015**, *6*, e84. [CrossRef] [PubMed]
5. Itzkowitz, S.H.; Yio, X. Inflammation and cancer IV. Colorectal cancer in inflammatory bowel disease: The role of inflammation. *Am. J. Physiol. Gastrointest. Liver Physiol.* **2004**, *287*, G7–G17. [CrossRef] [PubMed]
6. Winawer, S.J.; Zauber, A.G.; Ho, M.N.; O'Brien, M.J.; Gottlieb, L.S.; Sternberg, S.S.; Waye, J.D.; Schapiro, M.; Bond, J.H.; Panish, J.F.; et al. Prevention of colorectal cancer by colonoscopic polypectomy. The National Polyp Study Workgroup. *N. Engl. J. Med.* **1993**, *329*, 1977–1981. [CrossRef]
7. Zauber, A.G.; Winawer, S.J.; O'Brien, M.J.; Lansdorp-Vogelaar, I.; van Ballegooijen, M.; Hankey, B.F.; Shi, W.; Bond, J.H.; Schapiro, M.; Panish, J.F.; et al. Colonoscopic polypectomy and long-term prevention of colorectal-cancer deaths. *N. Engl. J. Med.* **2012**, *366*, 687–696. [CrossRef]
8. Kahi, C.J.; Imperiale, T.F.; Juliar, B.E.; Rex, D.K. Effect of screening colonoscopy on colorectal cancer incidence and mortality. *Clin. Gastroenterol. Hepatol.* **2009**, *7*, 770–775, quiz 711. [CrossRef]
9. Bretthauer, M.; Løberg, M.; Wieszczy, P.; Kalager, M.; Emilsson, L.; Garborg, K.; Rupinski, M.; Dekker, E.; Spaander, M.; Bugajski, M.; et al. Effect of Colonoscopy Screening on Risks of Colorectal Cancer and Related Death. *N. Engl. J. Med.* **2022**, *387*, 1547–1556. [CrossRef]
10. Rex, D.K. Have we defined best colonoscopic polypectomy practice in the United States? *Clin. Gastroenterol. Hepatol.* **2007**, *5*, 674–677. [CrossRef]
11. Gupta, N.; Bansal, A.; Rao, D.; Early, D.S.; Jonnalagadda, S.; Wani, S.B.; Edmundowicz, S.A.; Sharma, P.; Rastogi, A. Prevalence of advanced histological features in diminutive and small colon polyps. *Gastrointest. Endosc.* **2012**, *75*, 1022–1030. [CrossRef]
12. Repici, A.; Hassan, C.; Vitetta, E.; Ferrara, E.; Manes, G.; Gullotti, G.; Princiotta, A.; Dulbecco, P.; Gaffuri, N.; Bettoni, E.; et al. Safety of cold polypectomy for <10 mm polyps at colonoscopy: A prospective multicenter study. *Endoscopy* **2012**, *44*, 27–31. [PubMed]
13. Burgess, N.G.; Hourigan, L.F.; Zanati, S.A.; Brown, G.J.; Singh, R.; Williams, S.J.; Raftopoulos, S.C.; Ormonde, D.; Moss, A.; Byth, K.; et al. Risk Stratification for Covert Invasive Cancer Among Patients Referred for Colonic Endoscopic Mucosal Resection: A Large Multicenter Cohort. *Gastroenterology* **2017**, *153*, 732–742.e1. [CrossRef]

14. Holt, B.A.; Bourke, M.J. Wide field endoscopic resection for advanced colonic mucosal neoplasia: Current status and future directions. *Clin. Gastroenterol. Hepatol.* **2012**, *10*, 969–979. [CrossRef]
15. Vosko, S.; Shahidi, N.; Sidhu, M.; van Hattem, W.A.; Bar-Yishay, I.; Schoeman, S.; Tate, D.J.; Hourigan, L.F.; Singh, R.; Moss, A.; et al. Optical evaluation for predicting cancer in large non-pedunculated colorectal polyps is accurate for flat lesions. *Clin. Gastroenterol. Hepatol.* **2021**, *19*, 2425–2434. [CrossRef]
16. Shahidi, N.; Vosko, S.; Gupta, S.; Whitfield, A.; Cronin, O.; O'Sullivan, T.; van Hattem, W.A.; Sidhu, M.; Tate, D.J.; Lee, E.Y.T.; et al. A Rectum-Specific Selective Resection Algorithm Optimizes Oncologic Outcomes for Large Nonpedunculated Rectal Polyps. *Clin. Gastroenterol. Hepatol.* **2023**, *21*, 72–80.e2. [CrossRef]
17. Shahidi, N.; Bourke, M.J. How to Manage the Large Nonpedunculated Colorectal Polyp. *Gastroenterology* **2021**, *160*, 2239–2243.e2231. [CrossRef] [PubMed]
18. Kaltenbach, T.; Anderson, J.C.; Burke, C.A.; Dominitz, J.A.; Gupta, S.; Lieberman, D.; Robertson, D.J.; Shaukat, A.; Syngal, S.; Rex, D.K. Endoscopic Removal of Colorectal Lesions: Recommendations by the US Multi-Society Task Force on Colorectal Cancer. *Am. J. Gastroenterol.* **2020**, *115*, 435–464. [CrossRef]
19. Ferlitsch, M.; Moss, A.; Hassan, C.; Bhandari, P.; Dumonceau, J.M.; Paspatis, G.; Jover, R.; Langner, C.; Bronzwaer, M.; Nalankilli, K.; et al. Colorectal polypectomy and endoscopic mucosal resection (EMR): European Society of Gastrointestinal Endoscopy (ESGE) Clinical Guideline. *Endoscopy* **2017**, *49*, 270–297. [CrossRef] [PubMed]
20. Jayanna, M.; Burgess, N.G.; Singh, R.; Hourigan, L.F.; Brown, G.J.; Zanati, S.A.; Moss, A.; Lim, J.; Sonson, R.; Williams, S.J.; et al. Cost Analysis of Endoscopic Mucosal Resection vs Surgery for Large Laterally Spreading Colorectal Lesions. *Clin. Gastroenterol. Hepatol.* **2016**, *14*, 271–278.e2. [CrossRef]
21. Ahlenstiel, G.; Hourigan, L.F.; Brown, G.; Zanati, S.; Williams, S.J.; Singh, R.; Moss, A.; Sonson, R.; Bourke, M.J. Actual endoscopic versus predicted surgical mortality for treatment of advanced mucosal neoplasia of the colon. *Gastrointest. Endosc.* **2014**, *80*, 668–676. [CrossRef] [PubMed]
22. Jideh, B.; Bourke, M.J. How to Perform Wide-Field Endoscopic Mucosal Resection and Follow-up Examinations. *Gastrointest. Endosc. Clin. N. Am.* **2019**, *29*, 629–646. [CrossRef] [PubMed]
23. Holt, B.A.; Bassan, M.S.; Sexton, A.; Williams, S.J.; Bourke, M.J. Advanced mucosal neoplasia of the anorectal junction: Endoscopic resection technique and outcomes (with videos). *Gastrointest. Endosc.* **2014**, *79*, 119–126. [CrossRef] [PubMed]
24. Shahidi, N.; Sidhu, M.; Vosko, S.; van Hattem, W.A.; Bar-Yishay, I.; Schoeman, S.; Tate, D.J.; Holt, B.; Hourigan, L.F.; Lee, E.Y.; et al. Endoscopic mucosal resection is effective for laterally spreading lesions at the anorectal junction. *Gut* **2020**, *69*, 673–680. [CrossRef]
25. Bassan, M.S.; Holt, B.; Moss, A.; Williams, S.J.; Sonson, R.; Bourke, M.J. Carbon dioxide insufflation reduces number of postprocedure admissions after endoscopic resection of large colonic lesions: A prospective cohort study. *Gastrointest. Endosc.* **2013**, *77*, 90–95. [CrossRef] [PubMed]
26. Cronin, O.; Sidhu, M.; Shahidi, N.; Gupta, S.; O'Sullivan, T.; Whitfield, A.; Wang, H.; Kumar, P.; Hourigan, L.F.; Byth, K.; et al. Comparison of the morphology and histopathology of large nonpedunculated colorectal polyps in the rectum and colon: Implications for endoscopic treatment. *Gastrointest. Endosc.* **2022**, *96*, 118–124.
27. Kudo, S.; Tamura, S.; Nakajima, T.; Yamano, H.; Kusaka, H.; Watanabe, H. Diagnosis of colorectal tumorous lesions by magnifying endoscopy. *Gastrointest. Endosc.* **1996**, *44*, 8–14. [CrossRef]
28. Sano, Y.; Tanaka, S.; Kudo, S.E.; Saito, S.; Matsuda, T.; Wada, Y.; Fujii, T.; Ikematsu, H.; Uraoka, T.; Kobayashi, N.; et al. Narrow-band imaging (NBI) magnifying endoscopic classification of colorectal tumors proposed by the Japan NBI Expert Team. *Dig. Endosc.* **2016**, *28*, 526–533. [CrossRef]
29. Chang, L.C.; Shun, C.T.; Hsu, W.F.; Tu, C.H.; Chen, C.C.; Wu, M.S.; Chiu, H.M. Risk of delayed bleeding before and after implementation of cold snare polypectomy in a screening colonoscopy setting. *Endosc. Int. Open* **2019**, *7*, E232–E238. [CrossRef]
30. Tolliver, K.A.; Rex, D.K. Colonoscopic polypectomy. *Gastroenterol. Clin. N. Am.* **2008**, *37*, 229–251+ix. [CrossRef]
31. Rex, D.K.; Anderson, J.C.; Pohl, H.; Lahr, R.E.; Judd, S.; Antaki, F.; Lilley, K.; Castelluccio, P.F.; Vemulapalli, K.C. Cold versus hot snare resection with or without submucosal injection of 6- to 15-mm colorectal polyps: A randomized controlled trial. *Gastrointest. Endosc.* **2022**, *96*, 330–338. [PubMed]
32. Bourke, M.J.; Bhandari, P. How I remove polyps larger than 20 mm. *Gastrointest. Endosc.* **2019**, *90*, 877–880. [CrossRef] [PubMed]
33. Moss, A.; Bourke, M.J.; Metz, A.J. A randomized, double-blind trial of succinylated gelatin submucosal injection for endoscopic resection of large sessile polyps of the colon. *Am. J. Gastroenterol.* **2010**, *105*, 2375–2382. [CrossRef]
34. Klein, A.; Tate, D.J.; Jayasekeran, V.; Hourigan, L.; Singh, R.; Brown, G.; Bahin, F.F.; Burgess, N.; Williams, S.J.; Lee, E.; et al. Thermal Ablation of Mucosal Defect Margins Reduces Adenoma Recurrence After Colonic Endoscopic Mucosal Resection. *Gastroenterology* **2019**, *156*, 604–613.e3. [CrossRef]
35. Sidhu, M.; Shahidi, N.; Gupta, S.; Desomer, L.; Vosko, S.; Arnout van Hattem, W.; Hourigan, L.F.; Lee, E.Y.T.; Moss, A.; Raftopoulos, S.; et al. Outcomes of Thermal Ablation of the Mucosal Defect Margin After Endoscopic Mucosal Resection: A Prospective International, Multicenter Trial of 1000 Large Nonpedunculated Colorectal Polyps. *Gastroenterology* **2021**, *161*, 163–170.e163.

36. van Hattem, W.A.; Shahidi, N.; Vosko, S.; Hartley, I.; Britto, K.; Sidhu, M.; Bar-Yishay, I.; Schoeman, S.; Tate, D.J.; Byth, K.; et al. Piecemeal cold snare polypectomy versus conventional endoscopic mucosal resection for large sessile serrated lesions: A retrospective comparison across two successive periods. *Gut* **2021**, *70*, 1691–1697.
37. Nanda, K.S.; Tutticci, N.; Burgess, N.G.; Sonson, R.; Williams, S.J.; Bourke, M.J. Endoscopic mucosal resection of laterally spreading lesions involving the ileocecal valve: Technique, risk factors for failure, and outcomes. *Endoscopy* **2015**, *47*, 710–718. [PubMed]
38. Vosko, S.; Gupta, S.; Shahidi, N.; van Hattem, W.A.; Zahid, S.; McKay, O.; Whitfield, A.; Sidhu, M.; Tate, D.J.; Lee, E.Y.T.; et al. Impact of technical innovations in EMR in the treatment of large nonpedunculated polyps involving the ileocecal valve (with video). *Gastrointest. Endosc.* **2021**, *94*, 959–968.e2. [CrossRef]
39. Tate, D.J.; Desomer, L.; Awadie, H.; Goodrick, K.; Hourigan, L.; Singh, R.; Williams, S.J.; Bourke, M.J. EMR of laterally spreading lesions around or involving the appendiceal orifice: Technique, risk factors for failure, and outcomes of a tertiary referral cohort (with video). *Gastrointest. Endosc.* **2018**, *87*, 1279–1288.e2. [CrossRef]
40. Bar-Yishay, I.; Shahidi, N.; Gupta, S.; Vosko, S.; van Hattem, W.A.; Schoeman, S.; Sidhu, M.; Tate, D.J.; Hourigan, L.F.; Singh, R.; et al. Outcomes of Deep Mural Injury After Endoscopic Resection: An International Cohort of 3717 Large Non-Pedunculated Colorectal Polyps. *Clin. Gastroenterol. Hepatol.* **2022**, *20*, e139–e147.
41. Vennix, S.; Pelzers, L.; Bouvy, N.; Beets, G.L.; Pierie, J.P.; Wiggers, T.; Breukink, S. Laparoscopic versus open total mesorectal excision for rectal cancer. *Cochrane Database Syst. Rev.* **2014**, Cd005200. [CrossRef]
42. Duran, E.; Tanriseven, M.; Ersoz, N.; Oztas, M.; Ozerhan, I.H.; Kilbas, Z.; Demirbas, S. Urinary and sexual dysfunction rates and risk factors following rectal cancer surgery. *Int. J. Color. Dis.* **2015**, *30*, 1547–1555. [CrossRef] [PubMed]
43. Celentano, V.; Cohen, R.; Warusavitarne, J.; Faiz, O.; Chand, M. Sexual dysfunction following rectal cancer surgery. *Int. J. Color. Dis.* **2017**, *32*, 1523–1530. [CrossRef]
44. Climent, M.; Martin, S.T. Complications of laparoscopic rectal cancer surgery. *Mini-Invasive Surg.* **2018**, *2*, 45. [CrossRef]
45. Lindgren, R.; Hallböök, O.; Rutegård, J.; Sjödahl, R.; Matthiessen, P. What is the risk for a permanent stoma after low anterior resection of the rectum for cancer? A six-year follow-up of a multicenter trial. *Dis. Colon. Rectum* **2011**, *54*, 41–47. [CrossRef]
46. Cuccurullo, D.; Pirozzi, F.; Sciuto, A.; Bracale, U.; La Barbera, C.; Galante, F.; Corcione, F. Relaparoscopy for management of postoperative complications following colorectal surgery: Ten years experience in a single center. *Surg. Endosc.* **2015**, *29*, 1795–1803. [CrossRef]
47. Mulita, F.; Liolis, E.; Akinosoglou, K.; Tchabashvili, L.; Maroulis, I.; Kaplanis, C.; Vailas, M.; Panos, G. Postoperative sepsis after colorectal surgery: A prospective single-center observational study and review of the literature. *Prz. Gastroenterol.* **2022**, *17*, 47–51. [PubMed]
48. Gibson, D.J.; Sidhu, M.; Zanati, S.; Tate, D.J.; Mangira, D.; Moss, A.; Singh, R.; Hourigan, L.F.; Raftopoulos, S.; Pham, A.; et al. Oncological outcomes after piecemeal endoscopic mucosal resection of large non-pedunculated colorectal polyps with covert submucosal invasive cancer. *Gut* **2022**, *71*, 2481–2488. [CrossRef]
49. Shahidi, N.; Vosko, S.; Gupta, S.; van Hattem, W.A.; Sidhu, M.; Tate, D.J.; Williams, S.J.; Lee, E.Y.T.; Burgess, N.; Bourke, M.J. Previously Attempted Large Nonpedunculated Colorectal Polyps Are Effectively Managed by Endoscopic Mucosal Resection. *Am. J. Gastroenterol.* **2021**, *116*, 958–966. [CrossRef]
50. Burgess, N.G.; Bassan, M.S.; McLeod, D.; Williams, S.J.; Byth, K.; Bourke, M.J. Deep mural injury and perforation after colonic endoscopic mucosal resection: A new classification and analysis of risk factors. *Gut* **2017**, *66*, 1779–1789. [CrossRef]
51. Bahin, F.F.; Naidoo, M.; Williams, S.J.; Hourigan, L.F.; Ormonde, D.G.; Raftopoulos, S.C.; Holt, B.A.; Sonson, R.; Bourke, M.J. Prophylactic endoscopic coagulation to prevent bleeding after wide-field endoscopic mucosal resection of large sessile colon polyps. *Clin. Gastroenterol. Hepatol.* **2015**, *13*, e721–e722. [CrossRef] [PubMed]
52. Gupta, S.; Sidhu, M.; Shahidi, N.; Vosko, S.; McKay, O.; Bahin, F.F.; Zahid, S.; Whitfield, A.; Byth, K.; Brown, G.; et al. Effect of prophylactic endoscopic clip placement on clinically significant post-endoscopic mucosal resection bleeding in the right colon: A single-centre, randomised controlled trial. *Lancet Gastroenterol. Hepatol.* **2022**, *7*, 152–160. [PubMed]
53. Burgess, N.G.; Metz, A.J.; Williams, S.J.; Singh, R.; Tam, W.; Hourigan, L.F.; Zanati, S.A.; Brown, G.J.; Sonson, R.; Bourke, M.J. Risk factors for intraprocedural and clinically significant delayed bleeding after wide-field endoscopic mucosal resection of large colonic lesions. *Clin. Gastroenterol. Hepatol.* **2014**, *12*, 651–661.e3. [CrossRef]
54. Gupta, S.; Lieberman, D.; Anderson, J.C.; Burke, C.A.; Dominitz, J.A.; Kaltenbach, T.; Robertson, D.J.; Shaukat, A.; Syngal, S.; Rex, D.K. Recommendations for Follow-Up After Colonoscopy and Polypectomy: A Consensus Update by the US Multi-Society Task Force on Colorectal Cancer. *Gastrointest. Endosc.* **2020**, *91*, 463–485.e5.
55. Bick, B.L.; Ponugoti, P.L.; Rex, D.K. High yield of synchronous lesions in referred patients with large lateral spreading colorectal tumors. *Gastrointest. Endosc.* **2017**, *85*, 228–233. [CrossRef]
56. O'Sullivan, T.; Tate, D.; Sidhu, M.; Gupta, S.; Elhindi, J.; Byth, K.; Cronin, O.; Whitfield, A.; Craciun, A.; Singh, R.; et al. The Surface Morphology of Large Nonpedunculated Colonic Polyps Predicts Synchronous Large Lesions. *Clin. Gastroenterol. Hepatol.* **2023**, in press. [CrossRef]
57. Desomer, L.; Tutticci, N.; Tate, D.J.; Williams, S.J.; McLeod, D.; Bourke, M.J. A standardized imaging protocol is accurate in detecting recurrence after EMR. *Gastrointest. Endosc.* **2017**, *85*, 518–526. [CrossRef] [PubMed]

58. João, M.; Areia, M.; Pinto-Pais, T.; Gomes, L.C.; Saraiva, S.; Alves, S.; Elvas, L.; Brito, D.; Saraiva, S.; Teixeira-Pinto, A.; et al. Can white-light endoscopy or narrow-band imaging avoid biopsy of colorectal endoscopic mucosal resection scars? A multicenter randomized single-blind crossover trial. *Endoscopy* **2023**, *55*, 601–607. [CrossRef] [PubMed]
59. Burgess, N.; Bourke, M. Can we stop routine biopsy of post-endoscopic mucosal resection scars? *Endoscopy* **2023**, *55*, 608–610. [CrossRef]
60. Tate, D.J.; Bahin, F.F.; Desomer, L.; Sidhu, M.; Gupta, V.; Bourke, M.J. Cold-forceps avulsion with adjuvant snare-tip soft coagulation (CAST) is an effective and safe strategy for the management of non-lifting large laterally spreading colonic lesions. *Endoscopy* **2018**, *50*, 52–62. [CrossRef]

Disclaimer/Publisher's Note: The statements, opinions and data contained in all publications are solely those of the individual author(s) and contributor(s) and not of MDPI and/or the editor(s). MDPI and/or the editor(s) disclaim responsibility for any injury to people or property resulting from any ideas, methods, instructions or products referred to in the content.

Review

Endoscopic Management of Dysplastic Barrett's Oesophagus and Early Oesophageal Adenocarcinoma

Leonardo Henry Eusebi [1,2,*], Andrea Telese [3,4], Chiara Castellana [1,2], Rengin Melis Engin [1,2], Benjamin Norton [3,5,6], Apostolis Papaefthymiou [5], Rocco Maurizio Zagari [2,7] and Rehan Haidry [3,4,*]

1. Gastroenterology Unit, IRCCS Azienda Ospedaliero-Universitaria di Bologna, 40138 Bologna, Italy; chiara.castellana@outlook.it (C.C.); renginmelis.engin@studio.unibo.it (R.M.E.)
2. Department of Medical and Surgical Sciences, University of Bologna, 40138 Bologna, Italy; roccomaurizio.zagari@unibo.it
3. Digestive Disease and Surgery Institute Cleveland Clinic, London SW1X 7HY, UK; telesea@ccf.org (A.T.); benjamin.norton@nhs.net (B.N.)
4. Division of Surgery and Interventional Science, University College London, London NW1 2BU, UK
5. Department of Gastroenterology, University College London Hospital (UCLH), London NW1 2BU, UK; a.papaefthymiou@nhs.net
6. Centre for Obesity Research, Department of Medicine, Rayne Institute, University College London, London NW1 2BU, UK
7. Esophagus and Stomach Organic Diseases Unit, IRCCS Azienda Ospedaliero-Universitaria di Bologna, 40138 Bologna, Italy
* Correspondence: leonardo.eusebi@unibo.it (L.H.E.); haidryr@ccf.org (R.H.); Tel./Fax: +39-0512143338 (L.H.E.)

Simple Summary: Among individuals with gastro-esophageal reflux disease, the prevalence of histologically confirmed BO is around 7%, with variations according to different geographical regions. Since Barrett's oesophagus may progress to cancer through various stages of dysplasia, a correct diagnosis is pivotal in the management of patients with Barrett, made through accurate endoscopic examination and tissue sampling. The management of BO depends most strongly on the presence and severity of dysplasia, thus regular endoscopic surveillance and biopsies are required to monitor for neoplastic progression. In the presence of Barrett's-associated neoplasia, endoscopic treatments are utilised, including resection techniques and ablation therapies, and long-term data support their safety and efficacy. However, they are not without risk, and for the optimal management of BO-associated neoplasia, it is recommended that patients are referred to expert centres.

Abstract: Barrett's oesophagus is a pathological condition whereby the normal oesophageal squamous mucosa is replaced by specialised, intestinal-type metaplasia, which is strongly linked to chronic gastro-oesophageal reflux. A correct endoscopic and histological diagnosis is pivotal in the management of Barrett's oesophagus to identify patients who are at high risk of progression to neoplasia. The presence and grade of dysplasia and the characteristics of visible lesions within the mucosa of Barrett's oesophagus are both important to guide the most appropriate endoscopic therapy. In this review, we provide an overview on the management of Barrett's oesophagus, with a particular focus on recent advances in the diagnosis and recommendations for endoscopic therapy to reduce the risk of developing oesophageal adenocarcinoma.

Keywords: Barrett's oesophagus; endoscopic treatment; radiofrequency ablation

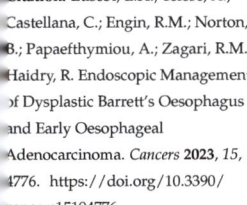

Citation: Eusebi, L.H.; Telese, A.; Castellana, C.; Engin, R.M.; Norton, B.; Papaefthymiou, A.; Zagari, R.M.; Haidry, R. Endoscopic Management of Dysplastic Barrett's Oesophagus and Early Oesophageal Adenocarcinoma. *Cancers* **2023**, *15*, 4776. https://doi.org/10.3390/cancers15194776

Academic Editor: B.P.L. (Bas) Wijnhoven

Received: 23 August 2023
Revised: 19 September 2023
Accepted: 20 September 2023
Published: 28 September 2023

Copyright: © 2023 by the authors. Licensee MDPI, Basel, Switzerland. This article is an open access article distributed under the terms and conditions of the Creative Commons Attribution (CC BY) license (https://creativecommons.org/licenses/by/4.0/).

1. Introduction

Barrett's oesophagus (BO) is a pathological condition that occurs due to metaplastic change that starts within the distal oesophagus. Here, the normal squamous mucosa is replaced by specialised intestinal-type metaplasia (SIM) that contains goblet cells [1,2].

Barrett's oesophagus is linked to gastro-oesophageal reflux disease (GORD) as an adaptive reaction to chronic reflux-induced damage [3]. The presence of reflux symptoms is significantly associated with an increased risk of BO, with the strongest association found between weekly reflux symptoms and long-segment BO [4]. Globally, about 15% of the general population experience reflux symptoms, with some areas reaching a prevalence of 50% [5,6]. However, among individuals with GORD, the prevalence of histologically confirmed BO is around 7%, with variations according to different geographical regions that range from 4% in Asian countries to 14% in North America [7,8].

Endoscopic screening may be considered in patients with chronic GORD symptoms who have at least three of the following risk factors: ≥50 years old, male sex, body mass index (BMI) ≥ 30 kg/m^2, and Caucasian ethnicity [7]. Also, the presence of hiatal hernia should be considered a significant risk factor for BO, whereas no association has been found between *Helicobacter pylori* infection and either endoscopically diagnosed or histologically confirmed BO [9,10]. Nevertheless, if a family history of Barrett's or oesophageal adenocarcinoma (OAC) is present, which should involve at least one first-degree relative, the threshold for screening is lower [11].

Barrett's oesophagus may progress to OAC through various stages: from non-dysplastic BO (NDBO) to low-grade dysplasia (LGD), to high-grade dysplasia (HGD), and eventually to OAC. The estimated annual risk of progression to OAC in NDBO is 0.3%, in LGD it is 1%, and the highest risk is in HGD, when it increases to 8% [12,13].

A correct diagnosis is pivotal in the management of patients with BO that is made through accurate endoscopic examination and tissue sampling. This both helps to establish a formal diagnosis of BO and identifies subjects at high risk of progression towards cancer. The management of BO depends most strongly on the presence and severity of dysplasia, although other individual variables are important. Initial treatment relies on lifestyle modification to reduce acid reflux with medications that suppress stomach acid production. This is combined with regular endoscopic surveillance and biopsies to monitor for neoplastic progression. In the presence of Barrett's-associated neoplasia, more invasive endoscopic treatments can be utilised, which include endoscopic mucosal resection (EMR) and endoscopic submucosal dissection (ESD) for the removal of visible dysplasia and endoscopic ablative techniques, including radiofrequency ablation (RFA) and cryotherapy, to eradicate non-visible dysplasia.

In this review, we discuss all aspects of the management of BO with a particular focus on recent advances in the diagnosis of SIM and the current evidence for endoscopic therapies to reduce the risk of progression towards OAC.

2. Diagnosis of Barrett's Oesophagus

Under normal conditions, the oesophagus is lined with a stratified squamous epithelium, which has a light-coloured and glossy appearance upon endoscopic inspection [14]. The squamocolumnar junction, also known as the Z-line, is a macroscopically visible line that marks the contact between the squamous and columnar epithelium. This is noticeable in the distal oesophagus at the level of the gastro-oesophageal junction (GOJ). In BO, there is proximal displacement of the squamocolumnar junction away from the GOJ [14]. The metaplastic columnar mucosa of BO is easily visible by its reddish colour (often described as 'salmon-coloured') and velvet-like texture compared to the pale and glossy squamous mucosa [15].

The diagnosis of BO requires a combination of endoscopic and histologic criteria. This involves the recognition of the abnormal distal oesophageal lining at the time of upper gastrointestinal endoscopy, which is then supported by histological evidence of a columnar-lined epithelium and an oesophageal SIM [14,16]. Consequently, oesophagogastroduodenoscopy (OGD) is considered the current gold standard for the diagnosis of BO [17].

An accurate diagnosis of BO relies on the precise delineation of the GOJ during endoscopy. This allows for determining whether there is a proximal migration of the squamocolumnar junction leading to a columnar-lined section of the epithelium within the lower oesophagus. The most straightforward landmark to define the GOJ, and the recommended minimum requirement, is the proximal limit of the longitudinal stomach folds with minimal air insufflation [18,19]. Thus, the initial endoscopic diagnosis of BO can be carried out when the endoscopist detects the presence of ≥1 centimetre of salmon-coloured mucosa extending proximally beyond the GOJ on the top of the gastric folds [20]. The use of incorrect landmarks for the GOJ can lead to a misclassification of BO, with a negative impact on the early diagnosis of neoplasia [21].

Barrett's oesophagus may appear during endoscopy as a lesion that extends segmentally or circumferentially. Once BO is identified, the Prague criteria are used to measure and classify its length, distinguishing between the circumferential extension (C) and the maximum longitudinal extension (M) of Barrett's metaplasia [22]. When the columnar epithelium rises at least 1 cm above the GOJ, interobserver agreement between endoscopists using the Prague criteria is excellent. However, the interobserver agreement was found to be poorer for shorter BO segments [14]. Moreover, short-segment BO is defined as having ≤3 cm of metaplastic epithelium, while long-segment BO is defined as having >3 cm of metaplastic epithelium above the GOJ [19]. An extension of the columnar epithelium < 1 cm, and in the absence of any confluent columnar-lined segment, should be considered as an irregular squamocolumnar junction rather than BO [11]. Indeed, patients with GORD are more likely to harbour an irregular Z-line, and up to 40% of biopsies taken from an irregular Z-line may contain intestinal metaplasia, but the relevance of this finding is still to be established [23,24]. Since the diagnosis of an irregular Z-line is subjective, and there is no accepted length cut-off to distinguish between an irregular Z-line and BO, it is suggested that 1 cm (M of the Prague criteria) should be the minimum length for an endoscopic diagnosis of BO. In general, biopsies are not recommended when an irregular Z-line is encountered, but they may be performed to aid the diagnosis depending on the degree of suspicion. If the biopsy specimens are taken within an irregular Z-line with no clear endoscopic evidence of BO, they should then be labelled as GOJ and not oesophageal biopsy samples [11].

However, the accuracy of a standard endoscopic examination with biopsy sampling for BO diagnosis can be limited by several factors, including the endoscopist's experience, the endoscopes definition, and the location of biopsies. Thus, to improve the diagnosis of oesophageal intestinal metaplasia and dysplasia, additional endoscopic procedures have been suggested. The ultimate objective of these newer methods is to enhance the endoscopic identification of curable Barrett's-associated neoplasia while lowering the procedure time, cost, and sampling error [25]. One of these procedures is chromoendoscopy, which has been increasingly used to improve the yield of SIM in BO [26]. As the name suggests, it involves the use of dyes, such as methylene blue, Lugol's iodine, indigo carmine, and acetic acid, which help to improve the detection rates by highlighting various features of the oesophageal mucosa [26,27]. Absorptive stains, such as Lugol's solution and methylene blue, identify specific epithelial cell types by preferential absorption or diffusion across the cell membrane. Contrast stains, such as indigo carmine, seep through mucosal crevices and highlight surface topography and mucosal irregularities [28]. Lugol's iodine is used for recognizing squamous tissue, squamous dysplasia, and squamous cell carcinomas because it preferentially stains the non-keratinised squamous epithelium [28]. This property makes it a good choice for staining oesophageal lesions as it is taken up by oesophageal squamous cells that contain glycogen [29]. However, it can be used to assess the success of the endoscopic therapy, as residual islands of Barrett's metaplasia are not stained by Lugol's iodine [29]. Methylene blue is a vital dye that may be used to detect BO because it is readily absorbed by columnar intestinal-type cells [30]. However, recent findings suggest that the detection of metaplasia by chromoendoscopy using methylene blue is not significantly different compared to the conventional four-quadrant biopsy technique, although the number of biopsies needed is significantly fewer [27,31]. Another factor to consider is the potential DNA-damaging effect of methylene blue on the Barrett's epithelium that may discourage its use [32]. The concomitant use of carmine dye or acetic acid staining with magnification endoscopy may enhance the recognition of different mucosal pit patterns in the columnar epithelium [29]. Chromoendoscopy with vital staining has been demonstrated to identify more patients with short-segment BO. Short segments are associated with a low yield of intestinal metaplasia (30–50%) when biopsy specimens are randomly acquired [33]. While chromoendoscopy significantly increases the detection of intestinal metaplasia and limits the number of biopsies required in short-segment BO, it does not appear to be beneficial in patients with an irregular Z-line (i.e., <1 cm of columnar mucosa in the distal oesophagus) [33].

In recent years, virtual chromoendoscopy has become available, enabling more practical chromoendoscopy without the use of dyes. Several virtual chromoendoscopy technologies have been developed, including narrow-band imaging (NBI; Olympus, Tokyo, Japan), blue light imaging (BLI; Fujifilm, Tokyo, Japan), and i-Scan (PENTAX Medical, Montvale, NJ, United States), based on light filters or post-image acquisition processing [11]. In comparison to standard resolution endoscopy, virtual chromoendoscopy allows for the better visualisation of mucosal glandular and vascular structures. Most evidence has been accumulated on the NBI system (Figure 1), including a prospective tandem study, which demonstrated that NBI led to a significantly higher rate of both the detection and grade of dysplasia with fewer biopsies [34]. Moreover, a recent meta-analysis of six studies reported a high diagnostic accuracy of NBI with targeted biopsies for detecting dysplasia of all grades compared to standard white light endoscopy with a standard biopsy protocol in a per-patient analysis. The authors reported a pooled sensitivity of NBI of 76% (95%CI: 0.61–0.91) and a pooled specificity of 99% (95%CI: 0.99–1.00) [35]. However, the interobserver agreement for the interpretation of virtual chromoendoscopy imaging is not always optimal, which may be a limitation in clinical practice [36]. Nevertheless, the increasing evidence for the potential benefits of virtual chromoendoscopy for the screening and surveillance of BO has led to its use being recommended when inspecting Barrett's segments [37].

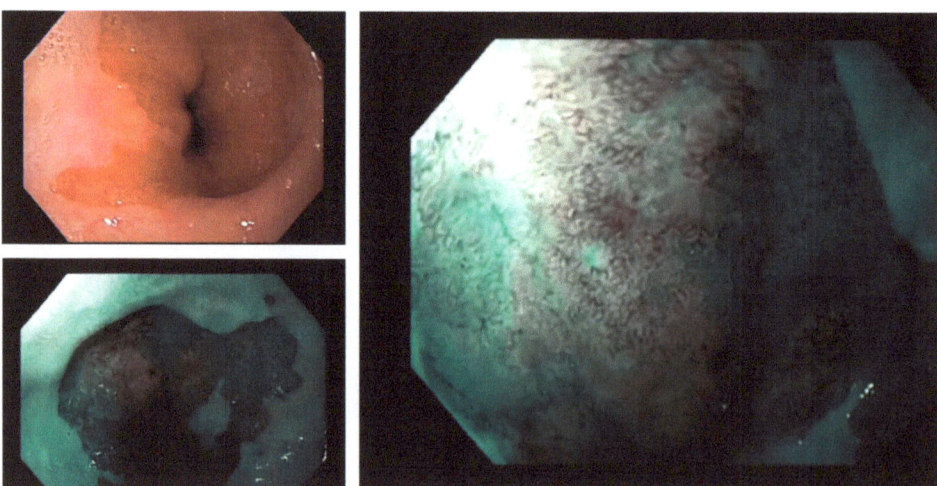

Figure 1. White light endoscopy, virtual chromoendoscopy (NBI), and virtual chromoendoscopy (NBI) with magnification images of Barrett's oesophagus.

Another recently developed diagnostic method is the confocal laser endomicroscopy (CLE) method. After an intravenous fluorescein injection, the oesophageal tissue is illuminated using a blue laser. This technique reproduces in vivo real-time imaging at a high magnification, allowing for the identification of suspicious lesions and for performing targeted biopsies [38]. A meta-analysis aiming to assess the accuracy of CLE for the diagnosis of neoplasia in BO, including more than 4000 lesions, showed a per-lesion pooled sensitivity and specificity of 77% (95%CI: 0.73–0.81) and 89% (95%CI: 0.87–0.90), respectively [39].

Thus, CLE appears to be very promising; however, its use is not currently recommended routinely but rather as an adjunctive imaging technique to identify dysplasia and cancer in select BO cases in expert centres [37].

When a diagnosis of BO is suspected during endoscopy, the endoscopist should perform biopsies following the "Seattle protocol". The presence of dysplasia within Barrett's mucosa is often patchy [40], which causes oesophageal biopsies to have a significant sampling error [41]. This protocol was designed to minimise the chance of missing a concealed lesion, which may be randomly distributed along the Barrett's epithelium. This protocol requires taking four-quadrant biopsy samples at every 1–2 cm intervals throughout the columnar-lined oesophagus. In addition, areas of any mucosal irregularity such as masses, nodules, and ulcerations must be sampled, as they are associated with a greater likelihood of harbouring dysplastic tissue [25]. The adherence to recommended surveillance procedures, such as the Seattle protocol, is associated with a higher dysplasia detection; however, it requires a lot of time, effort, and money and is still prone to sampling errors [41]. Therefore, it is unsurprising that adherence to such a protocol was found to be low among endoscopists, and adherence was inversely related to the length of the BO segment [41].

In Europe, the most widely used grading system for the histopathological diagnosis of BO-associated dysplasia is the revised Vienna classification. This original system was developed to standardise the terminology for the histological grading of gastrointestinal mucosal neoplasms. This was because of discrepancies in the grading systems used around the world for the categorisation of early neoplastic lesions. The system is versatile, dividing early mucosal lesions into one of five categories, and can be used for other gastrointestinal epithelial neoplastic or dysplastic lesions [42,43].

Three different types of columnar epithelia can be found in BO: a cardia-type epithelium almost completely composed of mucus-secreting cells; a gastric fundic-type epithelium with mucus-secreting cells, parietal cells, and chief cells; and an intestinal-type epithelium that is characterised by the presence goblet cells [44]. BO fundic- and cardia-type epithelia might morphologically look identical to the stomach's columnar epithelia. However, most of the professional guidelines concur that SIM is a necessary element for a formal diagnosis of BO [45].

Upon histological analysis, cells with BO-associated LGD exhibit mild architectural defects, an increased number of mitoses, and cytologic atypia, which includes an elevated nuclear/cytoplasmic ratio and nuclear elongation [46]. In the presence of LGD on random biopsies, the diagnosis should be confirmed by a second expert GI pathologist and referred to an expert centre [11,47–50]. It is also recommended to confirm the diagnosis of LGD after a surveillance interval of 6 months before offering endoscopic treatment [47–49]. Similarly, the diagnosis of HGD should be confirmed by a second expert GI pathologist and referred to a BO expert centre. Here, a repeat high-definition endoscopy should be completed to identify and treat all visible abnormalities in the dysplastic mucosa [48]. Therefore, the timing of endoscopic treatment is governed by the grade of dysplasia, and the visible characteristics of the dysplastic mucosa will guide to the most appropriate endoscopic therapy in the form of ablation or resection (Table 1).

Table 1. Guidelines on Barrett's oesophagus management.

	Scientific Society	Authors, Year [Ref]	ND-BO	LGD-BO	HGD-BO
European	BSG	Fitzgerald et al., 2014 [11] Di Pietro et al., 2018 [47]	• If maximum length < 3 cm, repeat OGD every 3–5 years. • If maximum length ≥ 3 cm, repeat OGD every 2–3 years.	• Repeat endoscopy in 6 months. • If LGD is confirmed, endoscopic ablation should be offered. • If ablation is not undertaken, 6-monthly surveillance.	• OGD in tertial referral centre. • If macroscopically visible lesion, endoscopic resection and RFA • If flat lining, RFA treatment.

Table 1. Cont.

	Scientific Society	Authors, Year [Ref]	ND-BO	LGD-BO	HGD-BO
European	ESGE	Weusten et al., 2017 [48]	• If columnar-lined oesophagus < 1 cm, no surveillance. • If BO ≥1 cm and < 3 cm, repeat OGD every 5 years. • If BO ≥ 3 cm and < 10 cm, repeat OGD every 3 years. • If BO ≥10 cm, refer to a BO expert centre.	• Repeat OGD at a BO expert centre in 6 months. • If no dysplasia is found, repeat after 1 year. After two subsequent endoscopies negative for dysplasia, follow standard surveillance for ND-BO. • If LGD is confirmed, endoscopic ablation should be offered.	• All visible abnormalities should be removed by endoscopic resection techniques. • If no suspicious visible lesions, take biopsies; if negative for dysplasia, repeat endoscopy at 3 months; if HGD confirmed, endoscopic ablation, preferably with RFA.
American	AGA	Sharma et al., 2020 [49]	• No endoscopic treatment indicated. • No indications on endoscopic surveillance.	• Repeat examination within 3–6 months to rule out visible lesions, which should prompt endoscopic resection. • Both endoscopic therapy and continued surveillance are reasonable options for the management of LGD-BO.	• Flat HGD should prompt a repeat HD-WLE (6–8 weeks) to evaluate for the presence of a visible lesion; these visible lesions should be removed by EMR. • Endoscopic therapy is the preferred treatment over oesophagectomy.
American	ACG	Shaheen et al., 2022 [50]	• If < 1 cm salmon-coloured mucosa or irregular Z-line, no biopsy. • If BO < 3 cm length, repeat OGD every 5 years. • If BO ≥ 3 cm length, repeat OGD every 3 years.	• Discuss risks and benefits of surveillance vs endoscopic therapy. • If surveillance, endoscopy every 6 months for one year, then annually. • If endoscopic therapy, resection of all visible lesions followed by ablation of the remaining BO.	• Endoscopic resection of all visible lesions followed by ablation of the remaining BO.
Asian Pacific	Asia-Pacific consensus	Fock et al., 2016 [51]	• No proven benefit in endoscopic surveillance of BO in the absence of dysplasia. • If surveillance, OGD every 3–5 years with biopsy protocol.	• Consider treatment or surveillance. • If treatment, resect visible lesions. In the absence of focal lesions, consider RFA. • If surveillance, repeat endoscopy in 6 months to confirm LGD.	• Endoscopic resection for BO with HGD and carcinoma in situ when visible lesions. • RFA to ablate all BO. • Surgery can be an alternative to endoscopic resection (with or without RFA).

BSG: British Society of Gastroenterology, ESGE: European Society of Gastrointestinal Endoscopy, AGA: American Gastroenterological Association, ACG: American College of Gastroenterology, ND-BO: non-dysplastic Barrett's oesophagus, LGD-BO: low-grade Barrett's oesophagus, HGD-BO: high-grade Barrett's oesophagus, OGD: oesophagogastroduodenoscopy, RFA: radiofrequency ablation

3. Endoscopic Resection Techniques for Dysplastic Barrett's and Early Oesophageal Adenocarcinoma

The main goal of an accurate endoscopic examination is to identify any suspicious lesions that require removal. For visible neoplasia, endoscopic resection (ER) is recommended. Elevated lesions are more likely to harbour neoplasia in comparison to flat lesions, but both require ER before ablation to increase the success of remission [52]. Moreover, one of the advantages of ER over ablation is that it allows for an accurate histological examination of the whole lesion. Indeed, changes in the histological stage after ER have been reported in up to a third of cases compared to initial biopsies, thus ensuring optimal management [17].

The two main ER techniques are EMR and ESD. The recent European Society of Gastrointestinal Endoscopy (ESGE) guidelines suggest using EMR for lesions ≤20 mm that have a low probability of submucosal invasion (i.e., Paris type 0-IIa, 0-IIb) and for larger or multifocal benign lesions. ESD should be performed for lesions suspicious for submucosal invasion (i.e., Paris type 0-Is, 0-IIc), for malignant lesions >20 mm, and for lesions in scarred or fibrotic areas [53]. Moreover, for deep excavated lesions (Paris type 0-III), ER is not recommended due to the high risk of deep invasion and lymph-node metastasis [54].

Early mucosal OAC, known as T1a, has a low risk of lymph node metastasis on systematic review (estimated around 2%), which means early, small lesions may be eligible for EMR [55]. A trial on 107 patients with BO-related lesions who were eligible for endoscopic treatment showed complete endoscopic eradication with EMR in 80% that included a T1a OAC in 36% [56].

With EMR, smaller lesions are usually resected en-bloc, whereas larger ones require multiple resections using the so-called piecemeal approach. Endoscopic mucosal resection can be performed using two main techniques: cap-snare and band-ligation. With the cap-snare technique, the lesion is first lifted and then drawn into the cap and resected by a snare. The band-ligation technique involves the release of an elastic band at the bottom of the lesion, which generates a pseudopolyp that can then be resected by a hot snare (Figure 2).

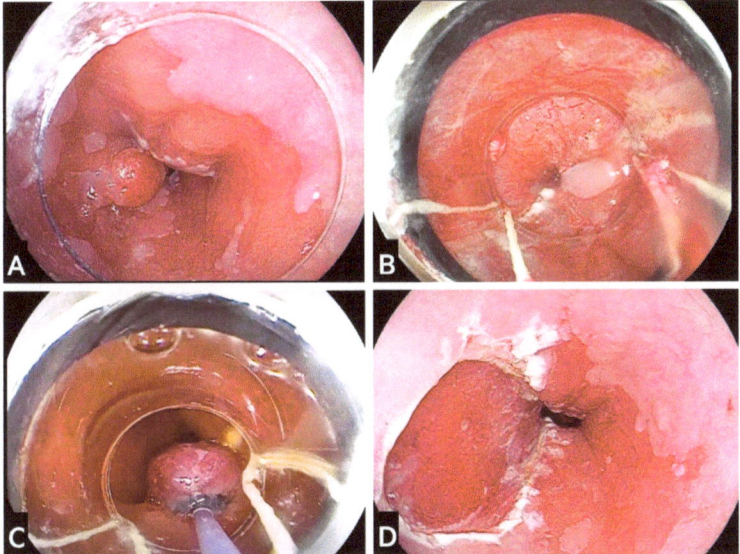

Figure 2. Endoscopic mucosal resection using multiband mucosectomy technique. (**A**) Evidence of a Paris Is nodule at 9 o'clock within a segment of Barrett's oesophagus; (**B**) plan for endoscopic mucosal resection using a multiband mucosectomy set that is applied over-the-scope; (**C**) the nodule is banded and a snare is placed around the lesion to enable resection; (**D**) final endoscopic view of the resection base.

The efficacy of both EMR techniques is comparable, although band-ligation is most commonly performed given its comparable ease and shorter time [20]. A comparative study between the two techniques has shown no significant differences in the maximum diameter of the resected specimen (approximately 16×11 mm for band-ligation versus 15×10 mm for cap-snare) nor any differences in the maximum diameter of the resected ulcer base after 24 h (approximately 21×14 mm for band-ligation versus 19×13 mm for cap-snare). In the same study, the overall complication rate was 2%, and the failure rate of ER was 7% with no significant differences between the two groups [57].

The ESD technique implies the use of an electrosurgical knife to dissect the submucosa underneath the lesion. Submucosal dissection is achieved after the injection of a lifting agent, which subsequently enables an en-bloc resection. This technique is particularly useful for the resection of large oesophageal lesions. Indeed, a higher percentage of en-bloc resections and subsequent lower rates of recurrence have been reported with ESD over traditional EMR [20,54].

Compared to surgery, ER appears effective for patients with dysplastic BO and early T1a OAC with a better safety profile [20]. In a meta-analysis of 7 studies that included 870 patients with HGD or T1a OAC, there was no significant difference in neoplastic remission or overall survival between surgery and endotherapy (ER and ablation). There was a higher rate of major adverse events in those undergoing surgery, whereas those having endotherapy had a higher rate of neoplasia recurrence, although most could be retreated successfully with endotherapy [58].

Endoscopic resection with ESD may be an alternative to surgery for T1b OAC, especially in patients who are poor surgical candidates. This is particularly true when the risk of lymph node metastasis is deemed low. Histopathological characteristics of T1b tumours associated with a low risk of lymph node metastasis include a tumour infiltration depth < 500 µm, the absence of poor differentiation, the absence of lymphovascular invasion, and clear deep resection margins (R0) [48]. The ability to achieve an R0 resection with ESD was determined in a retrospective cohort study that showed a rate of 87% for T1a OAC and 49% for T1b OAC [59].

A randomised clinical trial compared EMR to ESD for 40 patients with BO and HGD or early OAC, reporting that both ER techniques appeared to be highly effective in terms of the need for surgery, neoplasia remission, and recurrence [60]. Moreover, ESD achieved significantly higher en-bloc (R0) resection rates compared to EMR (59% vs. 12%), but the overall remission rates at 3 months were similar (94% vs. 94%); ESD was, however, more time consuming and caused severe AE more frequently [60].

A recent prospective study on 537 patients who underwent cap-assisted EMR or ESD followed by ablation showed that complete remission of dysplasia was higher in the patients treated with ESD compared to EMR at 2 years. Complete remission of intestinal metaplasia was similar in both groups, and there were no significant differences in complications [61].

The most common complication of ER is an oesophageal stricture, followed by bleeding and perforation. Strictures are related to the extension of the resected mucosa, with the risk increasing as the length of the circumferential resection increases. Perforation rates after EMR and ESD range between 0–5% and appear to be higher with ESD. Bleeding is common, but it is usually controlled with endoscopic haemostatic treatment. Admission for uncontrolled bleeding after ER is rare [62]. Konda et al., reported on the rates of complications among patients undergoing EMR. The rate of stricture formation was 41.5% (38% symptomatic), bleeding 3%, and perforation 19% [56].

After successful resection, recurrence rates up to 30% over 3 years have been observed for patients in whom the remaining BO is left untreated. Moreover, several studies have shown benefits in the outcomes of patients treated with RFA compared to endoscopic surveillance without any treatment, resulting in a reduced risk of neoplastic progression after RFA [63,64]. Therefore, endoscopic ablation is currently a guideline recommendation to achieve the complete eradication of all the remaining BO after ER of visible lesions [47–51].

Overall, straightforward cases should be managed following the current guidelines for BO follow-up and treatment (Figure 3). However, more complex cases require a case-by-case evaluation, taking into account several factors including institutional experience, the lesion's characteristics, and the patient's comorbidities and preferences, and they should therefore be discussed at MDT meetings in order to establish the most appropriate management.

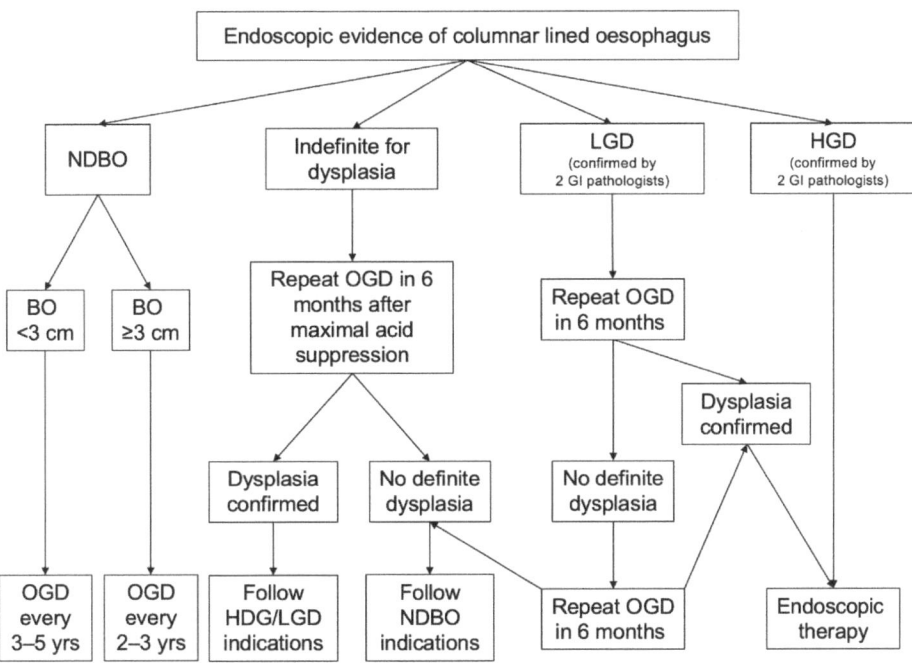

Figure 3. Endoscopic management of Barrett oesophagus based on the BSG guidelines [47].

4. Ablation Treatments for Barrett's Oesophagus

Endoscopic ablation of BO aims to destroy the abnormal mucosa to prevent further neoplastic progression. When discussing ablation techniques, it is crucial to emphasise that tissue disruption is limited to the mucosa. In addition, tissue coagulation prevents the acquisition of tissue for histological characterisation. This means that the endoscopist must be confident that the disease is limited to the mucosa to ensure complete eradication and avoid luminal or extraluminal recurrence.

Historically, several techniques have been used with different degrees of success. In the early 1990's, cases of BO ablation were reported using a neodymium-doped yttrium aluminium garnet (YAG) argon laser and photodynamic therapy [65–69]. These techniques were progressively abandoned in favour of argon plasma coagulation (APC) and multipolar electrocoagulation that were introduced later in the decade [70,71]. Their role in clinical practice remained controversial and not defined for almost a decade; in 2004, the American Gastroenterological Association workshop recognised the potential role of mucosal ablation in a subgroup of Barrett's patients. However, the selection criteria for patients that might benefit from mucosal ablation were not discussed by the working group [72]. More robust data were collected with the introduction of radiofrequency ablation (RFA), and the role of mucosal ablation was progressively recognised and advocated for by societal guidelines from the late 2000s and early 2010s [11,73,74].

At present, endoscopic ablation of BO is recommended for the treatment of residual BO following resection of any visible lesions and in patients with confirmed LGD to reduce the risk of progression toward more advanced neoplastic alterations, such as HGD and OAC [11,47,50].

As mentioned, several ablation techniques have been implemented over the last decades that can be carried out either by heating (e.g., RFA or APC) or freezing (e.g., cryoablation) to destroy the Barrett's epithelium. RFA is the most widely used in clinical practice; this technique uses a bipolar electrode in direct contact with the oesophageal mucosa to generate heat and induce a coagulative necrosis of the targeted mucosa. Among all the ablation techniques, RFA has the largest body of evidence, and its safety and efficacy has been evaluated in several studies including randomised trials and meta-analyses [75].

In 2009, Shaheen et al., published a randomised trial comparing ablation in BO-associated dysplasia compared to a control group using RFA. The rate of eradication in the patients with LGD was 90.5% compared with 23% in the control group, and the rate of eradication of HGD was 81% compared with 19% in the control group. The patients treated with RFA also had a significantly reduced disease progression (4% vs. 16%) and fewer cancers (1% vs. 9%), albeit with a stricture rate of 6% [76].

A subsequent large cohort study was published reporting the outcomes of 335 patients from the UK National Halo RFA Registry. This demonstrated eradication rates of 81% among all cases of BO-associated dysplasia after 12 months of treatment with a better response for short-segment BO [77]. The same authors showed an improvement in dysplasia and intestinal metaplasia clearance rates over the 6 years of observation from 77% to 92% and from 56% to 83%, respectively ($p < 0.0001$). In addition, the study demonstrated an increase in ER using EMR for visible lesions from 48% to 60% ($p = 0.013$) and a reduction in the rescue EMR following RFA from 13% to 2% ($p < 0.0001$). However, progression to OAC at 12 months remained statistically non-significant (3.6% vs. 2.1%, $p = 0.51$) [78].

Similarly, in 2014, Phoa et al., published the results of the SURF study showing that RFA is effective in treating LGD and eradicating BO. They achieved a complete eradication of dysplasia in 93% vs. 28% in the control group and a complete eradication of intestinal metaplasia in 88% vs. 0% in the control group ($p < 0.001$). This resulted in a reduced risk of progression to HGD or OAC by 25% and 7%, respectively. This study also showed that RFA presents an acceptable safety profile, with the most common adverse event being stricture formation that occurred in 12% of cases, which were treated with endoscopic dilatation [63]. The same cohort was analysed retrospectively after a median follow-up of 73 months in the study by Pouw et al. They reported a reduction in the absolute risk of BO progression following RFA of 32%, with only one case of progression to HGD/OAC in the RFA group (1.5%) during the follow-up compared to 23 cases in the surveillance group (34%). RFA achieved a complete BO clearance in 75 out of 83 patients, giving an eradication rate of 90%. Following RFA eradication, BO recurred in seven (9%) patients, of which three (4%) were diagnosed with LGD [64].

More recently, a prospective randomised study by Barret et al., showed a modest reduction in LGD and risk of progression at 3 years; indeed, the prevalence of LGD was 34% in the RFA group vs. 58% in the surveillance group (OR = 0.38; $p = 0.05$). Neoplastic progression was significantly higher in the surveillance group at 26% versus 12.5% in the RFA group ($p = 0.15$). A total of 22 adverse events were reported in the RFA group including bleeding or oesophageal stricture formation compared to no adverse events in the surveillance group. For this reason, the authors concluded that there was reduction in LGD prevalence and progression risk at three years The modest results in their study suggested that the risks and benefits of ablation should be weighed up carefully before proceeding to treat LGD dysplasia given the not-insignificant risk of complications after RFA [79].

In order to summarise these results, a recent meta-analysis by Shaheen [76], Phoa [63], and Barret [79] concluded that the pooled rate of progression of LGD to HGD or OAC was significantly lower in the RFA group than with endoscopic surveillance (RR 0.25; $p = 0.04$); however, the pooled risk of progression of LGD to OAC was slightly lower but not statistically significant (RR 0.56; $p = 0.65$). The patients in the RFA group also presented higher rates of complications including fever, bleeding, vomiting, nausea, and oesophageal strictures; hence, treatment options should be carefully weighed given the potential risk of oesophageal strictures following RFA treatment [80].

These data would suggest that, given the risk of potential complications and the heterogeneous rate of eradication, patient and centre selection remains paramount. In a retrospective analysis conducted in a French high-volume centre, including 96 consecutive patients with BO treated for dysplasia, there was a 59% rate of complete intestinal metaplasia eradication, a 79% rate of complete eradication of dysplasia, and a structure rate of 14% following RFA [81].

Catheter selection is another important aspect that is important to avoid overtreatment. There are different RFA catheters that can be used in practice depending on the clinical needs. In most scenarios, an over-the-scope RFA catheter can be used, which may include the Barrx 60, 90, and ultra-long catheters (Figure 4). These catheters differ in the dimension of the bipolar electrode, which includes a 15 mm long by 10 mm wide (Barrx 60), a 20 mm long by 13 mm wide (Barrx 90), and a 40 mm long by 13 mm wide (Barrx ultra long) electrode. A through-the-scope device is also available (Barrx channel 15.7 mm long by 7.5 mm wide) that has a flexible bipolar electrode, which is folded and passed into the working channel of a standard gastroscope. This device is particularly useful to target smaller Barrett's segments or treat patients with strictures that might not accommodate larger over-the-scope devices. For long circumferential segments, the use of a Barrx 360 express device can be considered. This lies separate to the scope and is inserted over a guidewire. It consists of a self-inflating and self-sizing balloon that is covered by a winding RFA electrode measuring 4 cm in length.

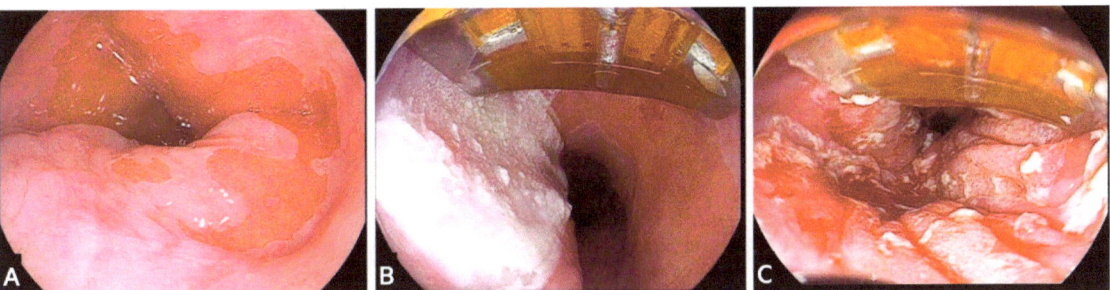

Figure 4. Radiofrequency ablation of dysplastic Barrett's segment using a focal catheter. (**A**) Evidence of dysplastic Barrett's segment undergoing ablation therapy; (**B**) over-the-scope ultra Barrx 90 focal RFA catheter with an area of ablated mucosa at 9 o'clock; (**C**) final endoscopic view following ablation of all Barrett's segments.

Among these, the Barrx 360 express might present the highest rate of patient-related and device-related adverse events. In a recent post-marketing surveillance data analysis from August 2011 to August 2021 from the Food and Drug Administration's Manufacturer and User Facility Device, a total of 87 patient-related adverse events including 15 strictures (17.2%), 13 mucosal laceration (14.9%), and 10 episodes of chest pain (11.4%) were reported in addition to 78 device-related malfunctions for RFA devices. The Barrx 360 express was involved in 61% of patient-related adverse events and 67% of device malfunction events; all the 15 oesophageal strictures secondary to treatment occurred with circumferential ablation devices [82].

Argon plasma coagulation is one of the first methods that was used to treat BO. Technically, APC is a non-contact thermal ablation technique that uses a through-the-scope catheter to deliver argon gas to the targeted mucosa. The gas is ionised when in contact with a high-voltage current on the tip of the catheter, and the resulting plasma causes thermal tissue coagulation. In 2006, the APBANEX prospective multicentre study showed a 77% rate of complete eradication of non-neoplastic BO treated with APC (90 W) in combination with esomeprazole 80 mg/day [83]. Following this, a randomised pilot study (the BRIDE study) has suggested that APC might have a similar efficacy and safety compared to RFA for the treatment of BO with HGD or OAC, with a more favourable cost difference [84].

APC can also be delivered following a submucosal injection of saline; this technique is known as hybrid APC (H-APC). The proposed advantage of the submucosal injection is to insulate and protect the subepithelial layers of the oesophagus, resulting in a lower stricture rate. An ex-vivo animal study showed that H-APC could reduce the coagulation depth compared to traditional APC and minimise thermal injury to the submucosal and muscular layers [85]. The same authors conduced a pilot study showing that H-APC achieved a complete macroscopical remission in 48 out of 50 treated patients (96%) after a median of 3.5 sessions (range 1–10). In this study, the histopathological eradication of BO was 78%, and the stricture rate was 2% [86].

Additional pilot studies and case series are available on the use of H-APC. The largest prospective study to date enrolled 146 patients and reported that, after 2 years, a total of 85 (66%) patients presented no recurrence of BO. In this study, H-APC showed an adverse event rate of 6%, with a stricture rate of 4% [87].

A recent systematic review and meta-analysis, pooling data from seven studies on H-APC, showed an overall complete remission rate of intestinal metaplasia of 91%, with an overall adverse event rate of 3%, including a stricture rate of 2%. However, as mentioned by the authors, the inclusion of non-controlled studies, retrospective cohorts, and case series might have lowered the overall quality of evidence [88]. Therefore, APC appears to be a safe and effective technique for the treatment of dysplastic Barrett's; however, its use might be limited when long segments of BO need to be eradicated. Therefore, this technique should be considered in cases of short or focal segments of metaplasia or those refractory to RFA [89].

Other methods based on cold ablation have also been tested. The most recent literature on cold ablation refers to a C2 cryoballoon ablation system (CbAS). This consists of a though-the-scope self-inflating, self-sizing balloon catheter, which is attached to a controller that regulates a flow of nitrous oxide into the balloon to freeze the oesophageal mucosa. Cellular disruption is obtained following intracellular ice crystal formation that alters the cellular architecture. This technique does not require energy generation and might be particularly suited to treat segments of BO in patients that have strictures which cannot be traversed with an RFA catheter. Initial evidence showed that CbAS is safe and effective for the treatment of short-segment BO, with a 95% rate of complete eradication of dysplasia and intestinal metaplasia [90]. In a larger study including a total of 120 patients with LGD, HGD, and intramucosal adenocarcinoma, CbAS achieved a complete eradication rate of dysplasia and intestinal metaplasia of 97% and 91%, respectively, with a stricture rate of 12.5% [91]. A systematic review and meta-analysis of 272 patients showed a pooled rate of compete eradication of intestinal metaplasia and dysplasia of 86% and 94%, respectively, with an adverse event rate of 12.5% [92]. A retrospective study showed a comparable outcome for dysplastic BO treatment compared to RFA, with a possible higher stricture rate (10.4% vs 4.4% $p = 0.04$) [93]. A large, prospective, European, multicentre study (EURO-COLDPLAY) is investigating the efficacy and safety of a focal cryoballoon for the treatment of BO, and an interim analysis has suggested using an 8 s rather than a standard 10 s duration of treatment. This is because the BO regression rates were similar, but there was a theoretical lower rate of stricture formation [94]. Novel cryoballoon devices should be implemented soon to treat larger segments of BO [95,96].

5. Failure of Therapy

Despite several advancements, the treatment of BO with a single technique might not achieve complete eradication. Therefore, a multimodal approach might be required for selected patients. Mittal et al., suggest considering further diagnostic tests and a multimodal approach in patients without a significant response after three sessions [97]. van Munster et al., retrospectively analysed the outcomes of patients undergoing RFA from the nationwide Dutch registry. In their study, 134 out of 1386 patients had poor mucosal healing following RFA that could be resolved with appropriate acid suppression and additional time; indeed, 67 of those 134 patients (50%) had normal squamous regeneration, achieving a complete eradication of BO in 97% of cases. These rates are similar to patients presenting with normal mucosal healing, whereas the remaining 67 patients (50%) had poor healing followed by poor squamous regeneration. A total of 74 out of 1386 patients (5%) with poor squamous epithelial regeneration had a higher risk of treatment failure (64% vs. 2%) and an increased risk of disease progression (15% vs. <1%) when compared to patients achieving normal mucosal regeneration. This study also identified risk factors independently associated with poor squamous regeneration such as higher body mass index, longer BO segments, reflux oesophagitis, and <50% squamous regeneration after baseline ER [98]. The same authors developed a model that identified other poor prognostic indicators associated with complex treatment courses such as those with a BO length ≥ 9 cm, the presence of HGD or OAC, and poor squamous epithelial regeneration [99].

6. Follow-up after Ablation

Long-term data from the UK National RFA registry show a risk of cancer at 10 years after ablation of 4% and a recurrence of dysplasia and intestinal metaplasia at 8 years of 6% and 19%, respectively. Nevertheless, most cases were treatable with the same modality [100].

Several post-RFA endoscopic surveillance intervals have been proposed; at present, the ACG guidelines are felt to be the most cost-effective strategy, suggesting endoscopic surveillance at 6 and 12 months followed by annual surveillance for patients with LGD, and a more intense surveillance at 3, 6, 9, 12, 18, and 24 months followed by annual surveillance for patients presenting with HGD [101].

van Munster et al., developed a prediction model of dysplasia recurrence after analysing data from 1154 patients during a mean follow-up of 4 years. During this time, a total of 38 patients developed recurrent disease (1% per person-year), and the authors identified some factors associated with recurrence such as the presence of new incident visible lesions during the treatment phase, a high number of endoscopic mucosal resections, male sex, an increased length of BO, the presence of HGD or OAC at baseline, and younger age [102]. This study could pave the road to further studies aiming to define predictors of recurrence for a more personalised surveillance strategy. This is particularly important because patients with BO-associated neoplasia are often frail and co-morbid and, following a successful endoscopic eradication therapy, are more likely to die from non-OAC causes [103].

7. Medical Management after the Endoscopic Treatment

Effective acid-suppression is considered to be an important condition for mucosal healing and squamous regeneration of BO following endoscopic therapy. The AGA expert review recommends the use of a proton pump inhibitor twice daily, and this was also further highlighted in a more recent paper by van Munster et al. [49,98]. H2-receptor agonist and sucralfate have also been used in European centres [104]; however, comparative studies on these drug regimens are lacking, and no definitive recommendation has been given.

Pain relief is another aspect to take into consideration following endoscopic intervention. It can usually be achieved with painkillers such as paracetamol, but there is no formal societal guidance on this issue. Similarly, no recommendations are available on diet; however, maintaining a liquid/soft diet in the days following the intervention might be reasonable to minimise the chances of traumatic injury.

Medical prevention of oesophageal strictures following the endoscopic treatment of BO is another aspect that has been investigated. A network meta-analysis conducted in 2019 showed that oral steroids might prevent postoperative strictures [105]. A more recent study evaluating the role of topical budesonide in patients undergoing oesophageal EMR or ESD showed conflicting results: no significant difference in stricture rates was seen in the patients taking topical budesonide compared to the patients not taking steroids (16% vs. 28%; $p = 0.23$); however, a logistic regression analysis taking into account potential confounders showed that the stricture rate was significantly lower (91%; 95%CI 0.0084–0.573; $p = 0.023$) in the budesonide cohort. The authors therefore suggest caution against concluding that budesonide is not effective, highlighting that, in their multivariate analysis, budesonide was associated with a lower stricture rate and concluding that budesonide might have a role in preventing stricture formation following oesophageal EMR and ESD [106].

Finally, with regards to the long-term management of patients with BO, other anti-reflux measures, such as surgical and endoscopic procedures, could be considered to address chronic gastroesophageal reflux insult, particularly in young patients who would require life-long endoscopic follow-up or for subjects who do not tolerate PPIs. However, anti-reflux surgery is not currently recommended by the ACG guidelines as an antineoplastic measure in patients with BO [50].

8. Conclusions and Future Directions

In the Western world, the incidence of BO is increasing alongside the rise in gastroesophageal reflux disease, its major risk factor. Although only a minority of patients with BO will progress to oesophageal adenocarcinoma, identifying high-risk individuals is pivotal, because oesophageal cancer is still associated with a high five-year mortality and associated care costs for healthcare systems. Future efforts should focus on improving endoscopists adherence to guidelines, with particular attention to the diagnosis of BO that includes the use of advanced endoscopic imaging and appropriate surveillance intervals.

Endoscopic endotherapy using both resection and ablation techniques have long-term data supporting their safety and efficacy; however, they are not without risk, and for the optimal management of BO-associated neoplasia, it is recommended that patients are referred to expert centres.

Author Contributions: Conceptualization, L.H.E., A.T. and R.H.; methodology, L.H.E. and A.T.; writing—original draft preparation, A.T., C.C. and R.M.E.; writing—review and editing, L.H.E., B.N and A.P.; supervision, L.H.E., R.M.Z. and R.H. All authors have read and agreed to the published version of the manuscript.

Funding: This article received no external funding.

Conflicts of Interest: R.H. has received educational grants to support their research infrastructure from Cook Endoscopy, Odin Vision, Pentax medical, Endogastric solutions, Apollo endosurgery, Medtronic.

References

1. Shaheen, N.J.; Falk, G.W.; Iyer, P.G.; Gerson, L.B. Acg Clinical Guideline: Diagnosis and Management of Barrett's Esophagus. *Am. J. Gastroenterol.* **2016**, *111*, 30–50. [CrossRef] [PubMed]
2. Burke, Z.D.; Tosh, D. Barrett's Metaplasia as a Paradigm for Understanding the Development of Cancer. *Curr. Opin. Genet. Dev.* **2012**, *22*, 494–499. [CrossRef] [PubMed]
3. Winters, C., Jr.; Spurling, T.J.; Chobanian, S.J.; Curtis, D.J.; Esposito, R.L.; Hacker, J.F., 3rd; Johnson, D.A.; Cruess, D.F.; Cotelingam, J.D.; Gurney, M.S.; et al. Barrett's Esophagus. A Prevalent, Occult Complication of Gastroesophageal Reflux Disease. *Gastroenterology* **1987**, *92*, 118–124. [CrossRef] [PubMed]
4. Eusebi, L.H.; Telese, A.; Cirota, G.G.; Haidry, R.; Zagari, R.M.; Bazzoli, F.; Ford, A.C. Effect of Gastro-Esophageal Reflux Symptoms on the Risk of Barrett's Esophagus: A Systematic Review and Meta-Analysis. *J. Gastroenterol. Hepatol.* **2022**, *37*, 1507–1516 [CrossRef]
5. El-Serag, H.B.; Sweet, S.; Winchester, C.C.; Dent, J. Update on the Epidemiology of Gastro-Oesophageal Reflux Disease: A Systematic Review. *Gut* **2014**, *63*, 871–880. [CrossRef]
6. Eusebi, L.H.; Ratnakumaran, R.; Yuan, Y.; Solaymani-Dodaran, M.; Bazzoli, F.; Ford, A.C. Global Prevalence of, and Risk Factors for, Gastro-Oesophageal Reflux Symptoms: A Meta-Analysis. *Gut* **2018**, *67*, 430–440. [CrossRef]
7. Qumseya, B.J.; Bukannan, A.; Gendy, S.; Ahemd, Y.; Sultan, S.; Bain, P.; Gross, S.A.; Iyer, P.; Wani, S. Systematic Review and Meta-Analysis of Prevalence and risk factors for Barrett's Esophagus. *Gastrointest. Endosc.* **2019**, *90*, 707–717.E1. [CrossRef]
8. Eusebi, L.H.; Cirota, G.G.; Zagari, R.M.; Ford, A.C. Global Prevalence of Barrett's Oesophagus and Oesophageal Cancer in Individuals with Gastro-Oesophageal Reflux: A Systematic Review and Meta-Analysis. *Gut* **2021**, *70*, 456–463. [CrossRef]
9. Zamani, M.; Alizadeh-Tabari, S.; Hasanpour, A.H.; Eusebi, L.H.; Ford, A.C. Systematic Review with Meta-Analysis: Association of Helicobacter Pylori Infection with Gastro-Oesophageal Reflux and Its Complications. *Aliment. Pharmacol. Ther.* **2021**, *54*, 988–998. [CrossRef]
10. Eusebi, L.H.; Telese, A.; Cirota, G.G.; Haidry, R.; Zagari, R.M.; Bazzoli, F.; Ford, A.C. Systematic Review with Meta-Analysis: Risk Factors for Barrett's Oesophagus in Individuals with Gastro-Oesophageal Reflux Symptoms. *Aliment. Pharmacol. Ther.* **2021**, *53*, 968–976. [CrossRef]
11. Fitzgerald, R.C.; Di Pietro, M.; Ragunath, K.; Ang, Y.; Kang, J.Y.; Watson, P.; Trudgill, N.; Patel, P.; Kaye, P.V.; Sanders, S.; et al British Society of Gastroenterology Guidelines on the Diagnosis and Management of Barrett's Oesophagus. *Gut* **2014**, *63*, 7–42 [CrossRef] [PubMed]

12. Singh, S.; Manickam, P.; Amin, A.V.; Samala, N.; Schouten, L.J.; Iyer, P.G.; Desai, T.K. Incidence of Esophageal Adenocarcinoma in Barrett's Esophagus with Low-Grade Dysplasia: A Systematic Review and Meta-Analysis. *Gastrointest. Endosc.* **2014**, *79*, 897–909.E4. [CrossRef]
13. Rastogi, A.; Puli, S.; El-Serag, H.B.; Bansal, A.; Wani, S.; Sharma, P. Incidence of Esophageal Adenocarcinoma in Patients with Barrett's Esophagus and High-Grade Dysplasia: A Meta-Analysis. *Gastrointest. Endosc.* **2008**, *67*, 394–398. [CrossRef] [PubMed]
14. Anand, O.; Wani, S.; Sharma, P. When and How to Grade Barrett's Columnar Metaplasia: The Prague System. *Best. Pract. Res. Clin. Gastroenterol.* **2008**, *22*, 661–669. [CrossRef] [PubMed]
15. Spechler, S.J. Barrett Esophagus and Risk of Esophageal Cancer: A Clinical Review. *JAMA* **2013**, *310*, 627–636. [CrossRef]
16. Sampliner, R.E. Updated Guidelines for the Diagnosis, Surveillance, and Therapy of Barrett's Esophagus. *Am. J. Gastroenterol.* **2002**, *97*, 1888–1895. [CrossRef] [PubMed]
17. Maione, F.; Chini, A.; Maione, R.; Manigrasso, M.; Marello, A.; Cassese, G.; Gennarelli, N.; Milone, M.; De Palma, G.D. Endoscopic Diagnosis and Management of Barrett's Esophagus with Low-Grade Dysplasia. *Diagnostics* **2022**, *12*, 1295. [CrossRef]
18. Mcclave, S.A.; Boyce, H.W., Jr.; Gottfried, M.R. Early Diagnosis of Columnar-Lined Esophagus: A New Endoscopic Diagnostic Criterion. *Gastrointest. Endosc.* **1987**, *33*, 413–416. [CrossRef]
19. Sharma, P.; Morales, T.G.; Sampliner, R.E. Short Segment Barrett's Esophagus—The Need for Standardization of the Definition and of Endoscopic Criteria. *Am. J. Gastroenterol.* **1998**, *93*, 1033–1036. [CrossRef]
20. Cotton, C.C.; Eluri, S.; Shaheen, N.J. Management of Dysplastic Barrett's Esophagus and Early Esophageal Adenocarcinoma. *Gastroenterol. Clin. N. Am.* **2022**, *51*, 485–500. [CrossRef]
21. Zagari, R.M.; Eusebi, L.H.; Galloro, G.; Rabitti, S.; Neri, M.; Pasquale, L.; Bazzoli, F. Attending Training Courses on Barrett's Esophagus Improves Adherence to Guidelines: A Survey from the Italian Society of Digestive Endoscopy. *Dig. Dis. Sci.* **2021**, *66*, 2888–2896. [CrossRef] [PubMed]
22. Sharma, P.; Dent, J.; Armstrong, D.; Bergman, J.J.; Gossner, L.; Hoshihara, Y.; Jankowski, J.A.; Junghard, O.; Lundell, L.; Tytgat, G.N.; et al. The Development and Validation of an Endoscopic Grading System for Barrett's Esophagus: The Prague C & M Criteria. *Gastroenterology* **2006**, *131*, 1392–1399. [CrossRef]
23. Kim, J.B.; Shin, S.R.; Shin, W.G.; Choi, M.H.; Jang, H.J.; Kim, K.O.; Park, C.H.; Baek, I.H.; Baik, G.H.; Kim, K.H.; et al. Prevalence of Minimal Change Lesions in Patients with Non-Erosive Reflux Disease: A Case-Control Study. *Digestion* **2012**, *85*, 288–294. [CrossRef] [PubMed]
24. Dickman, R.; Levi, Z.; Vilkin, A.; Zvidi, I.; Niv, Y. Predictors of Specialized Intestinal Metaplasia in Patients with an Incidental Irregular Z Line. *Eur. J. Gastroenterol. Hepatol.* **2010**, *22*, 135–138. [CrossRef] [PubMed]
25. Spechler, S.J.; Sharma, P.; Souza, R.F.; Inadomi, J.M.; Shaheen, N.J. American Gastroenterological Association Technical Review on the Management of Barrett's Esophagus. *Gastroenterology* **2011**, *140*, E18–E52. [CrossRef]
26. Canto, M.I.; Setrakian, S.; Willis, J.E.; Chak, A.; Petras, R.E.; Sivak, M.V. Methylene Blue Staining of Dysplastic and Nondysplastic Barrett's Esophagus: An in Vivo and Ex Vivo Study. *Endoscopy* **2001**, *33*, 391–400. [CrossRef]
27. Ngamruengphong, S.; Sharma, V.K.; Das, A. Diagnostic Yield of Methylene Blue Chromoendoscopy for Detecting Specialized Intestinal Metaplasia and Dysplasia in Barrett's Esophagus: A Meta-Analysis. *Gastrointest. Endosc.* **2009**, *69*, 1021–1028. [CrossRef]
28. Wong Kee Song, L.M.; Adler, D.G.; Chand, B.; Conway, J.D.; Croffie, J.M.; Disario, J.A.; Mishkin, D.S.; Shah, R.J.; Somogyi, L.; Tierney, W.M.; et al. Chromoendoscopy. *Gastrointest. Endosc.* **2007**, *66*, 639–649. [CrossRef]
29. Guelrud, M.; Herrera, I.; Essenfeld, H.; Castro, J. Enhanced Magnification Endoscopy: A New Technique to Identify Specialized Intestinal Metaplasia in Barrett's Esophagus. *Gastrointest. Endosc.* **2001**, *53*, 559–565. [CrossRef]
30. Canto, M.I. Vital Staining and Barrett's Esophagus. *Gastrointest. Endosc.* **1999**, *49*, S12–S16. [CrossRef]
31. Horwhat, J.D.; Maydonovitch, C.L.; Ramos, F.; Colina, R.; Gaertner, E.; Lee, H.; Wong, R.K. A Randomized Comparison of Methylene Blue-Directed Biopsy Versus Conventional Four-Quadrant Biopsy for the Detection of Intestinal Metaplasia and Dysplasia in Patients with Long-Segment Barrett's Esophagus. *Am. J. Gastroenterol.* **2008**, *103*, 546–554. [CrossRef] [PubMed]
32. Olliver, J.R.; Wild, C.P.; Sahay, P.; Dexter, S.; Hardie, L.J. Chromoendoscopy with Methylene Blue and Associated DNA Damage in Barrett's Oesophagus. *Lancet* **2003**, *362*, 373–374. [CrossRef]
33. Sharma, P.; Topalovski, M.; Mayo, M.S.; Weston, A.P. Methylene Blue Chromoendoscopy for Detection of Short-Segment Barrett's Esophagus. *Gastrointest. Endosc.* **2001**, *54*, 289–293. [CrossRef] [PubMed]
34. Wolfsen, H.C.; Crook, J.E.; Krishna, M.; Achem, S.R.; Devault, K.R.; Bouras, E.P.; Loeb, D.S.; Stark, M.E.; Woodward, T.A.; Hemminger, L.L.; et al. Prospective, Controlled Tandem Endoscopy Study of Narrow Band Imaging for Dysplasia Detection in Barrett's Esophagus. *Gastroenterology* **2008**, *135*, 24–31. [CrossRef]
35. Hajelssedig, O.E.; Zorron Cheng Tao Pu, L.; Thompson, J.Y.; Lord, A.; El Sayed, I.; Meyer, C.; Shaukat Ali, F.; Abdulazeem, H.M.; Kheir, A.O.; Siepmann, T.; et al. Diagnostic Accuracy of Narrow-Band Imaging Endoscopy with Targeted Biopsies Compared with Standard Endoscopy with Random Biopsies in Patients with Barrett's Esophagus: A Systematic Review and Meta-Analysis. *J. Gastroenterol. Hepatol.* **2021**, *36*, 2659–2671. [CrossRef]
36. Curvers, W.L.; Bohmer, C.J.; Mallant-Hent, R.C.; Naber, A.H.; Ponsioen, C.I.; Ragunath, K.; Singh, R.; Wallace, M.B.; Wolfsen, H.C.; Song, L.M.; et al. Mucosal Morphology in Barrett's Esophagus: Interobserver Agreement and Role of Narrow Band Imaging. *Endoscopy* **2008**, *40*, 799–805. [CrossRef] [PubMed]

37. Muthusamy, V.R.; Wani, S.; Gyawali, C.P.; Komanduri, S.; Participants, C.B.S.E.C.C. Aga Clinical Practice Update on New Technology and Innovation for Surveillance and Screening in Barrett's Esophagus: Expert Review. *Clin. Gastroenterol. Hepatol.* **2022**, *20*, 2696–2706.E1. [CrossRef] [PubMed]
38. Sharma, P.; Meining, A.R.; Coron, E.; Lightdale, C.J.; Wolfsen, H.C.; Bansal, A.; Bajbouj, M.; Galmiche, J.P.; Abrams, J.A.; Rastogi, A.; et al. Real-Time Increased Detection of Neoplastic Tissue in Barrett's Esophagus with Probe-Based Confocal Laser Endomicroscopy: Final Results of an International Multicenter, Prospective, Randomized, Controlled Trial. *Gastroint. Endosc.* **2011**, *74*, 465–472. [CrossRef]
39. Xiong, Y.Q.; Ma, S.J.; Zhou, J.H.; Zhong, X.S.; Chen, Q. A Meta-Analysis of Confocal Laser Endomicroscopy for the Detection of Neoplasia in Patients with Barrett's Esophagus. *J. Gastroenterol. Hepatol.* **2016**, *31*, 1102–1110. [CrossRef]
40. Mcardle, J.E.; Lewin, K.J.; Randall, G.; Weinstein, W. Distribution of Dysplasias and Early Invasive Carcinoma in Barrett's Esophagus. *Hum. Pathol.* **1992**, *23*, 479–482. [CrossRef]
41. Abrams, J.A.; Kapel, R.C.; Lindberg, G.M.; Saboorian, M.H.; Genta, R.M.; Neugut, A.I.; Lightdale, C.J. Adherence to Biopsy Guidelines for Barrett's Esophagus Surveillance in the Community Setting in the United States. *Clin. Gastroenterol. Hepatol.* **2009**, *7*, 736–742. [CrossRef] [PubMed]
42. Schlemper, R.J.; Riddell, R.H.; Kato, Y.; Borchard, F.; Cooper, H.S.; Dawsey, S.M.; Dixon, M.F.; Fenoglio-Preiser, C.M.; Fléjou, J.F.; Geboes, K.; et al. The Vienna Classification of Gastrointestinal Epithelial Neoplasia. *Gut* **2000**, *47*, 251–255. [CrossRef] [PubMed]
43. Dixon, M.F. Gastrointestinal Epithelial Neoplasia: Vienna Revisited. *Gut* **2002**, *51*, 130–131. [CrossRef]
44. Paull, A.; Trier, J.S.; Dalton, M.D.; Camp, R.C.; Loeb, P.; Goyal, R.K. The Histologic Spectrum of Barrett's Esophagus. *N. Engl. J. Med.* **1976**, *295*, 476–480. [CrossRef] [PubMed]
45. Spechler, S.J.; Souza, R.F. Barrett's Esophagus. *N. Engl. J. Med.* **2014**, *371*, 836–845. [CrossRef]
46. Levine, D.S.; Blount, P.L.; Rudolph, R.E.; Reid, B.J. Safety of a Systematic Endoscopic Biopsy Protocol in Patients with Barrett's Esophagus. *Am. J. Gastroenterol.* **2000**, *95*, 1152–1157. [CrossRef]
47. Di Pietro, M.; Fitzgerald, R.C. Revised British Society of Gastroenterology Recommendation on the Diagnosis and Management of Barrett's Oesophagus with Low-Grade Dysplasia. *Gut* **2018**, *67*, 392–393. [CrossRef]
48. Weusten, B.; Bisschops, R.; Coron, E.; Dinis-Ribeiro, M.; Dumonceau, J.M.; Esteban, J.M.; Hassan, C.; Pech, O.; Repici, A.; Bergman, J.; et al. Endoscopic Management of Barrett's Esophagus: European Society of Gastrointestinal Endoscopy (Esge) Position Statement. *Endoscopy* **2017**, *49*, 191–198. [CrossRef]
49. Sharma, P.; Shaheen, N.J.; Katzka, D.; Bergman, J. Aga Clinical Practice Update on Endoscopic Treatment of Barrett's Esophagus with Dysplasia and/or Early Cancer: Expert Review. *Gastroenterology* **2020**, *158*, 760–769. [CrossRef]
50. Shaheen, N.J.; Falk, G.W.; Iyer, P.G.; Souza, R.F.; Yadlapati, R.H.; Sauer, B.G.; Wani, S. Diagnosis and Management of Barrett's Esophagus: An Updated Acg Guideline. *Am. J. Gastroenterol.* **2022**, *117*, 559–587. [CrossRef]
51. Fock, K.M.; Talley, N.; Goh, K.L.; Sugano, K.; Katelaris, P.; Holtmann, G.; Pandolfino, J.E.; Sharma, P.; Ang, T.L.; Hongo, M.; et al. Asia-Pacific Consensus on the Management of Gastro-Oesophageal Reflux Disease: An Update Focusing on Refractory Reflux Disease and Barrett's Oesophagus. *Gut* **2016**, *65*, 1402–1415. [CrossRef]
52. De Matos, M.V.; Da Ponte-Neto, A.M.; De Moura, D.T.H.; Maahs, E.D.; Chaves, D.M.; Baba, E.R.; Ide, E.; Sallum, R.; Bernardo, W.M.; De Moura, E.G.H. Treatment of High-Grade Dysplasia and Intramucosal Carcinoma Using Radiofrequency Ablation or Endoscopic Mucosal Resection + Radiofrequency Ablation: Meta-Analysis and Systematic Review. *World J. Gastroint. Endosc.* **2019**, *11*, 239–248. [CrossRef] [PubMed]
53. Pimentel-Nunes, P.; Libânio, D.; Bastiaansen, B.A.J.; Bhandari, P.; Bisschops, R.; Bourke, M.J.; Esposito, G.; Lemmers, A.; Maselli, R.; Messmann, H.; et al. Endoscopic Submucosal Dissection for Superficial Gastrointestinal Lesions: European Society of Gastrointestinal Endoscopy (Esge) Guideline—Update 2022. *Endoscopy* **2022**, *54*, 591–622. [CrossRef] [PubMed]
54. Vantanasiri, K.; Iyer, P.G. State-of-the-Art Management of Dysplastic Barrett's Esophagus. *Gastroenterol. Rep.* **2022**, *10*, Goac068. [CrossRef] [PubMed]
55. Dunbar, K.B.; Spechler, S.J. The Risk of Lymph-Node Metastases in Patients with High-Grade Dysplasia or Intramucosal Carcinoma in Barrett's Esophagus: A Systematic Review. *Am. J. Gastroenterol.* **2012**, *107*, 850–862. [CrossRef] [PubMed]
56. Konda, V.J.; Gonzalez Haba Ruiz, M.; Koons, A.; Hart, J.; Xiao, S.Y.; Siddiqui, U.D.; Ferguson, M.K.; Posner, M.; Patti, M.G.; Waxman, I. Complete Endoscopic Mucosal Resection Is Effective and Durable Treatment for Barrett's-Associated Neoplasia. *Clin. Gastroenterol. Hepatol.* **2014**, *12*, 2002–2010.E2. [CrossRef]
57. May, A.; Gossner, L.; Behrens, A.; Kohnen, R.; Vieth, M.; Stolte, M.; Ell, C. A Prospective Randomized Trial of Two Different Endoscopic Resection Techniques for Early Stage Cancer of the Esophagus. *Gastrointest. Endosc.* **2003**, *58*, 167–175. [CrossRef]
58. Wu, J.; Pan, Y.M.; Wang, T.T.; Gao, D.J.; Hu, B. Endotherapy versus Surgery for Early Neoplasia in Barrett's Esophagus: A Meta-Analysis. *Gastrointest. Endosc.* **2014**, *79*, 233–241.E2. [CrossRef]
59. Van Munster, S.N.; Verheij, E.P.D.; Nieuwenhuis, E.A.; Offerhaus, J.; Meijer, S.L.; Brosens, L.A.A.; Weusten, B.; Alkhalaf, A.; Schenk, E.B.E.; Schoon, E.J.; et al. Extending Treatment Criteria for Barrett's Neoplasia: Results of a Nationwide Cohort of 138 Endoscopic Submucosal Dissection Procedures. *Endoscopy* **2022**, *54*, 531–541. [CrossRef]
60. Terheggen, G.; Horn, E.M.; Vieth, M.; Gabbert, H.; Enderle, M.; Neugebauer, A.; Schumacher, B.; Neuhaus, H. A Randomised Trial of Endoscopic Submucosal Dissection versus Endoscopic Mucosal Resection for Early Barrett's Neoplasia. *Gut* **2017**, *66*, 783–793. [CrossRef]

61. Codipilly, D.C.; Dhaliwal, L.; Oberoi, M.; Gandhi, P.; Johnson, M.L.; Lansing, R.M.; Harmsen, W.S.; Wang, K.K.; Iyer, P.G. Comparative Outcomes of Cap Assisted Endoscopic Resection and Endoscopic Submucosal Dissection in Dysplastic Barrett's Esophagus. *Clin. Gastroenterol. Hepatol.* **2022**, *20*, 65–73.E1. [CrossRef] [PubMed]
62. Zeki, S.S.; Bergman, J.J.; Dunn, J.M. Endoscopic Management of Dysplasia and Early Oesophageal Cancer. *Best. Pract. Res. Clin. Gastroenterol.* **2018**, *36–37*, 27–36. [CrossRef] [PubMed]
63. Phoa, K.N.; Van Vilsteren, F.G.; Weusten, B.L.; Bisschops, R.; Schoon, E.J.; Ragunath, K.; Fullarton, G.; Di Pietro, M.; Ravi, N.; Visser, M.; et al. Radiofrequency Ablation vs. Endoscopic Surveillance for Patients with Barrett Esophagus and Low-Grade Dysplasia: A Randomized Clinical Trial. *JAMA* **2014**, *311*, 1209–1217. [CrossRef] [PubMed]
64. Pouw, R.E.; Klaver, E.; Phoa, K.N.; Van Vilsteren, F.G.; Weusten, B.L.; Bisschops, R.; Schoon, E.J.; Pech, O.; Manner, H.; Ragunath, K.; et al. Radiofrequency Ablation for Low-Grade Dysplasia in Barrett's Esophagus: Long-Term Outcome of A Randomized Trial. *Gastrointest. Endosc.* **2020**, *92*, 569–574. [CrossRef]
65. Brandt, L.J.; Kauvar, D.R. Laser-Induced Transient Regression of Barrett's Epithelium. *Gastrointest. Endosc.* **1992**, *38*, 619–622. [CrossRef]
66. Sampliner, R.E.; Hixson, L.J.; Fennerty, M.B.; Garewal, H.S. Regression of Barrett's Esophagus by Laser Ablation in an Anacid Environment. *Dig. Dis. Sci.* **1993**, *38*, 365–368. [CrossRef]
67. Berenson, M.M.; Johnson, T.D.; Markowitz, N.R.; Buchi, K.N.; Samowitz, W.S. Restoration of Squamous Mucosa after Ablation of Barrett's Esophageal Epithelium. *Gastroenterology* **1993**, *104*, 1686–1691. [CrossRef]
68. Overholt, B.; Panjehpour, M.; Tefftellar, E.; Rose, M. Photodynamic Therapy for Treatment of Early Adenocarcinoma in Barrett's Esophagus. *Gastrointest. Endosc.* **1993**, *39*, 73–76. [CrossRef]
69. Overholt, B.F.; Panjehpour, M. Photodynamic Therapy for Barrett's Esophagus: Clinical Update. *Am. J. Gastroenterol.* **1996**, *91*, 1719–1723.
70. Dumoulin, F.L.; Terjung, B.; Neubrand, M.; Scheurlen, C.; Fischer, H.P.; Sauerbruch, T. Treatment of Barrett's Esophagus By Endoscopic Argon Plasma Coagulation. *Endoscopy* **1997**, *29*, 751–753. [CrossRef]
71. Sampliner, R.E.; Fennerty, B.; Garewal, H.S. Reversal of Barrett's Esophagus with Acid Suppression and Multipolar Electrocoagulation: Preliminary Results. *Gastrointest. Endosc.* **1996**, *44*, 532–535. [CrossRef] [PubMed]
72. Sharma, P.; Mcquaid, K.; Dent, J.; Fennerty, M.B.; Sampliner, R.; Spechler, S.; Cameron, A.; Corley, D.; Falk, G.; Goldblum, J.; et al. A Critical Review of the Diagnosis and Management of Barrett's Esophagus: The Aga Chicago Workshop. *Gastroenterology* **2004**, *127*, 310–330. [CrossRef] [PubMed]
73. Wang, K.K.; Sampliner, R.E.; Gastroenterology, P.P.C.O.T.A.C.O. Updated Guidelines 2008 for the Diagnosis, Surveillance and Therapy of Barrett's Esophagus. *Am. J. Gastroenterol.* **2008**, *103*, 788–797. [CrossRef] [PubMed]
74. Spechler, S.J.; Sharma, P.; Souza, R.F.; Inadomi, J.M.; Shaheen, N.J.; Association, A.G. American Gastroenterological Association Medical Position Statement on the Management of Barrett's Esophagus. *Gastroenterology* **2011**, *140*, 1084–1091. [CrossRef] [PubMed]
75. Luigiano, C.; Iabichino, G.; Eusebi, L.H.; Arena, M.; Consolo, P.; Morace, C.; Opocher, E.; Mangiavillano, B. Outcomes of Radiofrequency Ablation for Dysplastic Barrett's Esophagus: A Comprehensive Review. *Gastroenterol. Res. Pract.* **2016**, *2016*, 4249510. [CrossRef]
76. Shaheen, N.J.; Sharma, P.; Overholt, B.F.; Wolfsen, H.C.; Sampliner, R.E.; Wang, K.K.; Galanko, J.A.; Bronner, M.P.; Goldblum, J.R.; Bennett, A.E.; et al. Radiofrequency Ablation in Barrett's Esophagus with Dysplasia. *N. Engl. J. Med.* **2009**, *360*, 2277–2288. [CrossRef]
77. Haidry, R.J.; Dunn, J.M.; Butt, M.A.; Burnell, M.G.; Gupta, A.; Green, S.; Miah, H.; Smart, H.L.; Bhandari, P.; Smith, L.A.; et al. Radiofrequency Ablation and Endoscopic Mucosal Resection for Dysplastic Barrett's Esophagus and Early Esophageal Adenocarcinoma: Outcomes of the Uk National Halo Rfa Registry. *Gastroenterology* **2013**, *145*, 87–95. [CrossRef]
78. Haidry, R.J.; Butt, M.A.; Dunn, J.M.; Gupta, A.; Lipman, G.; Smart, H.L.; Bhandari, P.; Smith, L.; Willert, R.; Fullarton, G.; et al. Improvement over Time in Outcomes for Patients Undergoing Endoscopic Therapy for Barrett's Oesophagus-Related Neoplasia: 6-Year Experience from the First 500 Patients Treated in the Uk Patient Registry. *Gut* **2015**, *64*, 1192–1199. [CrossRef]
79. Barret, M.; Pioche, M.; Terris, B.; Ponchon, T.; Cholet, F.; Zerbib, F.; Chabrun, E.; Le Rhun, M.; Coron, E.; Giovannini, M.; et al. Endoscopic Radiofrequency Ablation or Surveillance in Patients with Barrett's Oesophagus with Confirmed Low-Grade Dysplasia: A Multicentre Randomised Trial. *Gut* **2021**, *70*, 1014–1022. [CrossRef]
80. Wang, Y.; Ma, B.; Yang, S.; Li, W.; Li, P. Efficacy and Safety of Radiofrequency Ablation vs. Endoscopic Surveillance for Barrett's Esophagus with Low-Grade Dysplasia: Meta-Analysis of Randomized Controlled Trials. *Front. Oncol.* **2022**, *12*, 801940. [CrossRef]
81. Benjamin, S.B.; Maher, K.A.; Cattau, E.L., Jr.; Collen, M.J.; Fleischer, D.E.; Lewis, J.H.; Ciarleglio, C.A.; Earll, J.M.; Schaffer, S.; Mirkin, K.; et al. Double-Blind Controlled Trial of the Garren-Edwards Gastric Bubble: An Adjunctive Treatment for Exogenous Obesity. *Gastroenterology* **1988**, *95*, 581–588. [CrossRef] [PubMed]
82. Dubrouskaya, K.; Hagenstein, L.; Ramai, D.; Adler, D.G. Clinical Adverse Events and Device Failures for the Barrx™ Radiofrequency Ablation Catheter System: A Maude Database Analysis. *Ann. Gastroenterol.* **2022**, *35*, 345–350. [CrossRef] [PubMed]
83. Manner, H.; May, A.; Miehlke, S.; Dertinger, S.; Wigginghaus, B.; Schimming, W.; Krämer, W.; Niemann, G.; Stolte, M.; Ell, C. Ablation of Nonneoplastic Barrett's Mucosa Using Argon Plasma Coagulation with Concomitant Esomeprazole Therapy (Apbanex): A Prospective Multicenter Evaluation. *Am. J. Gastroenterol.* **2006**, *101*, 1762–1769. [CrossRef] [PubMed]
84. Peerally, M.F.; Bhandari, P.; Ragunath, K.; Barr, H.; Stokes, C.; Haidry, R.; Lovat, L.; Smart, H.; Harrison, R.; Smith, K.; et al. Radiofrequency Ablation Compared with Argon Plasma Coagulation after Endoscopic Resection of High-Grade Dysplasia or Stage T1 Adenocarcinoma in Barrett's Esophagus: A Randomized Pilot Study (Bride). *Gastrointest. Endosc.* **2019**, *89*, 680–689. [CrossRef] [PubMed]

85. Manner, H.; Neugebauer, A.; Scharpf, M.; Braun, K.; May, A.; Ell, C.; Fend, F.; Enderle, M.D. The Tissue Effect of Argon-Plasma Coagulation with Prior Submucosal Injection (Hybrid-Apc) Versus Standard Apc: A Randomized Ex-Vivo Study. *United Eur. Gastroenterol. J.* **2014**, *2*, 383–390. [CrossRef]
86. Manner, H.; May, A.; Kouti, I.; Pech, O.; Vieth, M.; Ell, C. Efficacy and Safety of Hybrid-Apc for the Ablation of Barrett's Esophagus. *Surg. Endosc.* **2016**, *30*, 1364–1370. [CrossRef]
87. Knabe, M.; Beyna, T.; Rösch, T.; Bergman, J.; Manner, H.; May, A.; Schachschal, G.; Neuhaus, H.; Kandler, J.; Weusten, B.; et al. Hybrid Apc in Combination with Resection for the Endoscopic Treatment of Neoplastic Barrett's Esophagus: A Prospective, Multicenter Study. *Am. J. Gastroenterol.* **2022**, *117*, 110–119. [CrossRef]
88. Shah, S.N.; Chehade, N.E.H.; Tavangar, A.; Choi, A.; Monachese, M.; Chang, K.J.; Samarasena, J.B. Hybrid Argon Plasma Coagulation in Barrett's Esophagus: A Systematic Review and Meta-Analysis. *Clin. Endosc.* **2023**, *56*, 38–49. [CrossRef]
89. Trindade, A.J.; Wee, D.; Wander, P.; Stewart, M.; Lee, C.; Benias, P.C.; Mckinley, M.J. Successful Treatment of Refractory Barrett's Neoplasia with Hybrid Argon Plasma Coagulation: A Case Series. *Endoscopy* **2020**, *52*, 812–813. [CrossRef]
90. Künzli, H.T.; Schölvinck, D.W.; Meijer, S.L.; Seldenrijk, K.A.; Bergman, J.G.H.M.; Weusten, B.L.A.M. Efficacy of the Cryoballoon Focal Ablation System for the Eradication of Dysplastic Barrett's Esophagus Islands. *Endoscopy* **2017**, *49*, 169–175. [CrossRef]
91. Canto, M.I.; Trindade, A.J.; Abrams, J.; Rosenblum, M.; Dumot, J.; Chak, A.; Iyer, P.; Diehl, D.; Khara, H.S.; Corbett, F.S.; et al. Multifocal Cryoballoon Ablation for Eradication of Barrett's Esophagus-Related Neoplasia: A Prospective Multicenter Clinical Trial. *Am. J. Gastroenterol.* **2020**, *115*, 1879–1890. [CrossRef] [PubMed]
92. Westerveld, D.R.; Nguyen, K.; Banerjee, D.; Jacobs, C.; Kadle, N.; Draganov, P.V.; Yang, D. Safety and Effectiveness of Balloon Cryoablation for Treatment of Barrett's Associated Neoplasia: Systematic Review and Meta-Analysis. *Endosc. Int. Open* **2020**, *8*, E172–E178. [CrossRef] [PubMed]
93. Agarwal, S.; Alshelleh, M.; Scott, J.; Dhaliwal, L.; Codipilly, D.C.; Dierkhising, R.; Leggett, C.L.; Wang, K.K.; Otaki, F.A.; Trindade, A.J.; et al. Comparative Outcomes of Radiofrequency Ablation and Cryoballoon Ablation in Dysplastic Barrett's Esophagus: A Propensity Score-Matched Cohort Study. *Gastrointest. Endosc.* **2022**, *95*, 422–431.E2. [CrossRef] [PubMed]
94. Frederiks, C.N.; Overwater, A.; Alvarez Herrero, L.; Alkhalaf, A.; Schenk, E.; Repici, A.; Bergman, J.J.G.H.; Pouw, R.E.; Bisschops, R.; Haidry, R.J.; et al. Comparison of Focal Cryoballoon Ablation with 10- and 8-Second Doses for Treatment of Barrett's Esophagus-Related Neoplasia: Results from A Prospective European Multicenter Study (with Video). *Gastrointest. Endosc.* **2022**, *96*, 743–751.E4. [CrossRef] [PubMed]
95. Overwater, A.; Van Munster, S.N.; Nagengast, W.B.; Pouw, R.E.; Bergman, J.J.G.H.; Schoon, E.J.; Weusten, B.L.A.M. Novel Cryoballoon 180° Ablation System for Treatment of Barrett's Esophagus-Related Neoplasia: A First-in-Human Study. *Endoscopy* **2022**, *54*, 64–70. [CrossRef]
96. Van Munster, S.N.; Overwater, A.; Raicu, M.G.M.; Seldenrijk, K.C.A.; Nagengast, W.B.; Schoon, E.J.; Bergman, J.J.G.H.; Weusten, B.L.A.M. A Novel Cryoballoon Ablation System for Eradication of Dysplastic Barrett's Esophagus: A First-In-Human Feasibility Study. *Endoscopy* **2020**, *52*, 193–201. [CrossRef]
97. Mittal, C.; Muthusamy, V.R.; Simon, V.C.; Brauer, B.C.; Mullady, D.K.; Hollander, T.; Sloan, I.; Kushnir, V.; Early, D.; Rastogi, A.; et al. Threshold Evaluation for Optimal Number of Endoscopic Treatment Sessions to Achieve Complete Eradication of Barrett's Metaplasia. *Endoscopy* **2022**, *54*, 927–933. [CrossRef]
98. Van Munster, S.N.; Frederiks, C.N.; Nieuwenhuis, E.A.; Alvarez Herrero, L.; Bogte, A.; Alkhalaf, A.; Schenk, B.E.; Schoon, E.J.; Curvers, W.L.; Koch, A.D.; et al. Incidence and Outcomes of Poor Healing and Poor Squamous Regeneration after Radiofrequency Ablation Therapy for Early Barrett's Neoplasia. *Endoscopy* **2022**, *54*, 229–240. [CrossRef]
99. Van Munster, S.N.; Nieuwenhuis, E.; Bisschops, R.; Willekens, H.; Weusten, B.L.A.M.; Herrero, L.A.; Bogte, A.; Alkhalaf, A.; Schenk, E.B.E.; Schoon, E.J.; et al. Development and External Validation of A Model to Predict Complex Treatment after Radiofrequency Ablation for Barrett's Esophagus with Early Neoplasia. *Clin. Gastroenterol. Hepatol.* **2022**, *20*, 2495–2504.E5. [CrossRef]
100. Wolfson, P.; Ho, K.M.A.; Wilson, A.; Mcbain, H.; Hogan, A.; Lipman, G.; Dunn, J.; Haidry, R.; Novelli, M.; Olivo, A.; et al. Endoscopic Eradication Therapy for Barrett's Esophagus-Related Neoplasia: A Final 10-Year Report from the Uk National Halo Radiofrequency Ablation Registry. *Gastrointest. Endosc.* **2022**, *96*, 223–233. [CrossRef]
101. Menon, S.; Norman, R.; Mannath, J.; Iyer, P.G.; Ragunath, K. Comparative Cost-Effectiveness of Three Post-Radiofrequency Ablation Surveillance Intervals for Barrett's Esophagus. *Endosc. Int. Open* **2022**, *10*, E1053–E1064. [CrossRef] [PubMed]
102. Van Munster, S.N.; Nieuwenhuis, E.; Bisschops, R.; Willekens, H.; Weusten, B.L.A.M.; Herrero, L.A.; Bogte, A.; Alkhalaf, A.; Schenk, E.B.E.; Schoon, E.J.; et al. Dysplastic Recurrence after Successful Treatment for Early Barrett's Neoplasia: Development and Validation of a Prediction Model. *Gastroenterology* **2022**, *163*, 285–294. [CrossRef] [PubMed]
103. Verheij, E.P.D.; Van Munster, S.N.; Nieuwenhuis, E.; Cotton, C.C.; Weusten, B.L.A.M.; Alvarez Herrero, L.; Alkhalaf, A.; Schenk, B.E.; Schoon, E.J.; Curvers, W.; et al. All-Cause Mortality after Successful Endoscopic Eradication Therapy for Barrett's Related Neoplasia in a Nationwide Cohort of 1154 Patients. *Endoscopy* **2022**, *54*, S93.
104. Phoa, K.N.; Pouw, R.E.; Bisschops, R.; Pech, O.; Ragunath, K.; Weusten, B.L.; Schumacher, B.; Rembacken, B.; Meining, A.; Messmann, H.; et al. Multimodality Endoscopic Eradication for Neoplastic Barrett Oesophagus: Results of an European Multicentre Study (Euro-Ii). *Gut* **2016**, *65*, 555–562. [CrossRef]

105. Yang, J.; Wang, X.; Li, Y.; Lu, G.; Lu, X.; Guo, D.; Wang, W.; Liu, C.; Xiao, Y.; Han, N.; et al. Efficacy and Safety of Steroid in the Prevention of Esophageal Stricture after Endoscopic Submucosal Dissection: A Network Meta-Analysis. *J. Gastroenterol. Hepatol.* **2019**, *34*, 985–995. [CrossRef]
106. Bartel, M.J.; Mousa, O.Y.; Brahmbhatt, B.; Coffman, D.L.; Patel, K.; Repici, A.; Tokar, J.L.; Wolfsen, H.C.; Wallace, M.B. Impact of Topical Budesonide on Prevention of Esophageal Stricture after Mucosal Resection. *Gastrointest. Endosc.* **2021**, *93*, 1276–1282. [CrossRef]

Disclaimer/Publisher's Note: The statements, opinions and data contained in all publications are solely those of the individual author(s) and contributor(s) and not of MDPI and/or the editor(s). MDPI and/or the editor(s) disclaim responsibility for any injury to people or property resulting from any ideas, methods, instructions or products referred to in the content.

Review

The Role of Artificial Intelligence in Colorectal Cancer Screening: Lesion Detection and Lesion Characterization

Edward Young [1,*], Louisa Edwards [2] and Rajvinder Singh [1]

1. Faculty of Health and Medical Sciences, University of Adelaide, Lyell McEwin Hospital, Haydown Rd, Elizabeth Vale, SA 5112, Australia
2. Faculty of Health and Medical Sciences, University of Adelaide, Queen Elizabeth Hospital, Port Rd, Woodville South, SA 5011, Australia
* Correspondence: edward.young@sa.gov.au

Simple Summary: There has been an exponential rise in the availability of artificial intelligence systems in endoscopy in recent years. As a result, maintaining an informed understanding of the utility and efficacy of existing systems has become increasingly complex. This review aims to summarise the expanse of research in this area to guide proceduralists in making informed decisions regarding the use of artificial intelligence in colonoscopy. It focuses primarily on the application of artificial intelligence for the detection and characterisation of colorectal polyps in order to improve the efficacy of colorectal cancer screening and prevention.

Abstract: Colorectal cancer remains a leading cause of cancer-related morbidity and mortality worldwide, despite the widespread uptake of population surveillance strategies. This is in part due to the persistent development of 'interval colorectal cancers', where patients develop colorectal cancer despite appropriate surveillance intervals, implying pre-malignant polyps were not resected at a prior colonoscopy. Multiple techniques have been developed to improve the sensitivity and accuracy of lesion detection and characterisation in an effort to improve the efficacy of colorectal cancer screening, thereby reducing the incidence of interval colorectal cancers. This article presents a comprehensive review of the transformative role of artificial intelligence (AI), which has recently emerged as one such solution for improving the quality of screening and surveillance colonoscopy. Firstly, AI-driven algorithms demonstrate remarkable potential in addressing the challenge of overlooked polyps, particularly polyp subtypes infamous for escaping human detection because of their inconspicuous appearance. Secondly, AI empowers gastroenterologists without exhaustive training in advanced mucosal imaging to characterise polyps with accuracy similar to that of expert interventionalists, reducing the dependence on pathologic evaluation and guiding appropriate resection techniques or referrals for more complex resections. AI in colonoscopy holds the potential to advance the detection and characterisation of polyps, addressing current limitations and improving patient outcomes. The integration of AI technologies into routine colonoscopy represents a promising step towards more effective colorectal cancer screening and prevention.

Keywords: colonoscopy; artificial intelligence; polyp; adenoma; colorectal cancer

1. Introduction

In the context of modern healthcare, the integration of artificial intelligence (AI) has emerged as a transformative force, revolutionising various aspects of medical practice [1]. One promising application lies in the domain of colorectal cancer (CRC), where AI holds the potential to enhance the accuracy and efficiency of polyp detection and characterisation during colonoscopy—a pivotal procedure for early diagnosis and prevention. This review article delves into the dynamic intersection of AI and CRC management, with a specific focus on its application for polyp detection and characterisation during colonoscopy.

CRC accounts for more than 10% of cancer diagnoses and more than 9% of cancer-related mortality worldwide, necessitating effective screening and diagnostic strategies to curb its impact [2]. There is now compelling evidence that the implementation of population CRC screening in developed countries has led to a considerable reduction in its incidence and mortality [3,4]. Colonoscopy serves as the gold standard for both the detection and prevention of CRC, yet its efficacy is contingent on the skill and vigilance of the endoscopist [5]. Despite advances in endoscopic technology and improvement in adenoma detection, adenoma miss rates still remain as high as 26% in tandem colonoscopy studies [6]. Miss rates are particularly high for sessile serrated lesions (SSLs) (27%), proximal advanced adenomas (14%), and flat adenomas (34%) [6]. The integration of AI into colonoscopy holds the promise of augmenting human expertise, potentially reducing the miss rates of these inconspicuous polyps and thereby improving patient outcomes.

Drawing upon a plethora of studies, this review aims to dissect the methodologies and technological advancements that underpin AI-driven polyp detection and characterisation systems, with a particular focus on more recent real-world experiences with AI. By exploring the evolution, challenges, and outcomes associated with these technologies, we strive to provide insights into their potential to reshape CRC management paradigms. While this is not a formal systematic review, it has been based largely on a structured examination of published literature from Pubmed and Embase, with abstracts screened for relevance and reference lists reviewed for additional relevant studies.

2. Artificial Intelligence in Colonoscopy

Research in the field of AI-assisted colonoscopy has expanded exponentially in the last 5 years, with a wide range of AI systems now commercially available (Table 1). As a result, understanding the efficacy and accuracy of these individual systems has become increasingly complex while there are limited data available for direct comparison and no form of standardisation exists. Nevertheless, proponents of AI argue that the sophistication of deep learning and the vast datasets on which these systems are trained result in consistent accuracy at a high level. In the absence of standardisation, this review seeks to analyse the efficacy of the commercially available systems and the accuracy of this assertion.

Table 1. Commercially available artificial intelligence systems in colonoscopy.

Name	Company	Technique	Commercial Approval
EndoBRAIN	Cybernet Systems Corporation (Tokyo, Japan)	CADx	2018
GI Genius	Medtronic (Dublin, Ireland)	CADe	2019
EndoBRAIN-EYE	Cybernet Systems Corporation (Tokyo, Japan)	CADe	2020
DISCOVERY	Pentax Medical Company (Tokyo, Japan)	CADe	2020
ENDO-AID	Olympus Corporation (Tokyo, Japan)	CADe	2020
CAD EYE	Fujifilm (Tokyo, Japan)	CADe, CADx	2020
Wise Vision	NEC Corporation (Tokyo, Japan)	CADe	2020
EndoScreener	Wision A.I. (Shanghai, China)	CADe	2021

Machine learning involves the development of an algorithm based on a training dataset in order to predict the same pattern in unseen data. Initially, AI systems in endoscopy involved the manual introduction of polyp features to the machine learning algorithm for the program to recognise polyps; however, the accuracy of AI systems has catapulted with the introduction of deep learning. Deep learning is a type of machine learning characterised by self-learning, in that the program extracts data and recognises key features across multiple layers without any requirement for human input. It involves neural networks, imitating the complex interconnected networks of the human brain in order to analyse multiple increasingly complex layers of images. Convolutional neural networks (CNNs) are based on the principle of the visual cortex of the human brain for image processing. Using multiple filters, the CNN extracts key features from multiple versions of the same image before pooling layers to provide a final classification as the output based on learned polyp features. The key advantage of these systems is that the more data that is fed into the system, the more sophisticated the algorithm becomes, as the system is capable of continued independent learning. CNNs are a popular method for image recognition as they offer efficient performance, allowing for use in real-time video applications [7,8].

The number of AI systems developed or in development for upper and lower gastrointestinal endoscopy has expanded exponentially in recent years. Computer-aided detection (CADe) systems recognise characteristic features in order to discern the presence of a polyp within a still image or video. More recently, these systems have been integrated into real-time colonoscopy, alerting proceduralists to the presence of a polyp either with a coloured box around the entire display or a box around the polyp itself. Computer-aided diagnosis (CADx) systems are able to distinguish between polyp types and degrees of dysplasia, from benign hyperplastic polyps to advanced cancers, providing a real-time diagnosis to the proceduralist.

3. Polyp Detection

Since 2016, researchers have published deep learning algorithms for polyp detection (CADe) that have been tested in pre-clinical applications, such as polyp detection in still images or videos [9]. Only 3 years later, the first randomised controlled trials (RCTs) comparing CADe with existing standards were published [10]. Since then, there has been a vast amount of research published on real-time CADe systems, with strong support for their efficacy in polyp detection. Of the 15 RCTs reviewed here, 10 demonstrated a statistically significant increase in adenoma detection, although baseline and CADe adenoma detection rates (ADRs) are highly varied because of differing populations and study designs (Table 2) [10–24]. Although overall lesion detection is generally improved, many of these systems have been criticised for a lack of impact on the detection of advanced adenomas of heightened clinical significance. Many argue that these larger polyps are less likely to be missed by endoscopists, making the implementation of CADe systems less pivotal. While it may be true that larger polyps are less likely to be missed by endoscopists, the lack of demonstrable impact of CADe systems for advanced adenomas may simply reflect their reduced prevalence and, hence, the larger numbers required to adequately power these studies. For example, in the largest RCT by Xu et al., including 3059 patients, there was a statically significant increase in advanced adenoma (>10 mm, villous component or high-grade dysplasia) detection in the CADe group versus the control group (6.6% vs. 4.9%, $p = 0.041$) [12].

Table 2. Randomised controlled trials comparing artificial-intelligence-aided colonoscopy with control groups for adenoma detection.

Author, Year	CADe System	Control	Patients (n)	ADR (AI vs. Control)	Advanced ADR (AI vs. Control)
Nakashima et al., 2023 [11]	CAD EYE	HD-WLI	415	59.4% vs. 47.6% ($p = 0.018$)	7.2% vs. 7.7% ($p = 1$)
Xu et al., 2023 [12]	Eagle-Eye	HD-WLI	3059	39.9% vs. 32.4% ($p < 0.001$)	6.6% vs. 4.9% ($p = 0.041$)
Wang et al., 2023 [13]	EndoScreener	HD-WLI with second observer	1261	25.8% vs. 24.0% ($p = 0.464$)	0.314% vs. 0.39% ($p = 0.562$)

Table 2. Cont.

Author, Year	CADe System	Control	Patients (n)	ADR (AI vs. Control)	Advanced ADR (AI vs. Control)
Wei et al., 2023 [14]	EndoVigilant	HD-WLI	769	35.9% vs. 37.2% ($p = 0.774$)	N/A
Ahmad et al., 2022 [15]	GI Genius	HD-WLI	658	71.4% vs. 65.4% ($p = 0.09$)	N/A
Gimeno-Garcia et al., 2022 [16]	ENDO-AID	HD-WLI	370	55.1% vs. 43.8% ($p = 0.029$)	11.6% vs. 12.1% ($p = 0.89$)
Repici et al., 2022 [17]	GI Genius	HD-WLI	660	53.3% vs. 44.5% ($p < 0.02$)	12.7% vs. 12.7% ($p = 0.956$)
Rondonotti et al., 2022 [18]	CAD EYE	HD-WLI	800	53.6% vs. 45.3% (RR 1.18, 95% CI 1.03–1.36)	18.5% vs. 15.9% (RR 1.03, 95% CI 0.96–1.09)
Shaukat et al., 2022 [19]	SKOUT	HD-WLI	1359	47.8% vs. 43.9% ($p = 0.065$)	N/A
Luo et al., 2021 [20]	Xiamen Innovision	HD-WLI	150	PDR 38.7% vs. 34.0% ($p < 0.001$)	N/A
Xu et al., 2021 [21]	N/A	HD-WLI	2352	PDR 38.8% vs. 36.2% ($p = 0.183$)	N/A
Liu P et al., 2020 [22]	EndoScreener	HD-WLI	790	29.01% vs. 20.91% ($p = 0.009$)	1.43% vs. 3.92% ($p = 0.607$)
Liu W et al., 2020 [23]	Henan Xuanweitang Medical Information Technology Co.	HD-WLI	1026	39.1% vs. 23.89% ($p < 0.001$)	2.88% vs. 6.45% ($p = 0.821$)
Repici et al., 2020 [24]	GI-Genius	HD-WLI	685	54.8% vs. 40.4% (RR 1.30, 95% 1.14–1.45)	10.3% vs. 7.3% ($p = 0.769$)
Wang et al., 2019 [10]	EndoScreener	HD-WLI	1058	29.12% vs. 20.34% ($p < 0.001$)	3.41% vs. 5.95% ($p = 0.803$)

In an effort to synthesise the expanse of research in this area, multiple meta-analyses have been published comparing CADe with high-definition white light imaging (HD-WLI) control groups (Table 3). These studies have universally found an increase in ADR with CADe, with a 1.43–1.78 times increase in ADR versus HD-WLI [25–35]. The most significant difference has been in the detection of diminutive (<5 mm) adenomas. For larger polyps, the results have been varied, with four of the seven meta-analyses specifically analysing >10 mm adenomas finding a statistically significant improvement in detection. Interestingly, in their 2021 meta-analysis, Zhang et al. actually reported a reduction in the detection of advanced adenomas with CADe [34]. While this raises the possibility that the time and concentration consumed by higher diminutive polyp detection with CADe may detract from the detection of advanced lesions, this has not been borne out in other meta-analyses and was not the case in the largest RCT to date [12]. Sessile serrated lesions (SSLs) are a polyp subtype prone to being missed during colonoscopy because of their inconspicuous nature, as they are generally flat and difficult to differentiate from surrounding normal mucosa. For SSLs, RCTs have not been powered to demonstrate an effect as their incidence is considerably lower compared with adenomas. However, three meta-analyses assessed SSLs specifically, demonstrating a between 1.37- and 1.52-times increase in SSL detection with CADe, though one of these did not reach statistical significance [25,30,33].

Table 3. Meta-analyses comparing artificial-intelligence-aided colonoscopy with control groups for adenoma detection.

Author, Year	Studies (n)	Patients (n)	ADR (AI vs. Control)	≤5 mm Adenomas	≥10 mm Adenomas	Notes
Huang et al., 2022 [25]	10	6629	RR 1.43, $p < 0.001$	RR 1.71, $p < 0.001$	RR 1.73, $p < 0.001$	SSL per colonoscopy RR 1.53, $p < 0.001$
Sivananthan et al., 2022 [26]	7	5217	33.65% vs. 22.85%	0.691 adenomas per colonoscopy vs. 0.373 (pooled effect size 0.3, 95% CI 0.19–0.42)	N/A	91.7% higher detection of non-pedunculated adenomas
Ashat et al., 2021 [27]	6	5058	33.7% vs. 22.9% (OR 1.76, 95% CI 1.55–2.00)	OR 2.07, 95% CI 1.81–2.36, $p < 0.001$	OR 1.79, 95% CI 1.27–2.53, $p < 0.001$	
Barua et al., 2021 [28]	5	4311	29.6% vs. 19.3% (RR 1.52, 95% CI 1.31–1.77)	Mean difference, 0.15 (95% CI 0.12–0.28)	Mean difference 0.01, 95% CI 0.00–0.02	
Deliwala et al., 2021 [29]	6	4996	OR 1.77 (95% CI 1.57–2.08)	OR 1.33 (95% CI 1.12–1.59)	OR 1.24 (95% CI 0.87–1.78)	
Hassan et al., 2021 [30]	5	4354	36.6% vs. 25.2%, RR 1.44 (95% CI 1.27–1.62)	RR 1.69 (95% CI 1.48–1.84)	RR 1.46 (95% CI 1.04–2.06)	SSL per colonoscopy RR 1.52 (95% CI 1.14–2.02)
Li et al., 2021 [31]	5	4311	OR 1.75 (95% CI 1.52–2.01)	N/A	N/A	
Nazarian et al., 2021 [32]	8	5577	OR 1.53 (95% CI 1.32–1.77)	N/A	N/A	
Spadaccini et al., 2021 [33]	6	4996	OR 1.78 (95% CI 1.44–2.18)	N/A	OR 1.69 (95% CI 1.10–2.60)	No difference in SSL detection, OR 1.37 (95% CI 0.65–2.88)
Zhang et al., 2021 [34]	7	5427	OR 1.72 (95% CI 1.52–1.95)	OR 1.42 (95% CI 1.18–1.72)	OR 0.71 (95% CI 0.46–1.10)	Less advanced adenomas (OR 0.70, 95% CI 0.50–0.97) SSL OR 0.87 (95% CI 0.61–1.23)
Aziz et al., 2020 [35]	3	2815	32.9% vs. 20.8%, RR 1.58 (95% CI 1.39–1.80)	N/A	N/A	

N/A = variable not reported.

Overall, prospective studies into CADe for adenoma detection have been optimistic. Although many studies have not shown improved advanced adenoma detection, multiple meta-analyses and the largest RCT to date suggest that this is likely the case, and it has been conclusively demonstrated to improve the detection of diminutive adenomas. However, with the advent of commercially available CADe systems, data are now available in a real-world context, which may have greater generalisability than those conducted in a clinical trial setting. The largest of these, published by Ladabaum et al. in 2023, was a pragmatic real-world retrospective study whereby data were collected following the implementation of CADe in a single centre, compared with concurrent and historical controls [36]. In this study, the introduction of CADe resulted in no statistically significant difference in any detection metric, including ADR, adenomas per colonoscopy, or advanced adenoma detection. This was further supported by Levy et al., who demonstrated a reduction in ADR from 35.2% to 30.3% ($p < 0.001$) in their single-centre cohort study [37]. These studies highlighted the potential pitfalls of the use of CADe, including less thorough mucosal exposure due to a 'false sense of security' from the AI assistance; proceduralists dismissing lesions not highlighted by AI; and the cumulative effect of false positive detection and the resulting increase in withdrawal time. However, in two other large real-world propensity score-matched studies including a cumulative 2262 patients following the implementation of CADe, its introduction resulted in a 1.32–1.59-times higher ADR when compared with HD-WLI [38,39].

The differing results in these real-world implementation studies may relate in part to differences in the impact of AI on expert referral centres with already high ADR versus lower ADR proceduralists. Given the nature of the limited availability of CADe systems thus far, few studies have examined their impact on low-ADR endoscopists. As can be seen in Table 2, of the five studies not demonstrating a difference in ADR with CADe, only one study had a baseline ADR of less than 36% [38,39]. In this study by Wang et al., the control group included a second observer and was, therefore, not strictly a 'standard of care' control [13]. In one such study with a low baseline ADR, adenoma detection improved from 19.9% to 26.4% with the introduction of CADe [38]. Interestingly, in this study, proceduralists were stratified by experience, with experts defined as having performed more than 1000 colonoscopies, rather than by ADR. In doing so, they found no improvement in ADR in the 'non-expert' group. This raises the possibility that baseline ADR is of greater significance than procedural experience when determining the impact of CADe. This was also supported by Repici et al., who compared ADR with and without CADe across 660 colonoscopies performed by non-experts (<2000 colonoscopies) and found no correlation between examiner experience and the impact of AI on ADR [17]. In contrast, although not a controlled comparative study, Biscaglia et al. showed that with the assistance of CADe, trainee endoscopists (200–400 previous colonoscopies) could achieve the same ADR on tandem colonoscopy with expert, high-ADR endoscopists without AI assistance [40]. To the best of our knowledge, no studies have been published to date with stratification between endoscopists on baseline ADR in order to investigate this further.

While ADR is often used as a surrogate marker, the adenoma miss rate (AMR) is the most direct correlate with the potential for bowel cancer development despite surveillance colonoscopy. Few studies have directly examined the impact of CADe in this context. AMR refers to the number of adenomas 'missed' during a colonoscopy, generally based on tandem colonoscopy studies where an immediate repeat procedure detects additional adenomas. Three tandem colonoscopy studies (Table 4) have compared AMR for CADe versus HD-WLI, with a significant reduction when using CADe [41–43]. The SSL miss rate was higher in all three studies with HD-WLI, with two reaching statistical significance. In addition, non-polypoid and right-sided adenomas, both of which are frequently missed at colonoscopy, were less likely to be missed with the use of CADe. These are promising data for the potential of CADe to standardise the quality of colonoscopy by reducing miss rates for these more inconspicuous polyp subtypes.

Multiple previous studies have demonstrated the impact of fatigue on ADR, presumably because of a higher likelihood of human error. A 2009 retrospective study of 3619 colonoscopies found an ADR of 29.3% in the morning versus 25.3% in the afternoon ($p = 0.008$) [44]. This was reinforced by a prospective study that found that 27% more polyps were detected per patient during early morning cases, with an hour-by-hour decrease in adenoma detection as the day progressed [45]. Given CADe aims to reduce the likelihood of human error, two studies have assessed its role in preventing deterioration in ADR from physician fatigue. Lu et al. undertook a post hoc analysis of two prospective RCTs comparing CADe with HD-WLI, finding that while the ADR in morning sessions was higher in the control group, there was no longer any statistically significant difference in the CADe group [46]. In this cohort, the OR for adenoma detection during afternoon colonoscopy with CADe assistance versus without was 3.81 (95% CI 2.1–6.91) [46]. Similarly, Ritcher et al. performed a retrospective database analysis comparing ADR with CADe versus HD-WLI over the course of a day, demonstrating that while there was a statistically significant trend towards reduction in ADR throughout the day with HD-WLI ($p = 0.015$), this trend was no longer present in the CADe-assisted group ($p = 0.65$) [47].

Table 4. Tandem colonoscopies randomised to CADe or HD-WLI first.

Author, Year	Patients (n)	Adenoma Miss Rate (CADe vs. HD-WLI)	SSL Miss Rate (CADe vs. HD-WLI)	Non-Polypoid Adenoma Miss Rate	Right Colon Adenoma Miss Rate
Glissen-Brown et al., 2022 [41]	234	20.12% vs. 31.25% ($p = 0.0247$)	7.14% vs. 42.11% ($p = 0.0482$)	17.65% for CADe vs. 22.22% for HD-WLI ($p = 0.5872$)	Higher miss rate for HD-WLI in the right colon on multivariable analysis (OR 1.7865, $p = 0.0436$)
Wallace et al., 2022 [42]	230	15.5% vs. 32.4% ($p < 0.001$)	0% vs. 33.33% ($p = 0.455$)	Lower miss rate with CADe for nonpolypoid adenomas (OR 0.34, $p < 0.001$)	18.3% with CADe vs. 32.53% with HD-WLI ($p = 0.004$)
Kamba et al., 2021 [43]	346	13.8% vs. 36.7% ($p < 0.001$)	13% vs. 38.5% ($p = 0.0332$)	13.38% for CADe vs. 45.26% for HD-WLI ($p < 0.001$)	9.23% for CADe vs. 44.05% for HD-WLI ($p < 0.001$)

3.1. Criticisms of CADe

The two main criticisms of CADe are the impact on procedure time and the high rates of distracting false positive polyp identifications. In a 2022 ESGE position statement, the overwhelming consensus was that, for the use of CADe to become widespread, it would need to have an acceptable false-positive rate such that it does not significantly prolong procedure times [48].

Despite initial concerns from image- and video-based studies, the actual rates of false positives that have a meaningful impact on withdrawal time appear to be low, with 91% of false positives lasting less than half a second [49]. In their post hoc analysis of an RCT, Hassan et al. found that while overall false positive rates are high (27.3 per colonoscopy), only 5.7% of false positives required an additional exploration time of 4.8 s per false positive, adding a negligible 1% increase in total withdrawal time [50]. Nevertheless, although the majority of false positives are short-lived, they still have a considerable impact on proceduralist fatigue, with more than 80% of gastroenterologists reporting concerns regarding excessive false positive alerts in a 2023 survey assessing one commercially available CADe system [51]. These false positive alerts from CADe are most often related to bubbles or faeces falsely identified as polyps. As a result, Tang et al. examined whether this could be minimised using water exchange colonoscopy (where water is used rather than CO_2 insufflation during colonoscope insertion while, at the same time, fluid is suctioned to clear the lumen) in order to clear the field of view of the mucosa. In their 2022 study, they demonstrated a significant increase in the additional polyp detection rate with CADe versus HD-WLI after water exchange colonoscopy (30.1% vs. 12.3%, $p = 0.001$), with a lower rate of false positives related to faeces ($p = 0.007$) and bubbles ($p = 0.001$) due to the clearer field upon colonoscope withdrawal [52]. Techniques such as water exchange colonoscopy, therefore, stand to enhance the performance of CADe not only by improving mucosal visualisation but also by reducing rates of distracting false positives.

Regarding withdrawal times, it remains difficult to assess the true mucosal inspection time without this being impacted by the additional time spent on polyp assessment and resection. Though studies generally pause a stopwatch at the time of polypectomy, there are still delays when a polyp is found, for example, while the stopwatch is paused and restarted on each occasion. The most accurate assessment is, therefore, in the withdrawal time in patients where no polyps are found. Of the four meta-analyses from Table 3 directly examining withdrawal time, no study found any significant difference in withdrawal time in patients with no polyps, while three out of four found a slightly longer withdrawal time (up to a mean of 0.46 min) overall with CADe [25,27,29,33]. In all likelihood, despite false positives from CADe, the only meaningful difference in withdrawal times is in the impact on polyp detection.

3.2. Cost Effectiveness

There are controversies surrounding the cost-efficacy of implementing CADe-assisted colonoscopy in screening programs. Initially, the increase in adenoma detection will result in an increased healthcare burden because of requirements for pathological evaluation and a shortening of surveillance intervals. However, eventually, the reduction in adenoma miss rates may mean that surveillance guidelines are able to be adjusted, and there are significant cost savings if advanced colorectal cancers are able to be prevented. In 2022, Mori et al. investigated this further by performing a pooled analysis of RCTs, demonstrating that the proportion of patients who were recommended more intensive surveillance according to US guidelines increased from 8.4% in the control group to 11.3% in the CADe group (RR 1.35, 95% CI 1.16–1.57), which would place a significant burden on a strained healthcare system [53]. However, Areia et al. developed a microsimulation model in a hypothetic cohort to show that the implementation of CADe detection in a US population resulted in a yearly additional prevention of 7194 colorectal cancer cases and 2089 related deaths, with cost savings of USD 290 million [54]. This is aptly described in the World Endoscopy Organisation position statement on AI in colonoscopy in 2023, which states the following: 'In the short term, use of CADe is likely to increase health-care costs by detecting more adenomas', but 'the increased cost by CADe could be balanced by savings in costs related to cancer treatment due to CADe-related cancer prevention' [55].

3.3. Summary

CADe systems lead to improved adenoma detection, particularly for diminutive adenomas and polyp subgroups more likely to be missed because of human error, including non-polypoid adenomas, right-sided adenomas, and SSLs. While this has not yet been consistently supported by 'real-world' studies, the existing retrospective studies introduce forms of bias that may influence results. What has been demonstrated, however, is that, with the support of CADe, regular endoscopists can achieve equivalent performance in adenoma detection to expert high-ADR endoscopists in referral centres, standardising the quality of service provision. Given the dramatic increase in demand for colonoscopy with the implementation of population screening programs, not all patients will have access to expert referral centres for colonoscopy. CADe systems, therefore, have the capacity to make equality of healthcare provision a reality despite inevitable resource limitations. This sentiment is echoed by the European Society of Gastrointestinal Endoscopy (ESGE) 2022 position paper on AI in gastrointestinal endoscopy, stating that 'the task of AI is to lift the less experienced to the level of experienced endoscopists rather than to further increase the high ADR values of the high-detector experts' [48]. In this way, CADe is clearly meeting its objective.

4. Polyp Characterisation

In addition to lesion detection, the other primary focus of AI systems in colonoscopy has been on the characterisation of polyps (computer-aided diagnosis—CADx). Although expert interventional endoscopists with advanced mucosal imaging are able to achieve a high degree of accuracy in histology prediction, this requires specialised training, experience, and time that may not be available in the general endoscopy setting [56]. Accurate histology prediction is of particular importance in two commonly encountered settings in colonoscopy. For diminutive (<5 mm) polyps, accurate prediction facilitates the safe use of the 'resect and discard' and 'do not resect' strategies, as discussed below [57]. For larger polyps, the prediction of histology guides appropriate referral pathways for non-interventional endoscopists, either for endoscopic or surgical resection.

For overall histology prediction, multiple image-based studies and three meta-analyses have demonstrated the superiority of CADx compared with non-expert endoscopists [58–68]. However, in each of these meta-analyses, CADx has been unable to outperform expert endoscopists [58–60]. In addition, in existing real-time colonoscopy studies, CADx has not been shown to significantly improve the sensitivity or specificity of overall histology prediction. Barua et al. compared CADx with non-expert endoscopists (1–5 years of colonoscopy experience) across 518 patients with 892 polyps and demonstrated no significant difference in sensitivity (90.4% vs. 88.4%) or specificity (85.9% vs. 83.1%) [69]. When compared with expert endoscopists, Li et al. found CADx to be inferior in terms of both sensitivity (61.8% vs. 70.3%, $p < 0.001$) and overall accuracy (71.6% vs. 75.2%, $p = 0.023$) [70].

4.1. Diminutive Polyps

Despite a degree of variability in the evidence described above, there are certain circumstances where the accuracy of CADx has been more clearly established, including for the diagnosis of diminutive polyps. In this context, accurate histology prediction serves to avoid unnecessary and expensive pathologic evaluations. The Preservation and Incorporation of Valuable Endoscopic Innovations (PIVI) initiative is a program from the American Society for Gastrointestinal Endoscopy (ASGE) aiming to establish thresholds for endoscopic technologies aimed at addressing important clinical questions and needs in endoscopic diagnosis and intervention [57]. A key focus has been on two strategies to reduce the burden of the histopathological analysis of diminutive colorectal polyps. According to PIVI, diminutive polyps outside of the rectosigmoid colon should be resected but do not require pathological analysis provided endoscopic imaging-based histology prediction results in more than 90% agreement with pathology for surveillance intervals (the 'resect and discard' strategy). In addition, diminutive rectosigmoid polyps do not require resection if the endoscopic appearance is of a hyperplastic polyp, provided endoscopic imaging achieves a negative predictive value of more than 90% for adenomatous histology (the 'do not resect' strategy). In this context, CADx has been able to comprehensively surpass expectations.

Multiple image-based studies have shown CADx to be superior to non-expert endoscopists for diminutive polyps, with a 96–97% NPV and a sensitivity of 92.3–98.1% [71–75]. Once again, the accuracy of CADx has not outperformed expert endoscopists; however, the widespread adoption of CADx would allow endoscopists of all levels of expertise to employ the 'do not resect' or 'resect and discard' strategies, thereby improving the cost-effectiveness of colonoscopic screening programs. This was assessed in real-time colonoscopy by Rondonotti et al., including all patients with at least one diminutive rectosigmoid polyp assessed by an endoscopist with CADx assistance [76]. An AI-assisted high-confidence prediction was made in 92.3% of polyps, with NPVs of 91% and 97.4% agreement with ESGE surveillance intervals. Although the initial AI-assisted accuracy was significantly higher in expert (91.9%) versus non-expert (82.3%) endoscopists, there was a significant trend over time in non-experts, such that, for the final 50 polyps, there was no difference in NPV for non-experts (95.2%) versus experts (93.9%).

In fact, certain studies have argued that, for diminutive polyps, pathologic analysis can be misleading, and CADx systems may even outperform the gold standard. In 2019, Ponugoti et al. highlighted the significant discordance that exists between high-confidence expert endoscopist histology prediction and pathologic evaluation for ≤ 3 mm polyps, postulating that, for polyps of this size, there are frequently issues with processing and retrieval [77]. Subsequently, Shahidi et al. examined the accuracy of CADx diagnoses of 644 ≤ 3 mm polyps, with a discrepancy between endoscopic and pathological diagnoses in 28.9% of lesions [78]. CADx agreed with expert endoscopists in 90.3% of discordant cases, again highlighting the potential inaccuracy of pathology as the accepted gold standard for polyps of this size.

Critics of CADx argue that the histological predictions of these systems are significantly influenced by the dataset on which they are trained. For example, in datasets with an underrepresentation of SSLs, the CADx system may be less likely to report a lesion as such. To assess the consistency of these systems, Hassan et al. compared the histology predictions of two CADx systems trained on differing datasets: CAD-EYE and GI-Genius [79]. They found no difference in sensitivity or specificity for the two systems. For ≤ 5 mm rectosigmoid polyps, the negative predictive value well surpassed the PIVI threshold for both the CAD-

EYE (97%) and GI-Genius (97.7%) systems. Based on the ESGE surveillance guidelines, there was 98.3% agreement with guideline-recommended surveillance intervals with both systems. While datasets may impact the outputs of these systems, it is likely that the high volume of polyp images in the training sets is such that the accuracy is more than adequate to facilitate widespread use of the 'resect and discard' and 'do not resect' strategies.

4.2. Larger Polyps

For larger polyps, the potential benefit of CADx is in the identification of appropriate resection strategies or appropriate referral in the case of non-interventional endoscopists. Three studies have examined CADx specifically in larger polyps in comparison with endoscopists. Luo et al. trained a CADx system and tested this on a 1634-image dataset from 156 lesions with high-grade dysplasia or adenocarcinoma [80]. The polyps were stratified by the CADx system into 'P0' with a submucosal invasion depth of less than 1000 μm and, therefore, endoscopically resectable or 'P1' where there was at least deep submucosal invasion or more advanced cancer. In the testing set, the model had an overall accuracy of 91.1%, a sensitivity of 91.2%, and a specificity of 91.0%, with no significant difference in accuracy compared with experienced interventional endoscopists. When only early adenocarcinomas were included in the analysis, the CADx model was superior to experienced endoscopists (sensitivity 65.3% vs. 40.0%) for differentiating endoscopically resectable lesions, suggesting there may be surface signatures on polyps even with deep submucosal invasion that have not yet been identified by experts in advanced mucosal imaging. Nemoto et al. analysed 1513 early adenocarcinomas, from intramucosal to deep submucosal invasive cancer, comparing their CADx system with trainee and expert endoscopists [81]. CADx showed high specificity at 94.4% for deep submucosal invasion, although sensitivity was low at 59.8%. The AUROC was 85.1% and was equivalent to the two experts (88.2% and 85.9%) and superior to the trainees (77%, $p = 0.0076$ and 66.2%, $p < 0.001$). Yao et al. developed a CADx system trained on 339 large sessile polyps, differentiating malignant from non-malignant polyps [82]. The overall accuracy was 90.4%, which was comparable to expert endoscopists and superior to both senior and junior endoscopists [82]. In this study, with the assistance of CADx, the accuracy of junior endoscopists improved from 75.4% to 85.3% ($p = 0.002$).

While CADx systems are yet to convincingly outperform expert endoscopists in guiding resection strategies, the future of these systems may be in optimising appropriate referrals to experts in endoscopic resection. Additionally, they may obviate the need for a biopsy prior to referral. This is of particular importance as biopsies have been well established as a strong predictor of failed en bloc endoscopic submucosal dissection for colorectal polyps, increasing the odds of severe fibrosis by more than eight times [83].

For expert interventionalists, one role of CADx may be in combination with endocytoscopy systems. Endocytoscopy involves a device that can be either incorporated into the endoscope or as a separate probe-based system, utilising a high-power fixed-focus lens to achieve ultra-high magnification in excess of 450× [84]. This novel technology allows for in vivo visualisations of tissue at the cellular level in real time, with accuracy as high as 85.8–97% for detecting the depth of submucosal invasion [85–88]. However, these systems require significant training and experience to interpret images. This technology may become more accessible with the advent of AI systems, with EndoBRAIN and EndoBRAIN-Plus now commercially available for the interpretation of endocytoscopic images. Studies thus far have demonstrated a high degree of accuracy for endocytoscopy-based CADx systems, with specificity of up to 97.3–98.9% for differentiating invasive cancer from non-malignant adenoma [89,90]. Kudo et al. compared AI with both trainee and expert endoscopists for endocytoscopic interpretation, with superior accuracy (98% vs. 69% and 93.3%, $p < 0.001$), sensitivity (96.9% vs. 70.8% and 92.8%, $p < 0.001$), and specificity (100% vs. 65.7% and 94.3%, $p < 0.001$) [91]. While these studies demonstrate some benefit for even expert endoscopists in differentiating invasive cancers from non-malignant adenomas, the eventual goal of CADx with endocytoscopy would be to differentiate between depths of submucosal invasion in order to assess suitability for endoscopic resection techniques, a feat not able to be consistently achieved by even the most experienced interventionalists.

In addition, another area for further study that may impact expert endoscopists would be in the assessment of resection margins. To date, no endoscopic systems have been developed for this purpose; however, a recent study performed using hyperspectral imaging on surgical specimens showed high accuracy (AUC 97%) for classifying the components of resected tissue into cancer, adenomatous margins, and healthy mucosa [92]. While this is essentially a proof-of-concept study only, it has highlighted the potential for AI to analyse the completeness of large resections and, therefore, theoretically reduce adenoma recurrence rates.

4.3. Summary

CADx systems have been proven to be highly accurate in differentiating neoplastic from non-neoplastic polyps, as well as in recognising invasive cancers. Similar to CADe, these systems are yet to consistently outperform expert endoscopists. Nevertheless, their future may be in the elevation of the accuracy of regular endoscopists to nearing that of highly trained interventionalists in order to guide conservative strategies for diminutive polyps and appropriate referral strategies for larger polyps requiring advanced resection techniques.

5. Conclusions

This review provides compelling evidence of the transformative potential of artificial intelligence in the realm of polyp detection and characterisation during colonoscopy. The key findings underscore two crucial aspects that significantly impact healthcare provision, particularly in resource-constrained settings.

First and foremost, the evidence reviewed demonstrates that CADe enhances adenoma detection in studies with low baseline ADR and increases the detection of inconspicuous polyps more frequently missed by endoscopists. This outcome carries substantial implications for public health, as it promises to bolster the consistency of healthcare delivery. In regions or communities where access to highly trained interventionalists may be limited, AI can serve as a reliable and consistent ally in early polyp detection, potentially preventing the progression of colorectal cancer and improving patient outcomes. This democratisation of expertise through AI could bridge the gap in healthcare equality, ensuring that more individuals receive accurate and timely diagnoses, ultimately reducing the burden of colorectal cancer on health systems.

Secondly, this manuscript highlights how AI can elevate the accuracy of polyp characterisation when used by regular endoscopists to nearly that of highly trained expert interventionalists. This development holds significant promise for overburdened healthcare systems worldwide, where access to specialist interventionalists is often limited. AI's ability to assist in precise polyp characterisation can help mitigate the risk of misdiagnoses, reducing unnecessary treatments, and enhancing patient care quality. Moreover, by empowering non-experts with advanced AI tools, we can ensure that patients in underserved regions receive comprehensive care, irrespective of the available expertise.

In a world where resource limitations persist and not everyone has access to highly trained interventionalists, this manuscript's findings underscore the profound public health implications of AI in colonoscopy. AI's capacity to augment both adenoma detection and polyp characterisation in the hands of all proceduralists not only promises to enhance healthcare consistency but also signifies a crucial step towards healthcare equity. As we continue to harness the power of artificial intelligence in medicine, the potential to democratise expertise and improve the overall health outcomes of diverse populations becomes increasingly tangible and vital.

Author Contributions: E.Y.—literature review, manuscript writing, editing, review, and submission. L.E.—literature review, manuscript writing, editing, and review. R.S.—conceptualisation, review, editing, and submission. All authors have read and agreed to the published version of the manuscript.

Funding: This research received no external funding.

Conflicts of Interest: The authors declare no conflict of interest.

References

1. Rajpurkar, P.; Chen, E.; Banerjee, O.; Topol, E.J. AI in health and medicine. *Nat. Med.* **2022**, *28*, 31–38. [CrossRef]
2. Bray, F.; Ferlay, J.; Soerjomataram, I.; Siegel, R.L.; Torre, L.A.; Jemal, A. Global cancer statistics 2018: GLOBOCAN estimates of incidence and mortality worldwide for 36 cancers in 185 countries. *CA Cancer J. Clin.* **2018**, *68*, 394–424. [CrossRef]
3. Arnold, M.; Sierra, M.S.; Laversanne, M.; Soerjomataram, I.; Jemal, A.; Bray, F. Global patterns and trends in colorectal cancer incidence and mortality. *Gut* **2017**, *66*, 683–691. [CrossRef]
4. Cardoso, R.; Guo, F.; Heisser, T.; Hackl, M.; Ihle, P.; De Schutter, H.; Van Damme, N.; Valerianova, Z.; Atanasov, T.; Májek, O.; et al. Colorectal cancer incidence, mortality, and stage distribution in European countries in the colorectal cancer screening era: An international population-based study. *Lancet Oncol.* **2021**, *22*, 1002–1013. [CrossRef]
5. Corley, D.A.; Jensen, C.D.; Marks, A.R.; Zhao, W.K.; Lee, J.K.; Doubeni, C.A.; Zauber, A.G.; de Boer, J.; Fireman, B.H.; Schottinger, J.E.; et al. Adenoma detection rate and risk of colorectal cancer and death. *N. Engl. J. Med.* **2014**, *370*, 1298–1306. [CrossRef]
6. Zhao, S.; Wang, S.; Pan, P.; Xia, T.; Chang, X.; Yang, X.; Guo, L.; Meng, Q.; Yang, F.; Qian, W.; et al. Magnitude, Risk Factors, and Factors Associated with Adenoma Miss Rate of Tandem Colonoscopy: A Systematic Review and Meta-analysis. *Gastroenterology* **2019**, *156*, 1661–1674.e11. [CrossRef]
7. Choi, R.Y.; Coyner, A.S.; Kalpathy-Cramer, J.; Chiang, M.F.; Campbell, J.P. Introduction to Machine Learning, Neural Networks, and Deep Learning. *Transl. Vis. Sci. Technol.* **2020**, *9*, 14.
8. Min, J.K.; Kwak, M.S.; Cha, J.M. Overview of Deep Learning in Gastrointestinal Endoscopy. *Gut Liver* **2019**, *13*, 388–393. [CrossRef]
9. Fernández-Esparrach, G.; Bernal, J.; López-Cerón, M.; Córdova, H.; Sánchez-Montes, C.; Rodríguez de Miguel, C.; Sánchez, F.J. Exploring the clinical potential of an automatic colonic polyp detection method based on the creation of energy maps. *Endoscopy* **2016**, *48*, 837–842. [CrossRef]
10. Wang, P.; Berzin, T.M.; Glissen Brown, J.R.; Bharadwaj, S.; Becq, A.; Xiao, X.; Liu, P.; Li, L.; Song, Y.; Zhang, D.; et al. Real-time automatic detection system increases colonoscopic polyp and adenoma detection rates: A prospective randomised controlled study. *Gut* **2019**, *68*, 1813–1819. [CrossRef]
11. Nakashima, H.; Kitazawa, N.; Fukuyama, C.; Kawachi, H.; Kawahira, H.; Momma, K.; Sakaki, N. Clinical Evaluation of Computer-Aided Colorectal Neoplasia Detection Using a Novel Endoscopic Artificial Intelligence: A Single-Center Randomized Controlled Trial. *Digestion* **2023**, *104*, 193–201. [CrossRef]
12. Xu, H.; Tang, R.S.Y.; Lam, T.Y.T.; Zhao, G.; Lau, J.Y.W.; Liu, Y.; Wu, Q.; Rong, L.; Xu, W.; Li, X.; et al. Artificial Intelligence-Assisted Colonoscopy for Colorectal Cancer Screening: A Multicenter Randomized Controlled Trial. *Clin. Gastroenterol. Hepatol.* **2023**, *21*, 337–346.e3. [CrossRef]
13. Wang, P.; Liu, X.G.; Kang, M.; Peng, X.; Shu, M.L.; Zhou, G.Y.; Liu, P.X.; Xiong, F.; Deng, M.M.; Xia, H.F.; et al. Artificial intelligence empowers the second-observer strategy for colonoscopy: A randomized clinical trial. *Gastroenterol. Rep.* **2023**, *11*, goac081. [CrossRef]
14. Wei, M.T.; Shankar, U.; Parvin, R.; Hasan Abbas, S.; Chaudhary, S.; Friedlander, Y.; Friedland, S. Evaluation of computer aided detection during colonoscopy in the community (AI-SEE): A multicenter randomized clinical trial. *Am. J. Gastroenterol.* **2023**, *118*, 1841–1847. [CrossRef]
15. Ahmad, A.; Wilson, A.; Haycock, A.; Humphries, A.; Monahan, K.; Suzuki, N.; Thomas-Gibson, S.; Vance, M.; Bassett, P.; Thiruvilangam, K.; et al. Evaluation of a real-time computer-aided polyp detection system during screening colonoscopy: AI-DETECT study. *Endoscopy* **2022**, *55*, 313–319. [CrossRef]
16. Gimeno-García, A.Z.; Hernández Negrin, D.; Hernández, A.; Nicolás-Pérez, D.; Rodríguez, E.; Montesdeoca, C.; Alarcon, O.; Romero, R.; Baute Dorta, J.L.; Cedrés, Y.; et al. Usefulness of a novel computer-aided detection system for colorectal neoplasia: A randomized controlled trial. *Gastrointest. Endosc.* **2022**, *97*, 528–536. [CrossRef]
17. Repici, A.; Spadaccini, M.; Antonelli, G.; Correale, L.; Maselli, R.; Galtieri, P.A.; Pellegatta, G.; Capogreco, A.; Milluzzo, S.M.; Lollo, G.; et al. Artificial intelligence and colonoscopy experience: Lessons from two randomised trials. *Gut* **2022**, *71*, 757–765. [CrossRef]
18. Rondonotti, E.; Di Paolo, D.; Rizzotto, E.R.; Alvisi, C.; Buscarini, E.; Spadaccini, M.; Tamanini, G.; Paggi, S.; Amato, A.; Scardino, G.; et al. Efficacy of a computer-aided detection system in a fecal immunochemical test-based organized colorectal cancer screening program: A randomized controlled trial (AIFIT study). *Endoscopy* **2022**, *54*, 1171–1179. [CrossRef]
19. Shaukat, A.; Lichtenstein, D.R.; Somers, S.C.; Chung, D.C.; Perdue, D.G.; Gopal, M.; Colucci, D.R.; Phillips, S.A.; Marka, N.A.; Church, T.R.; et al. Computer-Aided Detection Improves Adenomas per Colonoscopy for Screening and Surveillance Colonoscopy: A Randomized Trial. *Gastroenterology* **2022**, *163*, 732–741. [CrossRef]

20. Luo, Y.; Zhang, Y.; Liu, M.; Lai, Y.; Liu, P.; Wang, Z.; Xing, T.; Huang, Y.; Li, Y.; Li, A.; et al. Artificial Intelligence-Assisted Colonoscopy for Detection of Colon Polyps: A Prospective, Randomized Cohort Study. *J. Gastrointest. Surg.* **2021**, *25*, 2011–2018. [CrossRef]
21. Xu, L.; He, X.; Zhou, J.; Zhang, J.; Mao, X.; Ye, G.; Chen, Q.; Xu, F.; Sang, J.; Wang, J.; et al. Artificial intelligence-assisted colonoscopy: A prospective, multicenter, randomized controlled trial of polyp detection. *Cancer Med.* **2021**, *10*, 7184–7193. [CrossRef]
22. Liu, P.; Wang, P.; Glissen Brown, J.R.; Berzin, T.M.; Zhou, G.; Liu, W.; Xiao, X.; Chen, Z.; Zhang, Z.; Zhou, C.; et al. The single-monitor trial: An embedded CADe system increased adenoma detection during colonoscopy: A prospective randomized study. *Therap Adv. Gastroenterol.* **2020**, *13*, 1756284820979165. [CrossRef]
23. Liu, W.N.; Zhang, Y.Y.; Bian, X.Q.; Wang, L.J.; Yang, Q.; Zhang, X.D.; Huang, J. Study on detection rate of polyps and adenomas in artificial-intelligence-aided colonoscopy. *Saudi J. Gastroenterol.* **2020**, *26*, 13–19. [CrossRef]
24. Repici, A.; Badalamenti, M.; Maselli, R.; Correale, L.; Radaelli, F.; Rondonotti, E.; Ferrara, E.; Spadaccini, M.; Alkandari, A.; Fugazza, A.; et al. Efficacy of Real-Time Computer-Aided Detection of Colorectal Neoplasia in a Randomized Trial. *Gastroenterology* **2020**, *159*, 512–520.e7. [CrossRef]
25. Huang, D.; Shen, J.; Hong, J.; Zhang, Y.; Dai, S.; Du, N.; Zhang, M.; Guo, D. Effect of artificial intelligence-aided colonoscopy for adenoma and polyp detection: A meta-analysis of randomized clinical trials. *Int. J. Colorectal Dis.* **2022**, *37*, 495–506. [CrossRef]
26. Sivananthan, A.; Nazarian, S.; Ayaru, L.; Patel, K.; Ashrafian, H.; Darzi, A.; Patel, N. Does computer-aided diagnostic endoscopy improve the detection of commonly missed polyps? A meta-analysis. *Clin. Endosc.* **2022**, *55*, 355–364. [CrossRef]
27. Ashat, M.; Klair, J.S.; Singh, D.; Murali, A.R.; Krishnamoorthi, R. Impact of real-time use of artificial intelligence in improving adenoma detection during colonoscopy: A systematic review and meta-analysis. *Endosc. Int. Open* **2021**, *9*, e513–e521. [CrossRef]
28. Barua, I.; Vinsard, D.G.; Jodal, H.C.; Løberg, M.; Kalager, M.; Holme, Ø.; Misawa, M.; Bretthauer, M.; Mori, Y. Artificial intelligence for polyp detection during colonoscopy: A systematic review and meta-analysis. *Endoscopy* **2021**, *53*, 277–284. [CrossRef]
29. Deliwala, S.S.; Hamid, K.; Barbarawi, M.; Lakshman, H.; Zayed, Y.; Kandel, P.; Malladi, S.; Singh, A.; Bachuwa, G.; Gurvits, G.E.; et al. Artificial intelligence (AI) real-time detection vs. routine colonoscopy for colorectal neoplasia: A meta-analysis and trial sequential analysis. *Int. J. Colorectal Dis.* **2021**, *36*, 2291–2303. [CrossRef]
30. Hassan, C.; Spadaccini, M.; Iannone, A.; Maselli, R.; Jovani, M.; Chandrasekar, V.T.; Antonelli, G.; Yu, H.; Areia, M.; Dinis-Ribeiro, M.; et al. Performance of artificial intelligence in colonoscopy for adenoma and polyp detection: A systematic review and meta-analysis. *Gastrointest. Endosc.* **2021**, *93*, 77–85.e6. [CrossRef]
31. Li, J.; Lu, J.; Yan, J.; Tan, Y.; Liu, D. Artificial intelligence can increase the detection rate of colorectal polyps and adenomas: A systematic review and meta-analysis. *Eur. J. Gastroenterol. Hepatol.* **2021**, *33*, 1041–1048. [CrossRef]
32. Nazarian, S.; Glover, B.; Ashrafian, H.; Darzi, A.; Teare, J. Diagnostic Accuracy of Artificial Intelligence and Computer-Aided Diagnosis for the Detection and Characterization of Colorectal Polyps: Systematic Review and Meta-analysis. *J. Med. Internet Res.* **2021**, *23*, e27370. [CrossRef]
33. Spadaccini, M.; Iannone, A.; Maselli, R.; Badalamenti, M.; Desai, M.; Chandrasekar, V.T.; Patel, H.K.; Fugazza, A.; Pellegatta, G.; Galtieri, P.A.; et al. Computer-aided detection versus advanced imaging for detection of colorectal neoplasia: A systematic review and network meta-analysis. *Lancet Gastroenterol. Hepatol.* **2021**, *6*, 793–802. [CrossRef]
34. Zhang, Y.; Zhang, X.; Wu, Q.; Gu, C.; Wang, Z. Artificial Intelligence-Aided Colonoscopy for Polyp Detection: A Systematic Review and Meta-Analysis of Randomized Clinical Trials. *J. Laparoendosc. Adv. Surg. Tech. A* **2021**, *31*, 1143–1149. [CrossRef]
35. Aziz, M.; Fatima, R.; Dong, C.; Lee-Smith, W.; Nawras, A. The impact of deep convolutional neural network-based artificial intelligence on colonoscopy outcomes: A systematic review with meta-analysis. *J. Gastroenterol. Hepatol.* **2020**, *35*, 1676–1683. [CrossRef]
36. Ladabaum, U.; Shepard, J.; Weng, Y.; Desai, M.; Singer, S.J.; Mannalithara, A. Computer-aided Detection of Polyps Does Not Improve Colonoscopist Performance in a Pragmatic Implementation Trial. *Gastroenterology* **2023**, *164*, 481–483.e6. [CrossRef]
37. Levy, I.; Bruckmayer, L.; Klang, E.; Ben-Horin, S.; Kopylov, U. Artificial Intelligence-Aided Colonoscopy Does Not Increase Adenoma Detection Rate in Routine Clinical Practice. *Am. J. Gastroenterol.* **2022**, *117*, 1871–1873. [CrossRef]
38. Ishiyama, M.; Kudo, S.E.; Misawa, M.; Mori, Y.; Maeda, Y.; Ichimasa, K.; Kudo, T.; Hayashi, T.; Wakamura, K.; Miyachi, H.; et al. Impact of the clinical use of artificial intelligence-assisted neoplasia detection for colonoscopy: A large-scale prospective, propensity score-matched study (with video). *Gastrointest. Endosc.* **2022**, *95*, 155–163. [CrossRef]
39. Schauer, C.; Chieng, M.; Wang, M.; Neave, M.; Watson, S.; Van Rijnsoever, M.; Walmsley, R.; Jafer, A. Artificial intelligence improves adenoma detection rate during colonoscopy. *N. Z. Med. J.* **2022**, *135*, 22–30.
40. Biscaglia, G.; Cocomazzi, F.; Gentile, M.; Loconte, I.; Mileti, A.; Paolillo, R.; Marra, A.; Castellana, S.; Mazza, T.; Di Leo, A.; et al. Real-time, computer-aided, detection-assisted colonoscopy eliminates differences in adenoma detection rate between trainee and experienced endoscopists. *Endosc. Int. Open* **2022**, *10*, e616–e621. [CrossRef]
41. Glissen Brown, J.R.; Mansour, N.M.; Wang, P.; Chuchuca, M.A.; Minchenberg, S.B.; Chandnani, M.; Liu, L.; Gross, S.A.; Sengupta, N.; Berzin, T.M. Deep Learning Computer-aided Polyp Detection Reduces Adenoma Miss Rate: A United States Multi-center Randomized Tandem Colonoscopy Study (CADeT-CS Trial). *Clin. Gastroenterol. Hepatol.* **2022**, *20*, 1499–1507.e4. [CrossRef]
42. Wallace, M.B.; Sharma, P.; Bhandari, P.; East, J.; Antonelli, G.; Lorenzetti, R.; Vieth, M.; Speranza, I.; Spadaccini, M.; Desai, M.; et al. Impact of Artificial Intelligence on Miss Rate of Colorectal Neoplasia. *Gastroenterology* **2022**, *163*, 295–304.c5. [CrossRef]

43. Kamba, S.; Tamai, N.; Saitoh, I.; Matsui, H.; Horiuchi, H.; Kobayashi, M.; Sakamoto, T.; Ego, M.; Fukuda, A.; Tonouchi, A.; et al. Reducing adenoma miss rate of colonoscopy assisted by artificial intelligence: A multicenter randomized controlled trial. *J. Gastroenterol.* **2021**, *56*, 746–757. [CrossRef]
44. Sanaka, M.R.; Deepinder, F.; Thota, P.N.; Lopez, R.; Burke, C.A. Adenomas are detected more often in morning than in afternoon colonoscopy. *Am. J. Gastroenterol.* **2009**, *104*, 1659–1664. [CrossRef]
45. Chan, M.Y.; Cohen, H.; Spiegel, B.M. Fewer polyps detected by colonoscopy as the day progresses at a Veteran's Administration teaching hospital. *Clin. Gastroenterol. Hepatol.* **2009**, *7*, 1217–1223. [CrossRef]
46. Lu, Z.; Zhang, L.; Yao, L.; Gong, D.; Wu, L.; Xia, M.; Zhang, J.; Zhou, W.; Huang, X.; He, C.; et al. Assessment of the Role of Artificial Intelligence in the Association Between Time of Day and Colonoscopy Quality. *JAMA Netw. Open* **2023**, *6*, e2253840. [CrossRef]
47. Richter, R.; Bruns, J.; Obst, W.; Keitel-Anselmino, V.; Weigt, J. Influence of artificial intelligence on the adenoma detection rate throughout the day. *Dig. Dis.* **2022**, *41*, 615–619. [CrossRef]
48. Messmann, H.; Bisschops, R.; Antonelli, G.; Libânio, D.; Sinonquel, P.; Abdelrahim, M.; Ahmad, O.F.; Areia, M.; Bergman, J.; Bhandari, P.; et al. Expected value of artificial intelligence in gastrointestinal endoscopy: European Society of Gastrointestinal Endoscopy (ESGE) Position Statement. *Endoscopy* **2022**, *54*, 1211–1231. [CrossRef]
49. Holzwanger, E.A.; Bilal, M.; Glissen Brown, J.R.; Singh, S.; Becq, A.; Ernest-Suarez, K.; Berzin, T.M. Benchmarking definitions of false-positive alerts during computer-aided polyp detection in colonoscopy. *Endoscopy* **2021**, *53*, 937–940. [CrossRef]
50. Hassan, C.; Badalamenti, M.; Maselli, R.; Correale, L.; Iannone, A.; Radaelli, F.; Rondonotti, E.; Ferrara, E.; Spadaccini, M.; Alkandari, A.; et al. Computer-aided detection-assisted colonoscopy: Classification and relevance of false positives. *Gastrointest. Endosc.* **2020**, *92*, 900–904.e4. [CrossRef]
51. Nehme, F.; Coronel, E.; Barringer, D.A.; Romero, L.; Shafi, M.A.; Ross, W.A.; Ge, P.S. Performance and Attitudes Toward Real-time Computer-aided Polyp Detection during Colonoscopy in a Large Tertiary Referral Center in the United States. *Gastrointest. Endosc.* **2023**, *98*, 100–109. [CrossRef]
52. Tang, C.P.; Lin, T.L.; Hsieh, Y.H.; Hsieh, C.H.; Tseng, C.W.; Leung, F.W. Polyp detection and false-positive rates by computer-aided analysis of withdrawal-phase videos of colonoscopy of the right-sided colon segment in a randomized controlled trial comparing water exchange and air insufflation. *Gastrointest. Endosc.* **2022**, *95*, 1198–1206.e6. [CrossRef]
53. Mori, Y.; Wang, P.; Løberg, M.; Misawa, M.; Repici, A.; Spadaccini, M.; Correale, L.; Antonelli, G.; Yu, H.; Gong, D.; et al. Impact of Artificial Intelligence on Colonoscopy Surveillance After Polyp Removal: A Pooled Analysis of Randomized Trials. *Clin. Gastroenterol. Hepatol.* **2023**, *21*, 949–959. [CrossRef]
54. Areia, M.; Mori, Y.; Correale, L.; Repici, A.; Bretthauer, M.; Sharma, P.; Taveira, F.; Spadaccini, M.; Antonelli, G.; Ebigbo, A.; et al. Cost-effectiveness of artificial intelligence for screening colonoscopy: A modelling study. *Lancet Digit. Health* **2022**, *4*, e436–e444. [CrossRef]
55. Mori, Y.; East, J.E.; Hassan, C.; Halvorsen, N.; Berzin, T.M.; Byrne, M.; von Renteln, D.; Hewett, D.; Repici, A.; Ramchandani, M.; et al. Benefits and Challenges in Implementation of Artificial Intelligence in Colonoscopy: World Endoscopy Organization Position Statement. *Dig. Endosc.* **2023**, *35*, 422–429. [CrossRef]
56. Young, E.J.; Rajandran, A.; Philpott, H.L.; Sathananthan, D.; Hoile, S.F.; Singh, R. Mucosal imaging in colon polyps: New advances and what the future may hold. *World J. Gastroenterol.* **2022**, *28*, 6632–6661. [CrossRef]
57. Rex, D.K.; Kahi, C.; O'Brien, M.; Levin, T.R.; Pohl, H.; Rastogi, A.; Burgart, L.; Imperiale, T.; Ladabaum, U.; Cohen, J.; et al. The American Society for Gastrointestinal Endoscopy PIVI (Preservation and Incorporation of Valuable Endoscopic Innovations) on real-time endoscopic assessment of the histology of diminutive colorectal polyps. *Gastrointest. Endosc.* **2011**, *73*, 419–422. [CrossRef]
58. Li, M.D.; Huang, Z.R.; Shan, Q.Y.; Chen, S.L.; Zhang, N.; Hu, H.T.; Wang, W. Performance and comparison of artificial intelligence and human experts in the detection and classification of colonic polyps. *BMC Gastroenterol.* **2022**, *22*, 517. [CrossRef]
59. Xu, Y.; Ding, W.; Wang, Y.; Tan, Y.; Xi, C.; Ye, N.; Wu, D.; Xu, X. Comparison of diagnostic performance between convolutional neural networks and human endoscopists for diagnosis of colorectal polyp: A systematic review and meta-analysis. *PLoS ONE* **2021**, *16*, e0246892. [CrossRef]
60. Lui, T.K.L.; Guo, C.G.; Leung, W.K. Accuracy of artificial intelligence on histology prediction and detection of colorectal polyps: A systematic review and meta-analysis. *Gastrointest. Endosc.* **2020**, *92*, 11–22.e6. [CrossRef]
61. Choi, S.J.; Kim, E.S.; Choi, K. Prediction of the histology of colorectal neoplasm in white light colonoscopic images using deep learning algorithms. *Sci. Rep.* **2021**, *11*, 5311. [CrossRef]
62. Meng, S.; Zheng, Y.; Wang, W.; Su, R.; Zhang, Y.; Zhang, Y.; Guo, B.; Han, Z.; Zhang, W.; Qin, W.; et al. A computer-aided diagnosis system using white-light endoscopy for the prediction of conventional adenoma with high grade dysplasia. *Dig. Liver Dis.* **2022**, *54*, 1202–1208. [CrossRef]
63. Hossain, E.; Abdelrahim, M.; Tanasescu, A.; Yamada, M.; Kondo, H.; Yamada, S.; Hamamoto, R.; Marugame, A.; Saito, Y.; Bhandari, P. Performance of a novel computer-aided diagnosis system in the characterization of colorectal polyps, and its role in meeting Preservation and Incorporation of Valuable Endoscopic Innovations standards set by the American Society of Gastrointestinal Endoscopy. *DEN Open* **2023**, *3*, e178. [CrossRef]

64. Jin, E.H.; Lee, D.; Bae, J.H.; Kang, H.Y.; Kwak, M.S.; Seo, J.Y.; Yang, J.I.; Yang, S.Y.; Lim, S.H.; Yim, J.Y.; et al. Improved Accuracy in Optical Diagnosis of Colorectal Polyps Using Convolutional Neural Networks with Visual Explanations. *Gastroenterology* **2020**, *158*, 2169–2179.e8. [CrossRef]
65. Ozawa, T.; Ishihara, S.; Fujishiro, M.; Kumagai, Y.; Shichijo, S.; Tada, T. Automated endoscopic detection and classification of colorectal polyps using convolutional neural networks. *Therap Adv. Gastroenterol.* **2020**, *13*, 1756284820910659. [CrossRef]
66. Song, E.M.; Park, B.; Ha, C.A.; Hwang, S.W.; Park, S.H.; Yang, D.H.; Ye, B.D.; Myung, S.J.; Yang, S.K.; Kim, N.; et al. Endoscopic diagnosis and treatment planning for colorectal polyps using a deep-learning model. *Sci. Rep.* **2020**, *10*, 30. [CrossRef]
67. Yang, Y.J.; Cho, B.J.; Lee, M.J.; Kim, J.H.; Lim, H.; Bang, C.S.; Jeong, H.M.; Hong, J.T.; Baik, G.H. Automated Classification of Colorectal Neoplasms in White-Light Colonoscopy Images via Deep Learning. *J. Clin. Med.* **2020**, *9*, 1593. [CrossRef]
68. Yamada, M.; Shino, R.; Kondo, H.; Yamada, S.; Takamaru, H.; Sakamoto, T.; Bhandari, P.; Imaoka, H.; Kuchiba, A.; Shibata, T.; et al. Robust automated prediction of the revised Vienna Classification in colonoscopy using deep learning: Development and initial external validation. *J. Gastroenterol.* **2022**, *57*, 879–889. [CrossRef]
69. Barua, I.; Wieszczy, P.; Kudo, S.-E.; Misawa, M.; Holme, Ø.; Gulati, S.; Williams, S.; Mori, K.; Itoh, H.; Takishima, K.; et al. Real-Time Artificial Intelligence–Based Optical Diagnosis of Neoplastic Polyps during Colonoscopy. *NEJM Evid.* **2022**, *1*, EVIDoa2200003. [CrossRef]
70. Li, J.W.; Wu, C.C.H.; Lee, J.W.J.; Liang, R.; Soon, G.S.T.; Wang, L.M.; Koh, X.H.; Koh, C.J.; Chew, W.D.; Lin, K.W.; et al. Real-World Validation of a Computer-Aided Diagnosis System for Prediction of Polyp Histology in Colonoscopy: A Prospective Multicenter Study. *Am. J. Gastroenterol.* **2023**, *118*, 1353–1364. [CrossRef]
71. Yoshida, N.; Inoue, K.; Tomita, Y.; Kobayashi, R.; Hashimoto, H.; Sugino, S.; Hirose, R.; Dohi, O.; Yasuda, H.; Morinaga, Y.; et al. An analysis about the function of a new artificial intelligence, CAD EYE with the lesion recognition and diagnosis for colorectal polyps in clinical practice. *Int. J. Colorectal Dis.* **2021**, *36*, 2237–2245. [CrossRef]
72. Byrne, M.F.; Chapados, N.; Soudan, F.; Oertel, C.; Linares Pérez, M.; Kelly, R.; Iqbal, N.; Chandelier, F.; Rex, D.K. Real-time differentiation of adenomatous and hyperplastic diminutive colorectal polyps during analysis of unaltered videos of standard colonoscopy using a deep learning model. *Gut* **2019**, *68*, 94–100. [CrossRef]
73. Mori, Y.; Kudo, S.E.; Misawa, M.; Saito, Y.; Ikematsu, H.; Hotta, K.; Ohtsuka, K.; Urushibara, F.; Kataoka, S.; Ogawa, Y.; et al. Real-Time Use of Artificial Intelligence in Identification of Diminutive Polyps During Colonoscopy: A Prospective Study. *Ann. Intern. Med.* **2018**, *169*, 357–366. [CrossRef]
74. Mori, Y.; Kudo, S.E.; Misawa, M.; Mori, K. Simultaneous detection and characterization of diminutive polyps with the use of artificial intelligence during colonoscopy. *VideoGIE* **2019**, *4*, 7–10. [CrossRef]
75. Zachariah, R.; Samarasena, J.; Luba, D.; Duh, E.; Dao, T.; Requa, J.; Ninh, A.; Karnes, W. Prediction of Polyp Pathology Using Convolutional Neural Networks Achieves "Resect and Discard" Thresholds. *Am. J. Gastroenterol.* **2020**, *115*, 138–144. [CrossRef]
76. Rondonotti, E.; Hassan, C.; Tamanini, G.; Antonelli, G.; Andrisani, G.; Leonetti, G.; Paggi, S.; Amato, A.; Scardino, G.; Di Paolo, D.; et al. Artificial intelligence-assisted optical diagnosis for the resect-and-discard strategy in clinical practice: The Artificial intelligence BLI Characterization (ABC) study. *Endoscopy* **2023**, *55*, 14–22. [CrossRef]
77. Ponugoti, P.; Rastogi, A.; Kaltenbach, T.; MacPhail, M.E.; Sullivan, A.W.; Thygesen, J.C.; Broadley, H.M.; Rex, D.K. Disagreement between high confidence endoscopic adenoma prediction and histopathological diagnosis in colonic lesions ≤ 3 mm in size. *Endoscopy* **2019**, *51*, 221–226. [CrossRef]
78. Shahidi, N.; Rex, D.K.; Kaltenbach, T.; Rastogi, A.; Ghalehjegh, S.H.; Byrne, M.F. Use of Endoscopic Impression, Artificial Intelligence, and Pathologist Interpretation to Resolve Discrepancies Between Endoscopy and Pathology Analyses of Diminutive Colorectal Polyps. *Gastroenterology* **2020**, *158*, 783–785.e1. [CrossRef]
79. Hassan, C.; Sharma, P.; Mori, Y.; Bretthauer, M.; Rex, D.K.; Repici, A. Comparative Performance of Artificial Intelligence Optical Diagnosis Systems for Leaving in Situ Colorectal Polyps. *Gastroenterology* **2023**, *164*, 467–469.e4. [CrossRef]
80. Luo, X.; Wang, J.; Han, Z.; Yu, Y.; Chen, Z.; Huang, F.; Xu, Y.; Cai, J.; Zhang, Q.; Qiao, W.; et al. Artificial intelligence-enhanced white-light colonoscopy with attention guidance predicts colorectal cancer invasion depth. *Gastrointest. Endosc.* **2021**, *94*, 627–638.e1. [CrossRef]
81. Nemoto, D.; Guo, Z.; Katsuki, S.; Takezawa, T.; Maemoto, R.; Kawasaki, K.; Inoue, K.; Akutagawa, T.; Tanaka, H.; Sato, K.; et al. Computer-Aided Diagnosis of Early-Stage Colorectal Cancer Using Non-Magnified Endoscopic White Light Images. *Gastrointest. Endosc.* **2023**, *98*, 90–99. [CrossRef]
82. Yao, L.; Lu, Z.; Yang, G.; Zhou, W.; Xu, Y.; Guo, M.; Huang, X.; He, C.; Zhou, R.; Deng, Y.; et al. Development and validation of an artificial intelligence-based system for predicting colorectal cancer invasion depth using multi-modal data. *Dig. Endosc.* **2023**, *35*, 625–635. [CrossRef]
83. Kuroha, M.; Shiga, H.; Kanazawa, Y.; Nagai, H.; Handa, T.; Ichikawa, R.; Onodera, M.; Naito, T.; Moroi, R.; Kimura, T.; et al. Factors Associated with Fibrosis during Colorectal Endoscopic Submucosal Dissection: Does Pretreatment Biopsy Potentially Elicit Submucosal Fibrosis and Affect Endoscopic Submucosal Dissection Outcomes? *Digestion* **2021**, *102*, 590–598. [CrossRef]
84. Singh, R.; Sathananthan, D.; Tam, W.; Ruszkiewicz, A. Endocytoscopy for diagnosis of gastrointestinal Neoplasia: The expert's approach. *Video J. Encycl. GI Endosc.* **2013**, *1*, 18–19. [CrossRef]
85. Kudo, T.; Kudo, S.E.; Mori, Y.; Wakamura, K.; Misawa, M.; Hayashi, T.; Miyachi, H.; Katagiri, A.; Ishida, F.; Inoue, H. Classification of nuclear morphology in endocytoscopy of colorectal neoplasms. *Gastrointest. Endosc.* **2017**, *85*, 628–638. [CrossRef]

86. Nakamura, H.; Kudo, S.E.; Misawa, M.; Kataoka, S.; Wakamura, K.; Hayashi, T.; Kudo, T.; Mori, Y.; Takeda, K.; Ichimasa, K.; et al. Evaluation of microvascular findings of deeply invasive colorectal cancer by endocytoscopy with narrow-band imaging. *Endosc. Int. Open* **2016**, *4*, E1280–E1285. [CrossRef]
87. Kudo, S.E.; Mori, Y.; Wakamura, K.; Ikehara, N.; Ichimasa, K.; Wada, Y.; Kutsukawa, M.; Misawa, M.; Kudo, T.; Hayashi, T.; et al. Endocytoscopy can provide additional diagnostic ability to magnifying chromoendoscopy for colorectal neoplasms. *J. Gastroenterol. Hepatol.* **2014**, *29*, 83–90. [CrossRef]
88. Kudo, T.; Kudo, S.-E.; Wakamura, K.; Mori, Y.; Misawa, M.; Hayashi, T.; Kutsukawa, M.; Ichimasa, K.; Miyachi, H.; Ishida, F.; et al. Diagnostic performance of endocytoscopy for evaluating the invasion depth of different morphological types of colorectal tumors. *Dig. Endosc.* **2015**, *27*, 755–762. [CrossRef]
89. Mori, Y.; Kudo, S.E.; Misawa, M.; Hotta, K.; Kazuo, O.; Saito, S.; Ikematsu, H.; Saito, Y.; Matsuda, T.; Kenichi, T.; et al. Artificial intelligence-assisted colonic endocytoscopy for cancer recognition: A multicenter study. *Endosc. Int. Open* **2021**, *9*, E1004–E1011. [CrossRef]
90. Takeda, K.; Kudo, S.E.; Mori, Y.; Misawa, M.; Kudo, T.; Wakamura, K.; Katagiri, A.; Baba, T.; Hidaka, E.; Ishida, F.; et al. Accuracy of diagnosing invasive colorectal cancer using computer-aided endocytoscopy. *Endoscopy* **2017**, *49*, 798–802. [CrossRef]
91. Kudo, S.E.; Misawa, M.; Mori, Y.; Hotta, K.; Ohtsuka, K.; Ikematsu, H.; Saito, Y.; Takeda, K.; Nakamura, H.; Ichimasa, K.; et al. Artificial Intelligence-assisted System Improves Endoscopic Identification of Colorectal Neoplasms. *Clin. Gastroenterol. Hepatol.* **2020**, *18*, 1874–1881.e2. [CrossRef]
92. Jansen-Winkeln, B.; Barberio, M.; Chalopin, C.; Schierle, K.; Diana, M.; Köhler, H.; Gockel, I.; Maktabi, M. Feedforward Artificial Neural Network-Based Colorectal Cancer Detection Using Hyperspectral Imaging: A Step towards Automatic Optical Biopsy. *Cancers* **2021**, *13*, 967. [CrossRef]

Disclaimer/Publisher's Note: The statements, opinions and data contained in all publications are solely those of the individual author(s) and contributor(s) and not of MDPI and/or the editor(s). MDPI and/or the editor(s) disclaim responsibility for any injury to people or property resulting from any ideas, methods, instructions or products referred to in the content.

 *cancers*

Review

The Role of Endoscopy in the Palliation of Pancreatico-Biliary Cancers: Biliary Drainage, Management of Gastrointestinal Obstruction, and Role in Relief of Oncologic Pain

Giacomo Emanuele Maria Rizzo [1,2], Lucio Carrozza [1], Gabriele Rancatore [1], Cecilia Binda [3], Carlo Fabbri [3], Andrea Anderloni [4] and Ilaria Tarantino [1,*]

[1] Endoscopy Unit, Department of Diagnostic and Therapeutic Services, IRCCS-ISMETT Palermo, 90127 Palermo, Italy; grizzo@ismett.edu (G.E.M.R.); lcarrozza@ismett.edu (L.C.); grancatore@ismett.edu (G.R.)
[2] Ph.D. Program, Department of Surgical, Oncological and Oral Sciences (Di.Chir.On.S.), University of Palermo, 90133 Palermo, Italy
[3] Gastroenterology and Digestive Endoscopy Unit, Forlì-Cesena Hospitals, AUSL Romagna, 48100 Forlì-Cesena, Italy; cecilia.binda@gmail.com (C.B.); carlo.fabbri@auslromagna.it (C.F.)
[4] Gastroenterology and Digestive Endoscopy Unit, Fondazione I.R.C.C.S. Policlinico San Matteo, Viale Camillo Golgi 19, 27100 Pavia, Italy; andrea_anderloni@hotmail.com
* Correspondence: itarantino@ismett.edu

Simple Summary: Palliative endoscopy has a fundamental role in the management of patients with advanced bilio-pancreatic cancers, which can involve the biliary tract and infiltrate the duodenal lumen or other close organs. Clinical presentations of these advanced cancers are mainly gastric outlet obstruction (GOO), obstructive jaundice, and unresponsive pain, which influence the patient's quality of life (QoL) and the oncologic management in terms of initiating or restarting systemic therapy. Our aim was to perform a literature review focusing on the role of endoscopy in the palliation of these advanced pancreatic and biliary cancers.

Abstract: Therapeutic endoscopy permits many and various treatments for cancer palliation in patients with bilio-pancreatic cancers, enabling different options, supporting patients during their route to oncologic treatments, and trying to improve their quality of life. Therefore, both endoscopic and endoscopic ultrasound (EUS)-guided techniques are performed in this scenario. We performed a literature review focusing on the role of endoscopy in the palliation of those advanced pancreatic and biliary cancers developing malignant biliary obstruction (MBO), gastric outlet obstruction (GOO), and pain unresponsive to medical therapies. Therefore, we explored and focused on the clinical outcomes of endoscopic procedures in this scenario. In fact, the endoscopic treatment is based on achieving biliary drainage in the case of MBO through endoscopic retrograde cholangiopancreatography (ERCP) or EUS-guided biliary drainage (EUS-BD), while GOO is endoscopically treated through the deployment of an enteral stent or the creation of EUS-guided gastro-entero-anastomosis (EUS-GEA). Furthermore, untreatable chronic abdominal pain is a major issue in patients unresponsive to high doses of painkillers, so EUS-guided celiac plexus neurolysis (CPN) or celiac ganglia neurolysis (CGN) helps to reduce dosage and have better pain control. Therefore, therapeutic endoscopy in the palliative setting is an effective and safe approach for managing most of the clinical manifestations of advanced biliopancreatic tumors.

Keywords: palliation; biliopancreatic cancer; endoscopy; biliary obstruction; pain; oncology

Citation: Rizzo, G.E.M.; Carrozza, L.; Rancatore, G.; Binda, C.; Fabbri, C.; Anderloni, A.; Tarantino, I. The Role of Endoscopy in the Palliation of Pancreatico-Biliary Cancers: Biliary Drainage, Management of Gastrointestinal Obstruction, and Role in Relief of Oncologic Pain. *Cancers* **2023**, *15*, 5367. https://doi.org/10.3390/cancers15225367

Academic Editors: Hajime Isomoto and Jan Willem B. de Groot

Received: 12 September 2023
Revised: 20 October 2023
Accepted: 7 November 2023
Published: 10 November 2023

Copyright: © 2023 by the authors. Licensee MDPI, Basel, Switzerland. This article is an open access article distributed under the terms and conditions of the Creative Commons Attribution (CC BY) license (https://creativecommons.org/licenses/by/4.0/).

1. Introduction

Endoscopy is the standard of care for the palliation of advanced cancers involving the gastrointestinal (GI) tract. The role of palliative endoscopy is variable and dependent on cancer advancement, which moves from the involvement of the biliary tract to the infiltration of the duodenal lumen or other close organs. Therapeutic endoscopy, including endoscopic ultrasound (EUS), has improved over the years to overcome the clinical symptoms of advanced neoplastic diseases, permitting different options, supporting patients during their route toward starting systemic chemotherapy, and even trying to improve their quality of life (QoL) [1].

We aimed to perform a literature review focusing on the role of endoscopy in the palliation of advanced pancreatic-biliary cancers, in order to highlight the technical and clinical aspects of those endoscopic procedures which are strengthening as first-line approaches in the case of cancer palliation.

2. Clinical Aspects of Advanced Pancreatic-Biliary Cancer

Pancreatic and biliary cancers are among the most aggressive cancers [2]. In the United States (US), researchers have estimated the average annual incidence rate (2015–2019) of pancreatic cancers at 13.2 per 100,000 inhabitants [2], so estimated new cases and deaths are 64,050 and 50,550 in 2023 [3]. Annual new cases of gallbladder and other biliary cancers, indeed, are estimated to be 12,220 in the US, while estimated deaths are 4510 [3]. Surely, even metastasis and neoplastic lymph nodes may involve the biliary tract and duodenal lumen, [4] creating the need for endoscopic treatments. Nowadays, the 5-year survival rate at the time of diagnosis is still dramatically low, being 10% for pancreatic cancer and 18% for localized/regional extrahepatic bile duct cancers (both hilar and distal) in the USA [3,5]. Therefore, palliation is the main aim in those advanced cases developing jaundice, oncologic pain, or vomiting, so the management of the latter conditions becomes of primary relevance. The trigger for mechanical obstruction is usually an infiltration or compression of the biliary and duodenal tract by the malignancy (Figure 1), which then clinically produces malignant biliary obstruction (MBO) or gastric outlet obstruction (GOO). On the other hand, both the malignancy itself and the involvement of nerves cause severe oncologic pain, which is arduous to resolve with a single intervention such as painkiller administration; nonetheless, alternative endoscopic therapies targeting the celiac plexus are available (Figure 1) [6]. However, MBO and GOO can be endoscopically treated, being caused by a mechanical obstruction, while cancer pain needs to be first treated by an expert in the field of pain therapy, even if EUS-guided therapeutic options may complement medical therapies [7,8].

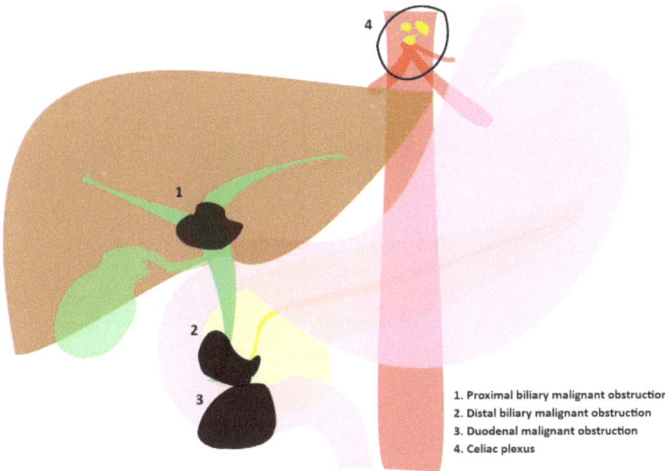

Figure 1. (1, 2, and 3) Sites of progression of advanced tumors involving the biliary and duodenal tract evolving in major clinical manifestations. (4) Celiac plexus as the "target" of endoscopic treatments in advanced tumors.

Materials and Methods

This is a comprehensive review of the role of endoscopy in the palliation of advanced pancreatic–biliary cancers. Considering the vastness of the topic, the search strategy, materials, and methods were adapted to each main topic of the review, and they are more deeply discussed in the Supplementary Materials [9,10]. Generally, the identification of the literature, the selection of sources, and the analysis, synthesis, and organization of the information were conducted by three researchers (G.E.M.R., L.C. and G.R.).

3. Endoscopic Treatments

Palliative endoscopic treatments in this scenario include procedures involving both endoscopic and EUS-guided techniques depending on the aim of the treatment and location of the issue (e.g., drainage, anastomosis creation, alcohol injection, ablation, debulking, and so on). In the case of MBO, which is one of the most common complications of malignancies involving the hepato-biliary-pancreatic system, the endoscopic treatment is based on achieving biliary drainage through endoscopic retrograde cholangiopancreatography (ERCP) or EUS-guided biliary drainage (EUS-BD). The resolution of jaundice reduces the risk of cholangitis and sepsis, and consequently improves QoL [11]. Furthermore, MBO can be divided based on its location into malignant distal biliary obstruction (dMBO) and malignant proximal biliary obstruction (pMBO). In the case of GOO, endoscopic treatment can include either duodenal stenting or EUS-guided gastro-enteroanastomosis (EUS-GEA). On the other hand, intractable oncologic pain in the case of bilio-pancreatic malignancies has been treated through celiac plexus neurolysis (CPN) over the years, firstly percutaneously, then through an EUS-guided approach, showing similar effectiveness and safety in randomized trials [12].

3.1. Malignant Biliary Obstruction (MBO)

3.1.1. Role of Endoscopic Retrograde Cholangiopancreatography (ERCP)

The transpapillary approach through ERCP is a milestone in the management of MBO with the advantages of avoiding external drainage [13], shorter hospitalization times, and lower rates of adverse events (8.6% vs. 12.3%, $p < 0.001$) compared to PTBD [14]. ERCP is also associated with lower rates of morbidity, peri- and post-procedural complications, and 30-day mortality (16.3% vs. 9.6%) when compared with the surgical approach, although surgical biliodigestive anastomosis showed a reduction in the rates of recurrent jaundice [15,16]. However, no interruption in the administration of oncological treatments is fundamental to achieving better oncological outcomes such as overall survival (OS) and progression-free survival (PFS), so the goal of ERCP is to permit BD in as many patients as possible.

3.1.2. Distal Malignant Biliary Obstruction (dMBO)

DMBO refers to malignant involvement of the distal part of the common bile duct (CBD) and it may be caused by intrinsic or extrinsic compression such as pancreatic head cancer, cholangiocarcinoma, ampullary cancer, or compression of metastatic lymph nodes [4,17]. Endoscopic treatments were historically based on ERCP, which is still considered the gold standard, even if EUS-guided approaches, which were initially used after ERCP failure, are becoming an alternative primary treatment, as suggested by recent studies and ongoing trials [18–20]. The European Society of Gastrointestinal Endoscopy (ESGE) guidelines recommend biliary self-expandable metal stent (SEMS) insertion for palliative drainage [21]. The choice of the type of stent to use is influenced by several factors such as the location of the stenosis, the patient's prognosis, and the availability of the prosthesis. There is enough evidence in the literature to suggest the choice of SEMS over a plastic prosthesis since remaining patent for longer improves patient outcomes. In the meta-analysis by Moole et al., where 11 studies with a total of 947 patients were selected, the pooled analysis of SEMS patency was 167 days, unlike the 73 days of the plastic stent [22]. Either covered SEMS (C-SEMS) or uncovered (U-SEMS) may be used, even if there is still a debate over which is the best due to conflicting results in the literature. In fact, C-SEMS seemed to prolong stent patency but had a higher migration rate [21] compared to U-SEMS, where tumor ingrowth through the metal mesh fixes the stent but reduces patency, even if a meta-analysis including nine randomized controlled trials (RCTs) found no difference in the length of stent patency [23]. Further meta-analyses evaluated the use of C-SEMS vs. U-SEMS without finding significant differences in clinical outcomes [24,25]. Regarding the safety and the rate of adverse events (AEs), the abovementioned meta-analysis did not demonstrate any higher risk of cholecystitis after C-SEMS insertion. Similarly, no differences in pancreatitis rate were shown between C-SEMS and U-SEMS. However, a novel type of stent was developed to counter stent ingrowth, the chemotherapy drug-eluting stent, but a meta-analysis of five studies comparing drug-eluting stents (197 patients) to SEMS (151 patients) reported a stent patency of 168 days vs. 149 days, respectively, with no major differences in the rates of cholecystitis (6.5% vs. 5.0%) or cholangitis (17% vs. 15%) [26]. Therefore, those stents have yet to receive receive FDA approval. Percutaneous biliary drainage (PTBD) has also been used as an alternative, showing similar efficacy with no significant differences in survival time or costs compared to endoscopic biliary drainage [27], but it needs an external approach and it could impact the QoL of patients. When jaundice secondary to pancreatic neoplasms is susceptible to neo-adjuvant chemotherapy, plastic biliary stent placement (of at least 10 Fr) was suggested until a few years ago, because the inflammatory reaction created by a SEMS made the surgical procedure more complex. A recent systematic review and meta-analysis by Du et al. conducted with the aim of comparing the clinical efficacy of metal stents versus plastic stents in patients undergoing neoadjuvant therapy included two randomized trials and six retrospective studies with a total of 316 patients, showing no significant differences in terms of operative and postoperative time, and the need for endoscopic reintervention and stent-related complications were significantly lower in the group treated with metal stents than in the one treated with plastic stents, respectively (18% vs. 80% and 15% vs. 44%) [28].

3.1.3. Proximal Malignant Biliary Obstruction (pMBO)

PMBO refers to malignant involvement of the proximal part of CBD caused by intrinsic obstruction or extrinsic compression by cancers, and it can involve the confluence of the hepatic ducts, often called 'Klatskin tumor', causing a malignant hilar biliary obstruction (hMBO) (Figure 2) [29,30]. Therefore, biliary drainage of the Klatskin tumors is strongly influenced by the extension of the neoplastic tissue, well-differentiated by the Bismuth classification (Supplementary Table S1) [31], because of the lower probability of concurrently draining through ERCP all of the hepatic segments when approaching a Bismuth type IV or III [21].

The retrograde approach is sometimes not the best option in the case of pMBO, especially when the tumor involves biliary confluence into both the right and left biliary ducts (type IV according to the Bismuth classification), because in these difficult cases sometimes it is not possible to drain both of the ducts, so patients do not resolve jaundice. Therefore, in the case of Klatskin tumor Bismuth IV or III it is extremely important to have a multidisciplinary approach together with an interventional radiologist in order to drain all the segments through a rendezvous or with an additional insertion of a PTBD [21]. A systematic review and meta-analysis of nine studies (n = 546 patients) showed a higher success rate with PTBD than ERCP in types III/IV, with comparable rates of adverse events and 30-day mortality [32]. On the other hand, Inamdar et al. reported that biliary drainage through ERCP showed a lower adverse event rate and shorter hospitalization when compared with PTBD [14]. Moreover, in a propensity score matching analysis, patients who underwent PTBD had lower overall survival and a higher risk for seeding metastasis when compared with ERCP [33]. Generally, PTBD is preferred when a patient has an altered gastro-duodenal anatomy, when the bile ducts to be drained are not accessible by ERCP, or when ERCP does not achieve adequate biliary drainage. Regardless of the method used, achieving ≥50% of total liver volume drainage is essential to relieve jaundice and reduce the risk of cholangitis. This was associated with longer overall survival particularly in the Bismuth III type [34]. Similarly, in their retrospective study, Takahashi et al. [35] correlated the percentage of liver volume to be drained with the patient's liver function and concluded that effective biliary drainage is achieved in patients with preserved liver function when >33% of the liver volume is drained, and in those with impaired liver function when >50% is drained. Anyway, regarding ERCP stenting, different meta-analyses comparing SEMS to plastic stents resulted in longer patient survival, lower risk of stent dysfunction and infection, and fewer reoperations when SEMS was deployed [36,37].

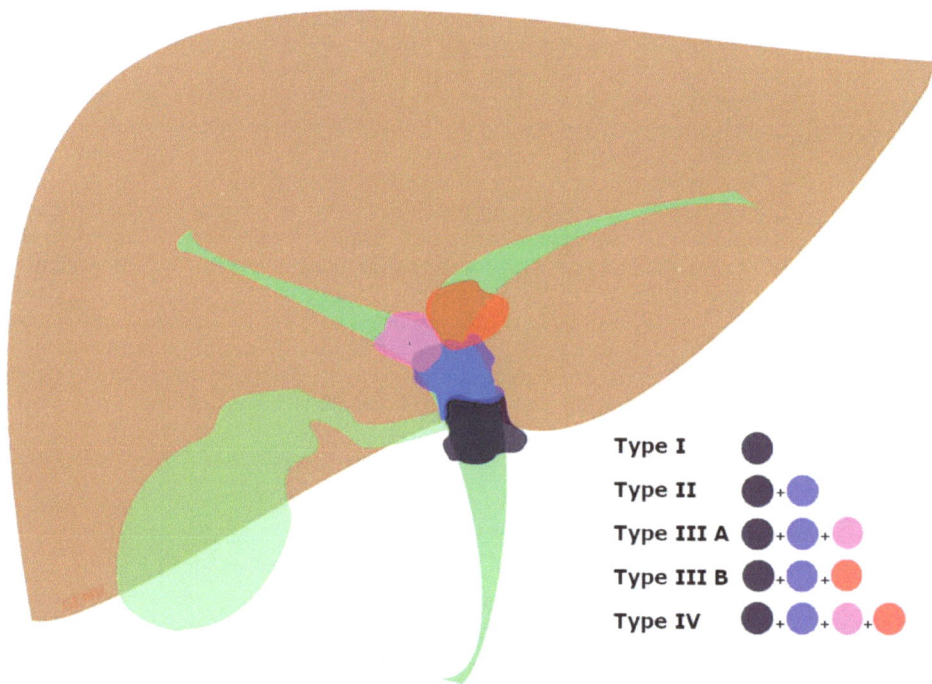

Figure 2. Graphical view of the Bismuth classification of Klatskin tumors [31]. Type (I) involving the common hepatic duct below the confluence; type (II) involving the biliary confluence; type (IIIA) involving the confluence and extending to the right hepatic duct; type (IIIB) involving the confluence and extending to the left hepatic duct; type (IV) involving the confluence and extending to both the right and left hepatic bile ducts.

3.1.4. Endoscopic Ultrasound Biliary Drainage (EUS-BD)

Although ERCP remains the gold standard in the treatment of dMBO, the international consensus statement for the management of malignant distal biliary stricture recommends that, when expertise is available, ultrasound endoscopic biliary drainage (EUS-BD) is an effective option in three situations: failed ERCP, difficult biliary cannulation, and postsurgical anatomy [13]. In fact, although PTBD has long been utilized, EUS-BD is a less invasive option with fewer procedure-related adverse events (8.80% vs. 31.22%, $p = 0.022$) and lower reintervention rates (0.34 vs. 0.93, $p = 0.02$) shown in a randomized open-label study [38], and recommended by European guidelines over PTBD [21]. Subsequently, these data were confirmed by a meta-analysis including 483 patients [39]. A systematic review of 42 studies including 1192 patients undergoing EUS-BD after ERCP failure reported a technical success rate of 94.7%, a clinical success rate of 91.6%, and an adverse event rate of 23%, which included bile leak (4.03%), bleeding (4.03%), pneumoperitoneum (3.02%), stent migration (2.68%), cholangitis (2.43%), abdominal pain (1.51%) and peritonitis (1.26%) [40]. Moreover,

EUS-BD techniques can be divided according to the anatomical location and the puncture site of the biliary access into choledochoduodenostomy (CDS), hepaticogastrostomy (HGS), rendezvous technique (RV), antegrade biliary stenting (AG), and gallbladder drainage (GBD) (Figure 3).

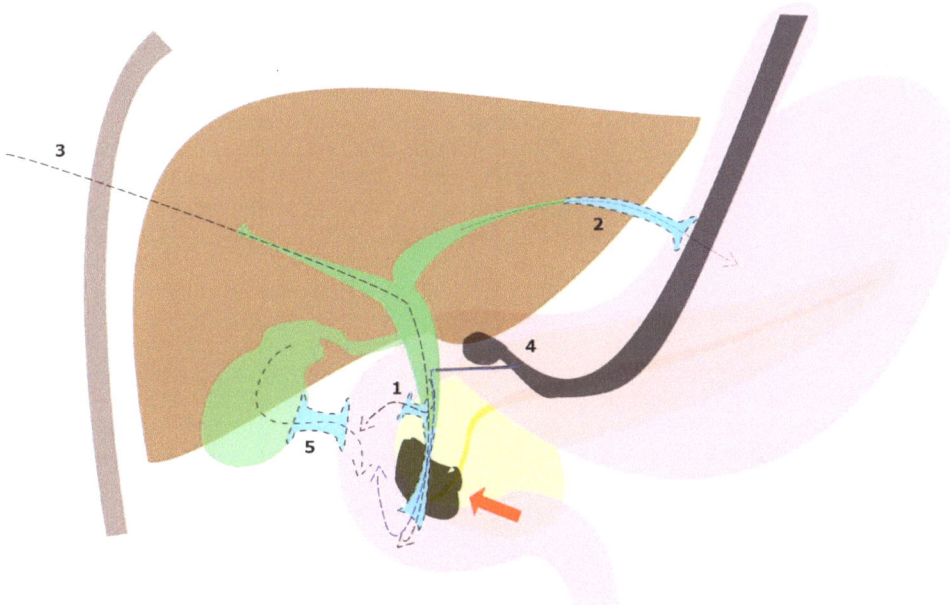

Figure 3. EUS-BD techniques for malignant biliary obstruction (red arrow) can be divided according to the anatomical location and the puncture site of the biliary access into (1) choledochoduodenostomy (CDS), (2) hepaticogastrostomy (HGS), (3) antegrade biliary stenting (AG), (4) rendezvous technique (RV) and (5) transduodenal gallbladder drainage (EUS-GBD).

In patients in whom ERCP fails, endoscopic ultrasound-guided choledochoduodenostomy (EUS-CDS) is considered the preferred choice for dMBO [41], as confirmed in a multicenter retrospective study comparing EUS-CDS to PTBD and demonstrating higher clinical success (84.6% vs. 62.1%, $p = 0.04$) for EUS-CDS with a significantly lower rate of reoperation (10.7% vs. 77.6%, $p < 0.001$) [42]. Biliary drainage through EUS-CDS permits direct access to the CBD from the duodenum creating a choledochoduodenostomy through the deployment of a plastic stent or fully covered metal stent, which is extremely useful and successful in the case of dMBO. Initially, FC-SEMS were preferred over plastic stents for CDS, as they have significantly lower rates of adverse events (13.0% vs. 42.8%, $p = 0.01$) and better stent patency [43,44], even if FC-SEMS theoretically increase the risk of stent migration. In this context, a fully-covered short metal stent with double flanges (lumen-apposing metal stent, LAMS) was developed for EUS-guided procedures about a decade ago [45] and it is on its way to becoming the preferred choice in the case of EUS-CDS. Furthermore, the application of the electrocautery-enhanced tip of the LAMS catheter has enabled a "free-hand", "single-step", and "exchange-free" procedure, making direct organ access possible without using further devices such as needles, guidewires, or dilator devices. A systematic review and meta-analysis containing seven studies including 284 patients who underwent EUS-BD using LAMS after ERCP failure showed high technical and clinical success rates (95.7% and 95.9%, respectively) with a 5.2% pooled rate of post-procedural adverse events and an 8.7% rate of recurrence [46]. Finally, those results were confirmed by a recent large multicenter study [47]. However, no differences in the technical and clinical success or post-procedure-related adverse events comparing LAMS vs. SEMS have been found so far [48,49], even if nowadays experts seem to prefer LAMS over FC-SEMS. On the other hand, EUS-guided hepatogastrostomy (EUS-HGS) is preferred in the case of hMBO, because it permits the creation of a fistulous duct between the gastric wall and the left intrahepatic duct, unlike EUS-CDS, which is indicated in dMBO. Moreover, when ERCP and/or PTBD do not achieve clinical success with adequate biliary drainage, ESGE suggests EUS-guided biliary drainage with EUS-HGS only for malignant inoperable hilar biliary obstruction with a dilated left hepatic duct [50]. However, current data on which is the best choice for MBO are conflicting, with some reports showing higher safety for the transduodenal route, while others have shown no such difference [51,52]. In a small randomized study comparing 25 patients who received EUS-HGS and 24 who received EUS-CDS, the clinical success of EUS-HGS was higher (91% versus 77%); however, adverse events were also slightly higher (20% vs. 12.5%), although neither outcome reached statistical significance [53]. A systematic review and meta-analysis of 10 studies by Uemura et al. comparing EUS-HGS ($n = 208$) and EUS-CDS ($n = 226$) found no difference in technical success (94.1% vs. 93.7%), clinical success (88.5% vs. 84.5%), or rates of adverse events [54]. Furthermore, a multicenter study on long-term patency of the two techniques conducted on 182 patients (95 EUS-HGS vs. 87 EUS-CDS) showed that EUS-CDS was associated with being 4.5 times more likely to achieve longer stent patency at the expense of a higher rate of adverse events [55]. Moreover, the EUS-guided gallbladder drainage appears a valid alternative as a rescue treatment after ERCP and EUS-CDS failure, showing adequate efficacy and safety for those patients who have dMBO and no involvement of the cystic duct. Therefore, a recent multicenter study involving 48 patients showed 100% and 81.3% technical and clinical success rates, respectively, with 10.4% of AEs [56]. Thus, the choice between these approaches is based on a combination of factors including procedural proficiency, risk of adverse events, and anatomical factors, such as the presence of a dilated bile duct or bile radicals, duodenal stenosis, and altered anatomy [57].

3.1.5. Comparison between ERCP and EUS-BD

The first study that compared ERPC vs. EUS in the drainage of biliary obstruction was a multicenter retrospective study demonstrating similar rates of technical success (94.23% for ERCP vs. 93.26% for EUS-BD, $p = 1.00$) and adverse events (8.65% for ERCP vs. 8.65% for EUS-BD); however, the ERCP was burdened by 4.8% of post-procedural pancreatitis [58]. Similar results were found in a meta-analysis showing that both techniques were equally effective in achieving biliary drainage (ERCP = 94.73%; EUS = 93.67; pooled odds ratio (OR): 1.20; 95% confidence interval (CI): 0.44–3.24) while there was no significant difference in adverse events (ERCP = 22.3%; EUS = 15.2%; OR 1.59; 95% CI 0.89–2.84), and furthermore, post-procedure pancreatitis (PEP) was significantly higher for ERCP (9.5% vs. EUS = 0; risk difference: 8%; 95% CI: 1–14%) [59]. Additionally, in cases of a gastroduodenal stent, the EUS-guided approach has been proven as technically and clinically superior when compared to ERCP [60], especially in the setting of concomitant double obstruction [61]. Finally, another systematic review and meta-analysis confirmed no significant differences in technical and clinical success between ERCP and EUS-BD, with lower rates of reintervention for EUS-BD [62].

3.2. Malignant Gastric Outlet Obstruction (mGOO)

The most frequent cause of mGOO in Western countries is pancreatic adenocarcinoma (between 15 and 25% of patients with pancreatic cancer develop MGOO during the course of the disease) [63]. Anyway, any other neoplasia occluding pylori or duodenum leads to mGOO, even if less frequently, as in the case of gastric cancer, neoplasms of the proximal duodenum and ampulla, local extension of advanced gallbladder carcinoma or cholangiocarcinoma, metastatic or primary malignancy in the duodenum, gastric carcinoid, or gastrointestinal stromal tumors/gastric leiomyosarcomas. The GOO-related clinical manifestations include abdominal pain, nausea and/or vomiting, early satiety and/or anorexia, bloating, and weight loss, which in the long term lead to cachexia. Furthermore, cancer progression increases these symptoms, also leading to dystrophy, general fatigue, dehydration, and electrolyte balance disorders [64,65]. By the way, the prognosis of these patients is related to tumor progression or an impaired general condition, so patient survival is also closely associated with the development of cachexia [66]. However, the lack of minimally invasive treatments in the past caused those patients with mGOO to undergo surgery to bypass the GI obstruction through a gastrojejunostomy, which was associated with a biliary shunt when occurring concurrently with biliary obstruction. However, those patients with advanced disease involving the GI tract are usually in poor condition and are not good candidates for surgery, so less invasive treatments have been developed over the years to rapidly and more safely treat this condition, improving many consequent outcomes, such as time to re-feeding, hospitalization time, and management costs. This goal was achieved with the development of endoscopic approaches such as enteral stent placement and more recently the creation of EUS-guided gastro-entero-anastomosis (EUS-GEA). Moreover, GOO-related symptoms were gathered into a score by Adler and colleagues, the gastric outlet obstruction scoring system (GOOSS score, Table 1), which is extremely helpful in easily following clinical outcomes after procedures through the improvement of patients' feeding [67]. In fact, clinical success is generally defined by remission of obstructive symptoms and resumption of oral feeding, when treating mGOO. Anyway, the choice between GEA and enteral stenting is dependent on different variables, so a prognostic scoring system was recently developed for patients with MGOO due to pancreatic adenocarcinoma in order to propose the best procedure depending on the survival predicted: a score between 0 and 1 indicates a better prognosis for the patient, so GEA should be preferred, while patients with a score between 2 and 4 have a worse prognosis and enteral stenting could be a better option [68].

Table 1. The gastric outlet obstruction scoring system (GOOSS).

Level of Oral Intake	GOOSS Score
No oral intake	0
Liquids only	1
Soft solids	2
Low-residue or full diet	3

3.2.1. Enteral Stenting

Endoscopic placement of an enteral stent was the first endoscopic option for treating mGOO as an alternative to surgical GEA [69], being extremely useful for those patients unfit for surgery, but it had a high rate of reintervention and low patency time compared to gastrojejunostomy [70]. Anyway, enteral stenting is alternatively used to re-establish channeling in patients with malignant gastrointestinal obstruction who are not eligible for surgery and with short life expectancy (less than 6 months) [71]. The first case reported in the literature of a self-expanding metallic stent for GOO dates back to 1992 [69]. Various studies have supported the efficacy and safety of enteral stenting in the management of unresectable mGOO since then [72–74]. Technical success, defined as the correct placement of the stent across the tumor stenosis, is frequently very high. In a systematic review with pooled analysis including 19 studies and 1281 patients, the overall pooled technical success rate was 97.3% and the clinical success rate was 85.7% [75]. According to the technique, a guidewire is placed beyond the duodenal stenosis over which the stent is then slid under radiological and endoscopic view (through-the-scope techniques). Finally, the injection of intraluminal contrast dye verifies both the regular flow through the SEMS after the obstruction site and the absence of any extra-luminal diffusion. Currently, we have three main types of enteral self-expandable metal stents (uncovered, partially covered, or fully covered) with different lengths, diameters, and radial expansive forces. In a systematic review including five trials with a total of 443 patients with MGOO, the authors compared the outcomes of covered SEMS vs. uncovered SEMS, showing that covered SEMS had a lower rate of stent occlusion (number-needed-to-treat, NNT, of 5) despite higher rates of stent migration compared with the uncovered SEMS (RD: 0.09, 95% CI [0.04, 0.14], I^2 9%, with a number-necessary-to-harm [NNH] of 11) [76]. In 2018, a systematic review confirmed that duodenal stenting had a faster return to oral intake, and shorter hospitalization time despite an increased recurrence of symptoms and increased reintervention rate when compared to surgical GEA [77]. As far as adverse events are concerned, the percentage varies between 0 and 30%, and they are strictly connected to the definition indicated in the study. Therefore, we have minor adverse events such as mild pain, nausea, and vomiting and major adverse events such as bleeding, perforation, and stent migration [78]. In the particular case where patients develop secondary MGOO and/or concurrent biliary obstruction, the positioning of the SEMS may increase the risk of biliary dysfunction. In the analysis by Hamada T et al., 410 patients with distal malignant biliary obstruction were enrolled and a duodenal SEMS was positioned in 33 (8%), 17 (52%) of whom developed biliary dysfunction with an average of 64 days after stent placement [79].

3.2.2. EUS-Guided Gastro-Entero-Anastomosis

EUS-guided GEA represents a novel and minimally invasive alternative to surgery and enteral stent for managing malignant GOO, and the literature shows increasing evidence in support of the advantages of EUS-guided anastomoses. In the past, surgery for gastrojejunostomy bypass was the most common option [80], but nowadays the development of EUS-GEA permits a less invasive option with similar efficacy and either fewer days of hospitalization or time to oral feeding for creating a GEA. Moreover, when malignancy causes concurrent biliary and duodenal obstruction, the EUS-guided approach may become the preferred one in the current era of EUS-guided procedures, [61] even if depending on the location of the obstruction, as indicated by the "bilioduodenal" classification [81]. The first EUS-guided method to create a GEA was reported in 2003 [82] in a porcine model but without adequate devices. Nowadays, the number of devices has increased but the technique is still not completely standardized, so some dedicated groups have worked on recognizing the differences among techniques and tertiary centers in order to better understand which is the best approach [83]. In general, the endosonographer firstly advances a catheter (or a double balloon/single balloon enteric tube) over a stiff guidewire through the gastric or duodenal stricture, and then saline is injected downstream of the stricture in order to fill the jejunal lumen. Finally, after EUS-identification of the enlarged enteral loop ("target"), the distal flange of the LAMS is deployed into the jejunal lumen (using the hands-free technique or through a guidewire previously placed through loop puncture with a fine-needle) and the proximal flange is deployed into the gastric lumen (with or without the intra-channel release technique). As described above, different variants of the technique have been developed over the years, changing the devices used for GEA creation or for loop enlargement, or in techniques for the target loop puncture. By the way, EUS-GEA for the treatment of gastric outlet obstruction (GOO) was initially performed only with one type of electrocautery lumen-apposing metal stents (EC-LAMS), especially thanks to the releasing system permitting the easy use of the wireless "free-hand" technique, but the use of another EC-LAMS was recently reported in the creation of a EUS-GEA [84,85], so further comparisons are expected in the near future. In general, therefore, the techniques for EUS-GEA can be summarized as direct EUS-GE, balloon-assisted EUS-GE, EUS-guided double-balloon-occluded gastrojejunostomy bypass (EPASS), and the wireless EUS-guided gastroenterostomy simplified technique (WEST), techniques that are described in-depth in other technical studies [85–90]. All in all, EUS-GEA is changing the approach to mGOO, moving toward becoming the standard of care in the future. In a meta-analysis including twelve studies and 290 patients the pooled technical success rate was 93.5% (95% CI, 89.7–6.0%; I^2 0%) and the pooled clinical success rate was 90.1% (95% CI 85.5–93.4%; I^2 0%), even if the studies included different techniques, mostly direct EUS-GE (68.2%), and indications were for mGOO only in 62.4% of cases [91]. Recently, a further and updated meta-analysis including 1493 patients with both benign and malignant GOO treated with EUS-GEA showed technical success and clinical success rates of 94%

and 89.9%, respectively. Furthermore, safety analysis showed a pooled rate of AEs of 13.1% [92]. Moreover, a recent multicenter retrospective study evaluated differences in treating mGOO with EUS-GEA ($n = 187$) vs. surgical gastrojejunostomy (SGJ, $n = 123$), showing significantly lower time to resumption of oral intake (1.40 vs. 4.06 days, $p < 0.001$) and a shorter length of stay (5.31 vs. 8.54 days, $p < 0.001$) comparing EUS-GEA with SGJ, with no differences in technical and clinical success between procedures (97.9% vs. 100% for TS and 94.1% vs. 94.3% for CS, respectively) [93]. In a matched comparison analysis of EUS-GEA vs. endoscopic stenting (ES), clinical success was, respectively, 100% vs. 75% ($p = 0.006$), with a lower recurrence rate (3.7% vs. 33.3%, $p = 0.02$) and a trend toward shorter time to chemotherapy [94]. However, a challenging scenario recently explored was the creation of EUS-GEA for GOO with peritoneal carcinomatosis, which showed slightly better outcomes when compared to SGJ, having comparable technical success (both 100%) and clinical success (88% vs. 85%, $p > 0.99$), but a lower rate of AEs (8% vs. 41%, $p = 0.01$, respectively). EUS-GEA is generally a safe technique, showing 12.9% of AEs in a prospective study evaluating 104 patients [94], which was similar to those pooled rates presented in the meta-analyses (13.1%), which was significantly low when compared to SGJ (13.4% vs. 33.3%, $p < 0.001$) [92,93]. Various comparative studies have been published in order to evaluate which is the most effective treatment for those patients developing mGOO, even if most of them have been retrospective so far. A recent meta-analysis of fifteen studies ($n = 1441$) showed higher pooled clinical success without recurrent GOO of EUS-GE when compared to ES or SGJ combined (OR, 2.60; 95% CI, 1.58–4.28) [95]. An overview of the outcomes, when comparing different techniques for mGOO, is shown in Table 2. Regarding the safety of these procedures, the table clearly shows some differences depending on the procedure, highlighting a generally better profile of the EUS-GE compared to a surgical approach, but even when compared to enteral stenting. Miller et al. showed a significantly lower pooled rate of AEs for the EUS-GE group compared to ES or SGJ grouped together (OR: 0.34; 95% CI: 0.20–0.58), or SGJ alone (OR: 0.17; 95% CI: 0.10–0.30) and no significant differences when compared to ES alone (OR: 0.57; 95% CI: 0.29–1.14) [95]. However, it is important to keep in mind that training for performing EUS-GE is still not well-established and these procedures are mainly performed by skilled and expert endosonographers, so this could be a bias in the context of real-world clinical practice.

Table 2. Differences between enteral stenting, surgical GJ, and EUS-GE in patients with mGOO.

Study	Study Design	N° Patients	Treatments Type	Technical Success, %	Clinical Success, %	Reintervention/Recurrence Rate	Hospitalization Median (Days)	Adverse Events
Jang S, 2019 [70]	Retrospective	183 ES; 127 SGJ	ES; SGJ	96% ES; 98% SGJ	79% ES; 80% SGJ	23% ES; 23% SGJ	4 ES; 9 SGJ	6% ES; 16% SGJ
Canakis, A. 2023 [93]	Retrospective	187 EUS-GE; 123 SGJ	EUS-GE; SGJ	97.8% EUS-GE; 100% SGJ	94% EUS-GE; 94% SGJ	15.5% EUS-GE; 1.63% SGJ	NA	11.9% EUS-GE; 17.9% SGJ
Jeurnick SM, 2010 [96]	Randomized Trial	18 SGJ; 21 ES	ES; SGJ	89% SGJ; 77% ES	NA	11% SGJ; 47% ES	15 SGJ; 7 ES	33% SGJ; 47% ES
Vanella, 2023 [94]	Prospective	28 EUS-GE; 28 ES *	EUS-GE; ES	100% ES; 96.4% EUS-GE	100% ES; 100% EUS-GE	33.3% ES; 3.7% EUS-GE	7 (4–21) ES; 6.5 (3–10.5) EUS-GE	25% ES; 7.1% EUS-GE
Van Wanrooij, 2022 [97]	Retrospective	88 EUS-GE; 88 ES **	EUS-GE; ES	EUS-GE 94%; ES 98%	EUS-GE 91%; ES 75%	1% EUS-GE; 26% ES	4 (2–10.8) EUS-GE; 4 (1–9.5) ES	10% EUS-GE; 21% ES
Ge, 2019 [98]	Retrospective	78 ES; 22 EUS-GE	EUS-GE; ES	100% EUS-GE; 100% ES	95.8% EUS-GE; 76.3% ES	8.3% EUS-GE; 32% ES	mean ± SD = 7.4 (9.1) EUS-GE; 7.9 (8.2) ES	20.8% EUS-GE; 40.2% ES
Chen, 2017 [99]	Retrospective	30 EUS-GE; 52 ES	EUS-GE; ES	86.7% EUS-GE; 94.2% ES	83.3% EUS-GE; 67.3% ES	4% EUS-GE; 28.6% ES	mean ± SD = 11.3 (6.6) EUS-GE; 9.5 (8.3) ES	16.7% EUS-GE; 11.5% ES
Bronswijk, 2021 [100]	Retrospective	37 EUS-GE; 37 SGJ **	EUS-GE; SGJ	94.6% EUS-GE; 100% SGJ	97.1% EUS-GE; 89.2% SGJ	0% EUS-GE; 5.4% SGJ	4 (2–8) EUS-GE; 8 (5.5–20) EUS-GE	2.7% EUS-GE; 27% SGJ
Khashab 2016 [101]	Retrospective	30 EUS-GE; 63 SGJ	EUS-GE; SGJ	87% EUS-GE; 100% SGJ	90% EUS-GE; 87% SGJ	3% EUS-GE; 14% SGJ	mean ± SD = 11.6 (6.6) EUS-GE; 12 (8.2) SGJ	16% EUS-GE; 25% SGJ
Perez-Miranda, 2017 [102]	Retrospective	25 EUS-GE; 29 SGJ	EUS-GE; SGJ	88% EUS-GE; 100% SGJ	84% EUS-GE; 90% SGJ	NA	mean: 9.4 EUS-GE; 8.9 SGJ	12% EUS-GE; 41% SGJ
Kouanda, 2021 [103]	Retrospective	40 EUS-GE; 26 SGJ	EUS-GE; SGJ	92.5% EUS-GE; 100% SGJ	92.5% EUS-GE; 100% SGJ	20% EUS-GE; 11.5% SGJ	5 EUS-GE; 14.5 SGJ	NA
Abbas, 2022 [104]	Retrospective	25 EUS-GE; 27 SGJ	EUS-GE; SGJ	100% EUS-GE; 100% SGJ	88% EUS-GE; 85% SGJ	NE	3.5 (2.5–9.5) EUS-GE; 9.5 (6–12) SGJ	8% EUS-GE; 41% SGJ

ES = enteral stenting; SGJ = surgical gastrojejunostomy; EUS-GE = endoscopic ultrasound-guided gastro-enterostomy; NE = not extractable. * EUS-GE and ES cohorts were matched according to baseline frailty and oncologic disease; ** after propensity score matching.

3.2.3. Natural Orifice Transluminal Endoscopic Surgery (NOTES)

Another endoscopic approach for palliation of tumors causing mGOO is the natural orifice transluminal endoscopic surgery (NOTES) for creating gastro-entero-anastomosis, which is still under development. Even if it has been proven to be as effective mostly in porcine models [105,106], it is still an option as a rescue therapy in the case of complete stent misdeployment during EUS-GEA [107,108]. Endoscopic access to the peritoneum was first described by Kalloo et al. [109] in a porcine model in 2004, changing our way of thinking about endoscopy and leading to the development of a new technique for creating EUS-GE or performing submucosal tunneling endoscopy. Various studies, mostly performed on animal models, demonstrated the feasibility and safety of performing NOTES-GE [106,110,111]. NOTES-GE includes several variations of the technique, but it generally starts in a similar way, with the identification of the small bowel segment closest to the gastric wall (usually corresponding to the ligament of Treitz) [112] through an echoendoscope. Then, a 19-gauge needle is used to insert a guidewire into the peritoneal space toward the ligament of Treitz, so the dilation of the tract permits the passage of a double channel forward-viewing endoscope into the peritoneal space. Under a direct endoscopic view, the small bowel distal to the obstruction is grasped by forceps while a 19-gauge needle punctures the small bowel inserting a guidewire. Therefore, an EC-LAMS is inserted under direct endoscopic visualization over the wire into the small bowel lumen, where the distal flange is deployed. Both the stent delivery catheter and the scope are finally pulled back within the gastric lumen, where the proximal flange is deployed creating the gastroenterostomy tract. Therefore, although animal data, as abovementioned, have demonstrated the feasibility of NOTES-GE, data on clinical settings are limited, and so the sample size is extremely small to apply it in clinical practice, despite the high technical and clinical success rates achieved in the described cases [112–114].

4. Pain Secondary to Bilio-Pancreatic Cancers

Patients with advanced bilio-pancreatic cancers may develop untreatable chronic abdominal pain, mainly due to the perineural invasion of tumor cells, and pain is present in 70–90% at diagnosis [115]. Pain management usually begins with medication titration in these oncological cases (i.e., progressing from nonsteroidal anti-inflammatory drugs to narcotics) but, unfortunately, they often are not able to fully relieve it despite adherence to the World Health Organization (WHO) analgesic ladder [116]. Moreover, celiac plexus neurolysis (CPN) also has a role in pain management in patients with advanced pancreatic cancer; in fact 16 trials have been published since 1997 evaluating its effectiveness in pain management and more than 50% of the patients enrolled had a reduction in pain intensity or decreased opioid consumption [117]. Therefore, alternative and additional therapeutic options to painkillers and opioids have been evaluated over the years, such as celiac plexus neurolysis (CPN) or celiac ganglia neurolysis (CGN) with various agents, administered either percutaneously or transgastrically [118]. CPN is the most widely used interventional procedure for the treatment of abdominal cancer pain, demonstrating efficacy for patients with both malignant and chronic non-malignant pain [119,120]. The celiac plexus is a dense network of autonomic fibers innervating visceral abdominal organs converging into the celiac ganglia, which are located in the retroperitoneum and adjacent to the origin of the celiac trunk. CPN may be able to reduce pain intensity and thus decrease systemic analgesic intake. Some authors have shown long-lasting pain relief for patients with pancreatic and intra-abdominal cancers with a benefit ranging from 50 days up to the time of death [121,122]. However, EUS technically permits performing CPN through the gastric wall, which allows for a safer and more effective procedure, as first described by Wiersema in 1996 and showing pain improvement in 79–88% of patients [123]. The safety profile is fundamental, because the EUS-guided transgastric approach allows direct access to the celiac plexus, leading to a reduction in the risk of injuries to the spinal nerve, diaphragm, or spinal artery.

EUS-Guided Neurolysis

Injection of substances into the celiac plexus is an established method for relieving pain in upper abdominal malignancies [124]. Absolute ethanol is commonly used after an injection of bupivacaine for performing CPN, while a combination of bupivacaine and triamcinolone is used in case of celiac plexus blockade. The safety of a combination of receiving 20 mL of 0.75% bupivacaine followed by 10 mL or 20 mL of alcohol for EUS-CPN was prospectively demonstrated in a cohort of 20 patients [125]. No major complications were seen in either group while minor self-limited AEs were seen in six (30%) subjects, including lightheadedness (5%), transient diarrhea (10%), and transient nausea and vomiting.

Technically, EUS-guided CPN consists of directly injecting substances in the two sides of the aorta at the level of origin of the celiac artery where the celiac ganglia are located, while maintaining the sagittal imaging of the aorta. Some authors inject 3 mL of 0.25% preservative-free bupivacaine followed by 10 mL of dehydrated 98% absolute ethanol into each side [126]. The result of the alcohol injection is an echogenic "cloud", which may cause discomfort after the procedure. In 2001, Gunaratnam et al. [127]. performed EUS-CPN in 58 patients with pancreatic cancer pain, reporting pain relief in 78% of them. Further initial data showed low efficacy (68.1% of patients with pain relief [128]), so predictive factors were also explored in order to enable rational selection of the therapeutic strategy. Therefore, in the first analysis in 2011, direct invasion of the celiac plexus and left-sided distribution of the injected ethanol were identified as significant predictors of a negative response to CPN [128]. Another evaluation of predictive factors in 2021 confirmed celiac plexus invasion (13.2 OR, 95% CI 3.02–46.27, $p = 0.003$) as significant negative independent pain response factors to EUS-CPN, also adding invisible ganglia (49 OR, 95% CI 2.25–17.91, $p = 0.011$) and presence of distant metastases (6.84 OR, 95% CI 2.34–19.15, $p = 0.022$) [129]. In 2008, a meta-analysis including eight studies with 283 oncologic patients undergoing EUS-CPN showed a pooled proportion of pain relief of 80.12% (95% CI 74.47–85.22) [130]. However, a meta-analysis evaluating the bilateral and unilateral EUS-CPN approaches and including 437 patients did not find a significant difference between the two approaches both in terms of short-term pain relief (SMD = 0.31, 95% CI (−0.20, 0.81), $p = 0.23$) and response to treatment (RR = 0.99, 95% CI (0.77, 1.41), $p = 0.97$), even if only the bilateral approach showed a significant reduction in the postoperative use of analgesics (RR = 0.66, 95% CI (0.47, 0.94), $p = 0.02$) compared to the unilateral approach [131]. However, on the other hand, a more specific variant of neurolysis consists of directly injecting agents into the ganglia, which are visualized as small and hypoechoic oval images at EUS-view. One of the first studies performing EUS-CGN with alcohol in 17 patients with pancreatic cancer and in 5 patients with chronic pancreatitis resulted in an improvement of pain scores in 94% and 80% of patients, respectively [132]. Later, a multicenter randomized trial comparing EUS-CGN and EUS-CPN showed a higher positive response rate at 7 postoperative days (POD) in the CGN group (73.5%) than in the CPN group (45.5%; $p = 0.026$), confirmed when evaluating the complete response rate (CGN group 50.0% vs. CPN group 18.2%; $p = 0.010$) [133]. A recent meta-analysis including 16 studies with 727 patients showed an overall response rate to EUS-CPN of 53% (95% CI 45–62%, I^2 68%, $p = 0.01$) at week four, regardless of the technique (central injection, bilateral injection, or CGN). Specifically, in subgroup analysis, EUS-CGN showed the highest proportion response, with 76% (95% CI, 71–82%; I^2 0.01%, $p = 0.38$) and 58% (95% CI, 48–69%; I^2 64.9%) at week two and four, respectively [117]. Recently, a multicenter prospective trial including 51 consecutive patients [134] evaluated the effectiveness of EUS-CPN in combination with EUS-CGN, defined as a decrease in the numerical rating scale (NRS) by ≥3 points 1 week after the procedure, which was 82.4%. However, complete pain relief, defined as NRS = 0 at 1 week after the procedure, was achieved only in 27.4% of patients.

5. Conclusions

In conclusion, endoscopy is an effective and safe approach for managing most of the clinical manifestations of advanced biliopancreatic tumors in the palliative setting (Figure 4).

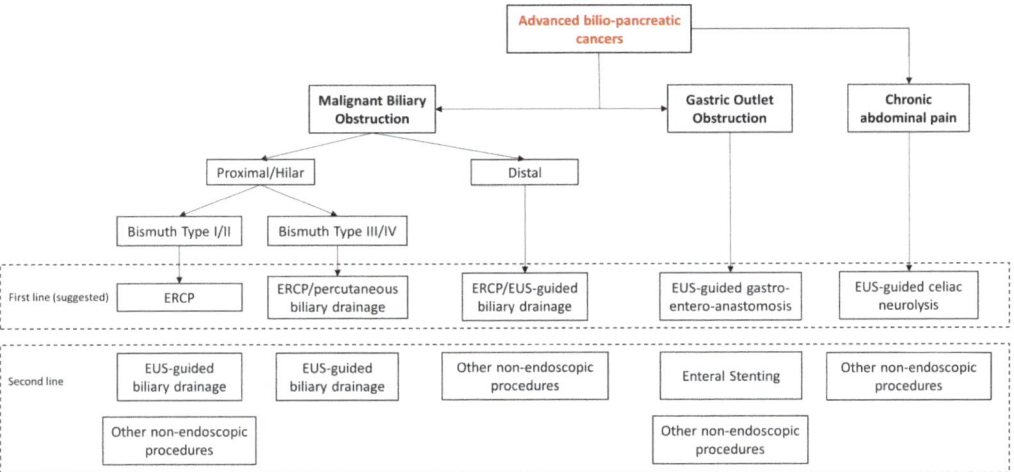

Figure 4. Algorithm of the management of clinical manifestations of advanced bilio-pancreatic tumors.

Furthermore, in addition to being a minimally invasive approach, which permits treating fragile and unfit-for-surgery patients, it has the advantage of treating many neoplastic clinical conditions during the same session, as in the case of MBO and mGOO, reducing anesthesiological risks and improving outcomes. However, palliative endoscopic advanced procedures require tertiary bilio-pancreatic centers due to the complexity of some techniques, mainly in the contest of EUS-guided therapeutic procedures. Moreover, tertiary centers guarantee the expertise of different specialists involved in the management of those patients with advanced bilio-pancreatic tumors, permitting them to propose the best options and manage the patient at 360 degrees, even in cases of the technical failure of endoscopic procedures.

Supplementary Materials: The following supporting information can be downloaded at: https://www.mdpi.com/article/10.3390/cancers15225367/s1, Table S1: Bismuth Classification of Klatskin tumors [31].

Author Contributions: G.E.M.R.: conceptualization, methodology, resources, writing—original draft, software, image creation, and visualization. L.C.: resources and writing—review and editing. G.R.: resources and writing—review and editing. C.B.: supervision, comments, and validation. C.F.: supervision and writing—review and editing. A.A.: supervision and writing—review and editing. I.T.: writing—original draft, review and editing, supervision, and validation. All authors have read and agreed to the published version of the manuscript.

Funding: This research received no external funding.

Conflicts of Interest: Cecilia Binda: lecturer for Fujifilm, Steris, Q3 Medical, and Boston Scientific. The other authors declare no conflict of interest.

References

1. Garcia-Alonso, F.J.; Chavarria, C.; Subtil, J.C.; Aparicio, J.R.; Busto Bea, V.; Martinez-Moreno, B.; Vila, J.J.; Martin-Alvarez, V.; Sanchez-Delgado, L.; De La Serna-Higuera, C.; et al. Prospective Multicenter Assessment of the Impact of Eus-Guided Gastroenterostomy on Patient Quality of Life in Unresectable Malignant Gastric Outlet Obstruction. *Gastrointest. Endosc.* **2023**, *98*, 28–35. [CrossRef] [PubMed]
2. Cancer Statistics Center. American Cancer Society. 2023. Available online: https://cancerstatisticscenter.cancer.org (accessed on 23 April 2023).
3. Siegel, R.L.; Miller, K.D.; Wagle, N.S.; Jemal, A. Cancer Statistics, 2023. *CA Cancer J. Clin.* **2023**, *73*, 17–48. [CrossRef]
4. Okamoto, T. Malignant Biliary Obstruction due to Metastatic Non-Hepato-Pancreato-Biliary Cancer. *World J. Gastroenterol.* **2022**, *28*, 985–1008. [CrossRef]
5. Survival Rates for Bile Duct Cancer. American Cancer Society. 2022. Available online: https://www.cancer.org/cancer/bile-duct-cancer/detection-diagnosis-staging/survival-by-stage.html (accessed on 23 April 2023).
6. Minaga, K.; Takenaka, M.; Kamata, K.; Yoshikawa, T.; Nakai, A.; Omoto, S.; Miyata, T.; Yamao, K.; Imai, H.; Sakamoto, H.; et al. Alleviating Pancreatic Cancer-Associated Pain Using Endoscopic Ultrasound-Guided Neurolysis. *Cancers* **2018**, *10*, 50. [CrossRef] [PubMed]
7. Wang, L.; Lu, M.; Wu, X.; Cheng, X.; Li, T.; Jiang, Z.; Shen, Y.; Liu, T.; Ma, Y. Contrast-Enhanced Ultrasound-Guided Celiac Plexus Neurolysis in Patients with Upper Abdominal Cancer Pain: Initial Experience. *Eur. Radiol.* **2020**, *30*, 4514–4523. [CrossRef] [PubMed]
8. Kaplan, J.; Khalid, A.; Cosgrove, N.; Soomro, A.; Mazhar, S.M.; Siddiqui, A.A. Endoscopic Ultrasound-Fine Needle Injection for Oncological Therapy. *World J. Gastrointest. Oncol.* **2015**, *7*, 466–472. [CrossRef]
9. Richardson, W.S.; Wilson, M.C.; Nishikawa, J.; Hayward, R.S. The well-built clinical question: A key to evidence-based decisions. *ACP J. Club* **1995**, *123*, A12–A13. [CrossRef]
10. Frandsen, T.F.; Nielsen, M.F.B.; Lindhardt, C.L.; Eriksen, M.B. Using the full PICO model as a search tool for systematic reviews resulted in lower recall for some PICO elements. *J. Clin. Epidemiol.* **2020**, *127*, 69–75. [CrossRef]
11. Boulay, B.R.; Birg, A. Malignant Biliary Obstruction: From Palliation to Treatment. *World J. Gastrointest. Oncol.* **2016**, *8*, 498–508. [CrossRef]
12. Yoon, W.J.; Oh, Y.; Yoo, C.; Jang, S.; Cho, S.S.; Suh, J.H.; Choi, S.S.; Park, D.H. Eus-Guided Versus Percutaneous Celiac Neurolysis for the Management of Intractable Pain due to Unresectable Pancreatic Cancer: A Randomized Clinical Trial. *J. Clin. Med.* **2020**, *9*, 1666. [CrossRef]
13. Nakai, Y.; Isayama, H.; Wang, H.P.; Rerknimitr, R.; Khor, C.; Yasuda, I.; Kogure, H.; Moon, J.H.; Lau, J.; Lakhtakia, S.; et al. International Consensus Statements for Endoscopic Management of Distal Biliary Stricture. *J. Gastroenterol. Hepatol.* **2020**, *35*, 967–979. [CrossRef] [PubMed]
14. Inamdar, S.; Slattery, E.; Bhalla, R.; Sejpal, D.V.; Trindade, A.J. Comparison of Adverse Events for Endoscopic vs. Percutaneous Biliary Drainage in the Treatment of Malignant Biliary Tract Obstruction in An Inpatient National Cohort. *JAMA Oncol.* **2016**, *2*, 112–117. [CrossRef] [PubMed]
15. Fernandez, Y.V.M.; Arvanitakis, M. Early Diagnosis and Management of Malignant Distal Biliary Obstruction: A Review on Current Recommendations and Guidelines. *Clin. Exp. Gastroenterol.* **2019**, *12*, 415–432. [CrossRef]
16. Distler, M.; Kersting, S.; Ruckert, F.; Dobrowolski, F.; Miehlke, S.; Grutzmann, R.; Saeger, H.D. Palliative Treatment of Obstructive Jaundice in Patients with Carcinoma of the Pancreatic Head or Distal Biliary Tree. Endoscopic Stent Placement vs. Hepaticojejunostomy. *JOP J. Pancreas* **2010**, *11*, 568–574. [CrossRef]
17. Valle, J.W.; Kelley, R.K.; Nervi, B.; Oh, D.Y.; Zhu, A.X. Biliary Tract Cancer. *Lancet* **2021**, *397*, 428–444. [CrossRef]
18. Paik, W.H.; Lee, T.H.; Park, D.H.; Choi, J.H.; Kim, S.O.; Jang, S.; Kim, D.U.; Shim, J.H.; Song, T.J.; Lee, S.S.; et al. Eus-Guided Biliary Drainage Versus Ercp for the Primary Palliation of Malignant Biliary Obstruction: A Multicenter Randomized Clinical Trial. *Am. J. Gastroenterol.* **2018**, *113*, 987–997. [CrossRef] [PubMed]
19. Park, J.K.; Woo, Y.S.; Noh, D.H.; Yang, J.I.; Bae, S.Y.; Yun, H.S.; Lee, J.K.; Lee, K.T.; Lee, K.H. Efficacy of Eus-Guided and Ercp-Guided Biliary Drainage for Malignant Biliary Obstruction: Prospective Randomized Controlled Study. *Gastrointest. Endosc.* **2018**, *88*, 277–282. [CrossRef]
20. Bang, J.Y.; Navaneethan, U.; Hasan, M.; Hawes, R.; Varadarajulu, S. Stent Placement by Eus or Ercp for Primary Biliary Decompression in Pancreatic Cancer: A Randomized Trial (with Videos). *Gastrointest. Endosc.* **2018**, *88*, 9–17. [CrossRef]
21. Dumonceau, J.M.; Tringali, A.; Papanikolaou, I.S.; Blero, D.; Mangiavillano, B.; Schmidt, A.; Vanbiervliet, G.; Costamagna, G.; Deviere, J.; Garcia-Cano, J.; et al. Endoscopic Biliary Stenting: Indications, Choice of Stents, and Results: European Society of Gastrointestinal Endoscopy (Esge) Clinical Guideline—Updated October 2017. *Endoscopy* **2018**, *50*, 910–930. [CrossRef]
22. Moole, H.; Jaeger, A.; Cashman, M.; Volmar, F.H.; Dhillon, S.; Bechtold, M.L.; Puli, S.R. Are Self-Expandable Metal Stents Superior to Plastic Stents in Palliating Malignant Distal Biliary Strictures? A Meta-Analysis and Systematic Review. *Med. J. Armed Forces India* **2017**, *73*, 42–48. [CrossRef]
23. Almadi, M.A.; Barkun, A.N.; Martel, M. No Benefit of Covered vs. Uncovered Self-Expandable Metal Stents in Patients with Malignant Distal Biliary Obstruction: A Meta-Analysis. *Clin. Gastroenterol. Hepatol.* **2013**, *11*, 27–37.e1. [CrossRef] [PubMed]
24. Li, J.; Li, T.; Sun, P.; Yu, Q.; Wang, K.; Chang, W.; Song, Z.; Zheng, Q. Covered Versus Uncovered Self-Expandable Metal Stents for Managing Malignant Distal Biliary Obstruction: A Meta-Analysis. *PLoS ONE* **2016**, *11*, e0149066. [CrossRef] [PubMed]

25. Saleem, A.; Leggett, C.L.; Murad, M.H.; Baron, T.H. Meta-Analysis of Randomized Trials Comparing the Patency of Covered and Uncovered Self-Expandable Metal Stents for Palliation of Distal Malignant Bile Duct Obstruction. *Gastrointest. Endosc.* **2011**, *74*, 321–327.e1–3. [CrossRef] [PubMed]
26. Mohan, B.P.; Canakis, A.; Khan, S.R.; Chandan, S.; Ponnada, S.; Mcdonough, S.; Adler, D.G. Drug Eluting Versus Covered Metal Stents in Malignant Biliary Strictures—Is There A Clinical Benefit?: A Systematic Review and Meta-Analysis. *J. Clin. Gastroenterol.* **2021**, *55*, 271–277. [CrossRef]
27. Sun, X.R.; Tang, C.W.; Lu, W.M.; Xu, Y.Q.; Feng, W.M.; Bao, Y.; Zheng, Y.Y. Endoscopic Biliary Stenting Versus Percutaneous Transhepatic Biliary Stenting in Advanced Malignant Biliary Obstruction: Cost-Effectiveness Analysis. *Hepatogastroenterology* **2014**, *61*, 563–566.
28. Du, J.; Gao, X.; Zhang, H.; Wan, Z.; Yu, H.; Wang, D. Stent Selection in Preoperative Biliary Drainage for Patients with Operable Pancreatic Cancer Receiving Neoadjuvant Therapy: A Meta-Analysis and Systematic Review. *Front. Surg.* **2022**, *9*, 875504. [CrossRef]
29. Blechacz, B.; Komuta, M.; Roskams, T.; Gores, G.J. Clinical Diagnosis and Staging of Cholangiocarcinoma. *Nat. Rev. Gastroenterol. Hepatol.* **2011**, *8*, 512–522. [CrossRef]
30. Klatskin, G. Adenocarcinoma of the Hepatic Duct at Its Bifurcation within the Porta Hepatis. An Unusual Tumor with Distinctive Clinical and Pathological Features. *Am. J. Med.* **1965**, *38*, 241–256. [CrossRef]
31. Bismuth, H.; Nakache, R.; Diamond, T. Management Strategies in Resection for Hilar Cholangiocarcinoma. *Ann. Surg.* **1992**, *215*, 31–38. [CrossRef]
32. Moole, H.; Dharmapuri, S.; Duvvuri, A.; Dharmapuri, S.; Boddireddy, R.; Moole, V.; Yedama, P.; Bondalapati, N.; Uppu, A.; Yerasi, C. Endoscopic Versus Percutaneous Biliary Drainage in Palliation of Advanced Malignant Hilar Obstruction: A Meta-Analysis and Systematic Review. *Can. J. Gastroenterol. Hepatol.* **2016**, *2016*, 4726078. [CrossRef]
33. Komaya, K.; Ebata, T.; Fukami, Y.; Sakamoto, E.; Miyake, H.; Takara, D.; Wakai, K.; Nagino, M.; Nagoya Surgical Oncology, G. Percutaneous Biliary Drainage Is Oncologically Inferior to Endoscopic Drainage: A Propensity Score Matching Analysis in Resectable Distal Cholangiocarcinoma. *J. Gastroenterol.* **2016**, *51*, 608–619. [CrossRef] [PubMed]
34. Vienne, A.; Hobeika, E.; Gouya, H.; Lapidus, N.; Fritsch, J.; Choury, A.D.; Chryssostalis, A.; Gaudric, M.; Pelletier, G.; Buffet, C.; et al. Prediction of Drainage Effectiveness During Endoscopic Stenting of Malignant Hilar Strictures: The Role of Liver Volume Assessment. *Gastrointest. Endosc.* **2010**, *72*, 728–735. [CrossRef] [PubMed]
35. Takahashi, E.; Fukasawa, M.; Sato, T.; Takano, S.; Kadokura, M.; Shindo, H.; Yokota, Y.; Enomoto, N. Biliary Drainage Strategy of Unresectable Malignant Hilar Strictures by Computed Tomography Volumetry. *World J. Gastroenterol.* **2015**, *21*, 4946–4953. [CrossRef]
36. Zorron Pu, L.; De Moura, E.G.; Bernardo, W.M.; Baracat, F.I.; Mendonca, E.Q.; Kondo, A.; Luz, G.O.; Furuya Junior, C.K.; Artifon, E.L. Endoscopic Stenting for Inoperable Malignant Biliary Obstruction: A Systematic Review and Meta-Analysis. *World J. Gastroenterol.* **2015**, *21*, 13374–13385. [CrossRef] [PubMed]
37. Almadi, M.A.; Barkun, A.; Martel, M. Plastic vs. Self-Expandable Metal Stents for Palliation in Malignant Biliary Obstruction: A Series of Meta-Analyses. *Am. J. Gastroenterol.* **2017**, *112*, 260–273. [CrossRef] [PubMed]
38. Lee, T.H.; Choi, J.H.; Park Do, H.; Song, T.J.; Kim, D.U.; Paik, W.H.; Hwangbo, Y.; Lee, S.S.; Seo, D.W.; Lee, S.K.; et al. Similar Efficacies of Endoscopic Ultrasound-Guided Transmural and Percutaneous Drainage for Malignant Distal Biliary Obstruction. *Clin. Gastroenterol. Hepatol.* **2016**, *14*, 1011–1019.e3. [CrossRef]
39. Sharaiha, R.Z.; Khan, M.A.; Kamal, F.; Tyberg, A.; Tombazzi, C.R.; Ali, B.; Tombazzi, C.; Kahaleh, M. Efficacy and Safety of Eus-Guided Biliary Drainage in Comparison with Percutaneous Biliary Drainage When Ercp Fails: A Systematic Review and Meta-Analysis. *Gastrointest. Endosc.* **2017**, *85*, 904–914. [CrossRef]
40. Wang, K.; Zhu, J.; Xing, L.; Wang, Y.; Jin, Z.; Li, Z. Assessment of Efficacy and Safety of Eus-Guided Biliary Drainage: A Systematic Review. *Gastrointest. Endosc.* **2016**, *83*, 1218–1227. [CrossRef]
41. Pawa, R.; Pleasant, T.; Tom, C.; Pawa, S. Endoscopic Ultrasound-Guided Biliary Drainage: Are We There Yet? *World J. Gastroint. Endosc.* **2021**, *13*, 302–318. [CrossRef]
42. Sawas, T.; Bailey, N.J.; Yeung, K.; James, T.W.; Reddy, S.; Fleming, C.J.; Marya, N.B.; Storm, A.C.; Abu Dayyeh, B.K.; Petersen, B.T.; et al. Comparison of Eus-Guided Choledochoduodenostomy and Percutaneous Drainage for Distal Biliary Obstruction: A Multicenter Cohort Study. *Endosc. Ultrasound* **2022**, *11*, 223–230. [CrossRef]
43. Gupta, K.; Perez-Miranda, M.; Kahaleh, M.; Artifon, E.L.; Itoi, T.; Freeman, M.L.; De-Serna, C.; Sauer, B.; Giovannini, M.; In, E.S.G. Endoscopic Ultrasound-Assisted Bile Duct Access and Drainage: Multicenter, Long-Term Analysis of Approach, Outcomes, and Complications of a Technique in Evolution. *J. Clin. Gastroenterol.* **2014**, *48*, 80–87. [CrossRef] [PubMed]
44. Schmidt, A.; Riecken, B.; Rische, S.; Klinger, C.; Jakobs, R.; Bechtler, M.; Kahler, G.; Dormann, A.; Caca, K. Wing-Shaped Plastic Stents vs. Self-Expandable Metal Stents for Palliative Drainage of Malignant Distal Biliary Obstruction: A Randomized Multicenter Study. *Endoscopy* **2015**, *47*, 430–436. [CrossRef] [PubMed]
45. Binmoeller, K.F.; Shah, J. A Novel Lumen-Apposing Stent for Transluminal Drainage of Nonadherent Extraintestinal Fluid Collections. *Endoscopy* **2011**, *43*, 337–342. [CrossRef] [PubMed]
46. Krishnamoorthi, R.; Dasari, C.S.; Thoguluva Chandrasekar, V.; Priyan, H.; Jayaraj, M.; Law, J.; Larsen, M.; Kozarek, R.; Ross, A.; Irani, S. Effectiveness and Safety of Eus-Guided Choledochoduodenostomy Using Lumen-Apposing Metal Stents (Lams): A Systematic Review and Meta-Analysis. *Surg. Endosc.* **2020**, *34*, 2866–2877. [CrossRef] [PubMed]

47. On, W.; Paranandi, B.; Smith, A.M.; Venkatachalapathy, S.V.; James, M.W.; Aithal, G.P.; Varbobitis, I.; Cheriyan, D.; Mcdonald, C.; Leeds, J.S.; et al. Eus-Guided Choledochoduodenostomy with Electrocautery-Enhanced Lumen-Apposing Metal Stents in Patients with Malignant Distal Biliary Obstruction: Multicenter Collaboration from the United Kingdom and Ireland. *Gastrointest. Endosc.* **2022**, *95*, 432–442. [CrossRef]
48. Amato, A.; Sinagra, E.; Celsa, C.; Enea, M.; Buda, A.; Vieceli, F.; Scaramella, L.; Belletrutti, P.; Fugazza, A.; Camma, C.; et al. Efficacy of Lumen-Apposing Metal Stents or Self-Expandable Metal Stents for Endoscopic Ultrasound-Guided Choledochoduodenostomy: A Systematic Review and Meta-Analysis. *Endoscopy* **2021**, *53*, 1037–1047. [CrossRef]
49. De Benito Sanz, M.; Najera-Munoz, R.; De La Serna-Higuera, C.; Fuentes-Valenzuela, E.; Fanjul, I.; Chavarria, C.; Garcia-Alonso, F.J.; Sanchez-Ocana, R.; Carbajo, A.Y.; Bazaga, S.; et al. Lumen Apposing Metal Stents Versus Tubular Self-Expandable Metal Stents for Endoscopic Ultrasound-Guided Choledochoduodenostomy in Malignant Biliary Obstruction. *Surg. Endosc.* **2021**, *35*, 6754–6762. [CrossRef]
50. Van Der Merwe, S.W.; Van Wanrooij, R.L.J.; Bronswijk, M.; Everett, S.; Lakhtakia, S.; Rimbas, M.; Hucl, T.; Kunda, R.; Badaoui, A.; Law, R.; et al. Therapeutic Endoscopic Ultrasound: European Society of Gastrointestinal Endoscopy (Esge) Guideline. *Endoscopy* **2022**, *54*, 185–205. [CrossRef]
51. Khashab, M.A.; El Zein, M.H.; Sharzehi, K.; Marson, F.P.; Haluszka, O.; Small, A.J.; Nakai, Y.; Park, D.H.; Kunda, R.; Teoh, A.Y.; et al. Eus-Guided Biliary Drainage or Enteroscopy-Assisted Ercp in Patients with Surgical Anatomy and Biliary Obstruction: An International Comparative Study. *Endosc. Int. Open* **2016**, *4*, E1322–E1327. [CrossRef]
52. Khashab, M.A.; Messallam, A.A.; Penas, I.; Nakai, Y.; Modayil, R.J.; De La Serna, C.; Hara, K.; El Zein, M.; Stavropoulos, S.N.; Perez-Miranda, M.; et al. International Multicenter Comparative Trial of Transluminal Eus-Guided Biliary Drainage Via Hepatogastrostomy vs. Choledochoduodenostomy Approaches. *Endosc. Int. Open* **2016**, *4*, E175–E181. [CrossRef]
53. Artifon, E.L.; Marson, F.P.; Gaidhane, M.; Kahaleh, M.; Otoch, J.P. Hepaticogastrostomy or Choledochoduodenostomy for Distal Malignant Biliary Obstruction after Failed Ercp: Is There Any Difference? *Gastrointest. Endosc.* **2015**, *81*, 950–959. [CrossRef] [PubMed]
54. Uemura, R.S.; Khan, M.A.; Otoch, J.P.; Kahaleh, M.; Montero, E.F.; Artifon, E.L.A. Eus-Guided Choledochoduodenostomy Versus Hepaticogastrostomy: A Systematic Review and Meta-Analysis. *J. Clin. Gastroenterol.* **2018**, *52*, 123–130. [CrossRef] [PubMed]
55. Tyberg, A.; Napoleon, B.; Robles-Medranda, C.; Shah, J.N.; Bories, E.; Kumta, N.A.; Yague, A.S.; Vazquez-Sequeiros, E.; Lakhtakia, S.; El Chafic, A.H.; et al. Hepaticogastrostomy Versus Choledochoduodenostomy: An International Multicenter Study on Their Long-Term Patency. *Endosc. Ultrasound* **2022**, *11*, 38–43. [CrossRef] [PubMed]
56. Binda, C.; Anderloni, A.; Fugazza, A.; Amato, A.; De Nucci, G.; Redaelli, A.; Di Mitri, R.; Cugia, L.; Pollino, V.; Macchiarelli, R.; et al. Eus-Guided Gallbladder Drainage Using a Lumen-Apposing Metal Stent As Rescue Treatment for Malignant Distal Biliary Obstruction: A Large Multicenter Experience. *Gastrointest. Endosc.* **2023**, *98*, 765–773. [CrossRef]
57. Bang, J.Y.; Hawes, R.; Varadarajulu, S. Endoscopic Biliary Drainage for Malignant Distal Biliary Obstruction: Which Is Better—Endoscopic Retrograde Cholangiopancreatography or Endoscopic Ultrasound? *Dig. Endosc.* **2022**, *34*, 317–324. [CrossRef]
58. Dhir, V.; Itoi, T.; Khashab, M.A.; Park, D.H.; Yuen Bun Teoh, A.; Attam, R.; Messallam, A.; Varadarajulu, S.; Maydeo, A. Multicenter Comparative Evaluation of Endoscopic Placement of Expandable Metal Stents for Malignant Distal Common Bile Duct Obstruction by Ercp or Eus-Guided Approach. *Gastrointest. Endosc.* **2015**, *81*, 913–923. [CrossRef]
59. Kakked, G.; Salameh, H.; Cheesman, A.R.; Kumta, N.A.; Nagula, S.; Dimaio, C.J. Primary Eus-Guided Biliary Drainage Versus Ercp Drainage for the Management of Malignant Biliary Obstruction: A Systematic Review and Meta-Analysis. *Endosc. Ultrasound* **2020**, *9*, 298–307. [CrossRef]
60. Yamao, K.; Kitano, M.; Takenaka, M.; Minaga, K.; Sakurai, T.; Watanabe, T.; Kayahara, T.; Yoshikawa, T.; Yamashita, Y.; Asada, M.; et al. Outcomes of Endoscopic Biliary Drainage in Pancreatic Cancer Patients with An Indwelling Gastroduodenal Stent: A Multicenter Cohort Study in West Japan. *Gastrointest. Endosc.* **2018**, *88*, 66–75.e2. [CrossRef]
61. Rizzo, G.E.M.; Carrozza, L.; Quintini, D.; Ligresti, D.; Traina, M.; Tarantino, I. A Systematic Review of Endoscopic Treatments for Concomitant Malignant Biliary Obstruction and Malignant Gastric Outlet Obstruction and the Outstanding Role of Endoscopic Ultrasound-Guided Therapies. *Cancers* **2023**, *15*, 2585. [CrossRef]
62. Lyu, Y.; Li, T.; Cheng, Y.; Wang, B.; Cao, Y.; Wang, Y. Endoscopic Ultrasound-Guided vs. Ercp-Guided Biliary Drainage for Malignant Biliary Obstruction: A Up-To-Date Meta-Analysis and Systematic Review. *Dig. Liver Dis.* **2021**, *53*, 1247–1253. [CrossRef]
63. Shone, D.N.; Nikoomanesh, P.; Smith-Meek, M.M.; Bender, J.S. Malignancy Is the Most Common Cause of Gastric Outlet Obstruction in the Era of H2 Blockers. *Am. J. Gastroenterol.* **1995**, *90*, 1769–1770. [PubMed]
64. Van Hooft, J.E.; Van Montfoort, M.L.; Jeurnink, S.M.; Bruno, M.J.; Dijkgraaf, M.G.; Siersema, P.D.; Fockens, P. Safety and Efficacy of A New Non-Foreshortening Nitinol Stent in Malignant Gastric Outlet Obstruction (Duoniti Study): A Prospective, Multicenter Study. *Endoscopy* **2011**, *43*, 671–675. [CrossRef] [PubMed]
65. Storm, A.C.; Ryou, M. Advances in the Endoscopic Management of Gastric Outflow Disorders. *Curr. Opin. Gastroenterol.* **2017**, *33*, 455–460. [CrossRef]
66. Roxburgh, C.S.; Mcmillan, D.C. Role of Systemic Inflammatory Response in Predicting Survival in Patients with Primary Operable Cancer. *Future Oncol.* **2010**, *6*, 149–163. [CrossRef] [PubMed]
67. Adler, D.G.; Baron, T.H. Endoscopic Palliation of Malignant Gastric Outlet Obstruction Using Self-Expanding Metal Stents: Experience in 36 Patients. *Am. J. Gastroenterol.* **2002**, *97*, 72–78. [CrossRef]

68. Sugiura, T.; Okamura, Y.; Ito, T.; Yamamoto, Y.; Ashida, R.; Yoshida, Y.; Tanaka, M.; Uesaka, K. Prognostic Scoring System for Patients Who Present with A Gastric Outlet Obstruction Caused by Advanced Pancreatic Adenocarcinoma. *World J. Surg.* **2017**, *41*, 2619–2624. [CrossRef] [PubMed]
69. Topazian, M.; Ring, E.; Grendell, J. Palliation of Obstructing Gastric Cancer with Steel Mesh, Self-Expanding Endoprostheses. *Gastrointest. Endosc.* **1992**, *38*, 58–60. [CrossRef] [PubMed]
70. Jang, S.; Stevens, T.; Lopez, R.; Bhatt, A.; Vargo, J.J. Superiority of Gastrojejunostomy Over Endoscopic Stenting for Palliation of Malignant Gastric Outlet Obstruction. *Clin. Gastroenterol. Hepatol.* **2019**, *17*, 1295–1302.e1. [CrossRef]
71. Committee, A.S.O.P.; Jue, T.L.; Storm, A.C.; Naveed, M.; Fishman, D.S.; Qumseya, B.J.; Mcree, A.J.; Truty, M.J.; Khashab, M.A.; Agrawal, D.; et al. Asge Guideline on the Role of Endoscopy in the Management of Benign and Malignant Gastroduodenal Obstruction. *Gastrointest. Endosc.* **2021**, *93*, 309–322.e4. [CrossRef]
72. Baron, T.H. Expandable Metal Stents for the Treatment of Cancerous Obstruction of the Gastrointestinal Tract. *N. Engl. J. Med.* **2001**, *344*, 1681–1687. [CrossRef]
73. Yim, H.B.; Jacobson, B.C.; Saltzman, J.R.; Johannes, R.S.; Bounds, B.C.; Lee, J.H.; Shields, S.J.; Ruymann, F.W.; Van Dam, J.; Carr-Locke, D.L. Clinical Outcome of the Use of Enteral Stents for Palliation of Patients with Malignant Upper Gi Obstruction. *Gastrointest. Endosc.* **2001**, *53*, 329–332. [CrossRef] [PubMed]
74. Hori, Y.; Naitoh, I.; Hayashi, K.; Ban, T.; Natsume, M.; Okumura, F.; Nakazawa, T.; Takada, H.; Hirano, A.; Jinno, N.; et al. Predictors of Outcomes in Patients Undergoing Covered and Uncovered Self-Expandable Metal Stent Placement for Malignant Gastric Outlet Obstruction: A Multicenter Study. *Gastrointest. Endosc.* **2017**, *85*, 340–348.e1. [CrossRef] [PubMed]
75. Van Halsema, E.E.; Rauws, E.A.; Fockens, P.; Van Hooft, J.E. Self-Expandable Metal Stents for Malignant Gastric Outlet Obstruction: A Pooled Analysis of Prospective Literature. *World J. Gastroenterol.* **2015**, *21*, 12468–12481. [CrossRef] [PubMed]
76. Minata, M.K.; Bernardo, W.M.; Rocha, R.S.; Morita, F.H.; Aquino, J.C.; Cheng, S.; Zilberstein, B.; Sakai, P.; De Moura, E.G. Stents and Surgical Interventions in the Palliation of Gastric Outlet Obstruction: A Systematic Review. *Endosc. Int. Open* **2016**, *4*, E1158–E1170. [CrossRef]
77. Upchurch, E.; Ragusa, M.; Cirocchi, R. Stent Placement Versus Surgical Palliation for Adults with Malignant Gastric Outlet Obstruction. *Cochrane Database Syst. Rev.* **2018**, *5*, CD012506. [CrossRef]
78. Troncone, E.; Fugazza, A.; Cappello, A.; Del Vecchio Blanco, G.; Monteleone, G.; Repici, A.; Teoh, A.Y.B.; Anderloni, A. Malignant Gastric Outlet Obstruction: Which Is the Best Therapeutic Option? *World J. Gastroenterol.* **2020**, *26*, 1847–1860. [CrossRef]
79. Hamada, T.; Nakai, Y.; Isayama, H.; Sasaki, T.; Kogure, H.; Kawakubo, K.; Sasahira, N.; Yamamoto, N.; Togawa, O.; Mizuno, S.; et al. Duodenal Metal Stent Placement Is A Risk Factor for Biliary Metal Stent Dysfunction: An Analysis Using A Time-Dependent Covariate. *Surg. Endosc.* **2013**, *27*, 1243–1248. [CrossRef]
80. Denley, S.M.; Moug, S.J.; Carter, C.R.; Mckay, C.J. The Outcome of Laparoscopic Gastrojejunostomy in Malignant Gastric Outlet Obstruction. *Int. J. Gastrointest. Cancer* **2005**, *35*, 165–169. [CrossRef]
81. Mutignani, M.; Tringali, A.; Shah, S.G.; Perri, V.; Familiari, P.; Iacopini, F.; Spada, C.; Costamagna, G. Combined Endoscopic Stent Insertion in Malignant Biliary and Duodenal Obstruction. *Endoscopy* **2007**, *39*, 440–447. [CrossRef]
82. Fritscher-Ravens, A.; Mosse, C.A.; Mukherjee, D.; Mills, T.; Park, P.O.; Swain, C.P. Transluminal Endosurgery: Single Lumen Access Anastomotic Device for Flexible Endoscopy. *Gastrointest. Endosc.* **2003**, *58*, 585–591. [CrossRef]
83. Fabbri, C.; Coluccio, C.; Binda, C.; Fugazza, A.; Anderloni, A.; Tarantino, I.; i-EUS Group. Lumen-Apposing Metal Stents: How Far Are We from Standardization? An Italian Survey. *Endosc. Ultrasound* **2022**, *11*, 59–67. [CrossRef] [PubMed]
84. Mangiavillano, B.; Larghi, A.; Vargas-Madrigal, J.; Facciorusso, A.; Di Matteo, F.; Crino, S.F.; Pham, K.D.; Moon, J.H.; Auriemma, F.; Camellini, L.; et al. Eus-Guided Gastroenterostomy Using A Novel Electrocautery Lumen Apposing Metal Stent for Treatment of Gastric Outlet Obstruction (with Video). *Dig. Liver Dis.* **2023**, *55*, 644–648. [CrossRef]
85. Bronswijk, M.; Van Malenstein, H.; Laleman, W.; Van Der Merwe, S.; Vanella, G.; Petrone, M.C.; Arcidiacono, P.G. Eus-Guided Gastroenterostomy: Less Is More! the Wireless Eus-Guided Gastroenterostomy Simplified Technique. *VideoGIE* **2020**, *5*, 442. [CrossRef] [PubMed]
86. Miller, C.S.; Chen, Y.I.; Haito Chavez, Y.; Alghamdi, A.; Zogopoulos, G.; Bessissow, A. Double-Balloon Endoscopic Ultrasound-Guided Gastroenterostomy: Simplifying A Complex Technique Towards Widespread Use. *Endoscopy* **2020**, *52*, 151–152. [CrossRef] [PubMed]
87. Marino, A.; Bessissow, A.; Miller, C.; Valenti, D.; Boucher, L.; Chaudhury, P.; Barkun, J.; Forbes, N.; Khashab, M.A.; Martel, M.; et al. Modified Endoscopic Ultrasound-Guided Double-Balloon-Occluded Gastroenterostomy Bypass (M-Epass): A Pilot Study. *Endoscopy* **2022**, *54*, 170–172. [CrossRef] [PubMed]
88. Chen, Y.I.; Kunda, R.; Storm, A.C.; Aridi, H.D.; Thompson, C.C.; Nieto, J.; James, T.; Irani, S.; Bukhari, M.; Gutierrez, O.B.; et al. Eus-Guided Gastroenterostomy: A Multicenter Study Comparing the Direct and Balloon-Assisted Techniques. *Gastrointest. Endosc.* **2018**, *87*, 1215–1221. [CrossRef]
89. Itoi, T.; Ishii, K.; Ikeuchi, N.; Sofuni, A.; Gotoda, T.; Moriyasu, F.; Dhir, V.; Teoh, A.Y.; Binmoeller, K.F. Prospective Evaluation of Endoscopic Ultrasonography-Guided Double-Balloon-Occluded Gastrojejunostomy Bypass (Epass) for Malignant Gastric Outlet Obstruction. *Gut* **2016**, *65*, 193–195. [CrossRef]
90. Chen, Y.I.; Khashab, M.A. Endoscopic Approach to Gastrointestinal Bypass in Malignant Gastric Outlet Obstruction. *Curr. Opin. Gastroenterol.* **2016**, *32*, 365–373. [CrossRef]

91. Antonelli, G.; Kovacevic, B.; Karstensen, J.G.; Kalaitzakis, E.; Vanella, G.; Hassan, C.; Vilmann, P. Endoscopic Ultrasound-Guided Gastro-Enteric Anastomosis: A Systematic Review and Meta-Analysis. *Dig. Liver Dis.* **2020**, *52*, 1294–1301. [CrossRef]
92. Li, J.S.; Lin, K.; Tang, J.; Liu, F.; Fang, J. Eus-Guided Gastroenterostomy for Gastric Outlet Obstruction: A Comprehensive Meta-Analysis. *Minim. Invasive Ther. Allied Technol.* **2023**. ahead-of-print. [CrossRef]
93. Canakis, A.; Bomman, S.; Lee, D.U.; Ross, A.; Larsen, M.; Krishnamoorthi, R.; Alseidi, A.A.; Adam, M.A.; Kouanda, A.; Sharaiha, R.Z.; et al. Benefits of Eus-Guided Gastroenterostomy Over Surgical Gastrojejunostomy in the Palliation of Malignant Gastric Outlet Obstruction: A Large Multicenter Experience. *Gastrointest. Endosc.* **2023**, *98*, 348–359.e30. [CrossRef] [PubMed]
94. Vanella, G.; Dell'anna, G.; Capurso, G.; Maisonneuve, P.; Bronswijk, M.; Crippa, S.; Tamburrino, D.; Macchini, M.; Orsi, G.; Casadei-Gardini, A.; et al. Eus-Guided Gastroenterostomy for Management of Malignant Gastric Outlet Obstruction: A Prospective Cohort Study with Matched Comparison with Enteral Stenting. *Gastrointest. Endosc.* **2023**, *98*, 337–347.e5. [CrossRef] [PubMed]
95. Miller, C.; Benchaya, J.A.; Martel, M.; Barkun, A.; Wyse, J.M.; Ferri, L.; Chen, Y.I. Eus-Guided Gastroenterostomy vs. Surgical Gastrojejunostomy and Enteral Stenting for Malignant Gastric Outlet Obstruction: A Meta-Analysis. *Endosc. Int. Open* **2023**, *11*, E660–E672. [CrossRef] [PubMed]
96. Jeurnink, S.M.; Steyerberg, E.W.; Van Hooft, J.E.; Van Eijck, C.H.; Schwartz, M.P.; Vleggaar, F.P.; Kuipers, E.J.; Siersema, P.D.; Dutch, S.S.G. Surgical Gastrojejunostomy or Endoscopic Stent Placement for the Palliation of Malignant Gastric Outlet Obstruction (Sustent Study): A Multicenter Randomized Trial. *Gastrointest. Endosc.* **2010**, *71*, 490–499. [CrossRef] [PubMed]
97. Van Wanrooij, R.L.J.; Vanella, G.; Bronswijk, M.; De Gooyer, P.; Laleman, W.; Van Malenstein, H.; Mandarino, F.V.; Dell'anna, G.; Fockens, P.; Arcidiacono, P.G.; et al. Endoscopic Ultrasound-Guided Gastroenterostomy Versus Duodenal Stenting for Malignant Gastric Outlet Obstruction: An International, Multicenter, Propensity Score-Matched Comparison. *Endoscopy* **2022**, *54*, 1023–1031. [CrossRef]
98. Ge, P.S.; Young, J.Y.; Dong, W.; Thompson, C.C. Eus-Guided Gastroenterostomy Versus Enteral Stent Placement for Palliation of Malignant Gastric Outlet Obstruction. *Surg. Endosc.* **2019**, *33*, 3404–3411. [CrossRef]
99. Chen, Y.I.; Itoi, T.; Baron, T.H.; Nieto, J.; Haito-Chavez, Y.; Grimm, I.S.; Ismail, A.; Ngamruengphong, S.; Bukhari, M.; Hajiyeva, G.; et al. Eus-Guided Gastroenterostomy Is Comparable to Enteral Stenting with Fewer Re-Interventions in Malignant Gastric Outlet Obstruction. *Surg. Endosc.* **2017**, *31*, 2946–2952. [CrossRef]
100. Bronswijk, M.; Vanella, G.; Van Malenstein, H.; Laleman, W.; Jaekers, J.; Topal, B.; Daams, F.; Besselink, M.G.; Arcidiacono, P.G.; Voermans, R.P.; et al. Laparoscopic Versus Eus-Guided Gastroenterostomy for Gastric Outlet Obstruction: An International Multicenter Propensity Score-Matched Comparison (With Video). *Gastrointest. Endosc.* **2021**, *94*, 526–536.e2. [CrossRef]
101. Khashab, M.A.; Bukhari, M.; Baron, T.H.; Nieto, J.; El Zein, M.; Chen, Y.I.; Chavez, Y.H.; Ngamruengphong, S.; Alawad, A.S.; Kumbhari, V.; et al. International Multicenter Comparative Trial of Endoscopic Ultrasonography-Guided Gastroenterostomy Versus Surgical Gastrojejunostomy for the Treatment of Malignant Gastric Outlet Obstruction. *Endosc. Int. Open* **2017**, *5*, E275–E281. [CrossRef]
102. Perez-Miranda, M.; Tyberg, A.; Poletto, D.; Toscano, E.; Gaidhane, M.; Desai, A.P.; Kumta, N.A.; Fayad, L.; Nieto, J.; Barthet, M.; et al. Eus-Guided Gastrojejunostomy Versus Laparoscopic Gastrojejunostomy: An International Collaborative Study. *J. Clin. Gastroenterol.* **2017**, *51*, 896–899. [CrossRef]
103. Kouanda, A.; Binmoeller, K.; Hamerski, C.; Nett, A.; Bernabe, J.; Watson, R. Endoscopic Ultrasound-Guided Gastroenterostomy Versus Open Surgical Gastrojejunostomy: Clinical Outcomes and Cost Effectiveness Analysis. *Surg. Endosc.* **2021**, *35*, 7058–7067. [CrossRef] [PubMed]
104. Abbas, A.; Dolan, R.D.; Bazarbashi, A.N.; Thompson, C.C. Endoscopic Ultrasound-Guided Gastroenterostomy Versus Surgical Gastrojejunostomy for the Palliation of Gastric Outlet Obstruction in Patients with Peritoneal Carcinomatosis. *Endoscopy* **2022**, *54*, 671–679. [CrossRef] [PubMed]
105. Chiu, P.W.; Wai Ng, E.K.; Teoh, A.Y.; Lam, C.C.; Lau, J.Y.; Sung, J.J. Transgastric Endoluminal Gastrojejunostomy: Technical Development from Bench to Animal Study (with Video). *Gastrointest. Endosc.* **2010**, *71*, 390–393. [CrossRef] [PubMed]
106. Song, T.J.; Seo, D.W.; Kim, S.H.; Park, D.H.; Lee, S.S.; Lee, S.K.; Kim, M.H. Endoscopic Gastrojejunostomy with A Natural Orifice Transluminal Endoscopic Surgery Technique. *World J. Gastroenterol.* **2013**, *19*, 3447–3452. [CrossRef]
107. Rizzo, G.E.M.; Carrozza, L.; Tammaro, S.; Ligresti, D.; Traina, M.; Tarantino, I. Complete Intraperitoneal Maldeployment of A Lumen-Apposing Metal Stent During Eus-Guided Gastroenteroanastomosis for Malignant Gastric Outlet Obstruction: Rescue Retrieval with Peritoneoscopy Through Natural Orifice Transluminal Endoscopic Surgery. *VideoGIE* **2023**, *8*, 310–312. [CrossRef]
108. Gornals, J.B.; Consiglieri, C.F.; Maisterra, S.; Garcia-Sumalla, A.; Velasquez-Rodriguez, J.G.; Loras, C. Helpful Technical Notes for Intraperitoneal Natural Orifice Transluminal Endoscopic Surgery (Notes) Salvage in A Failed Eus-Guided Gastroenterostomy Scenario. *Endoscopy* **2022**, *54*, E287–E289. [CrossRef]
109. Kalloo, A.N.; Singh, V.K.; Jagannath, S.B.; Niiyama, H.; Hill, S.L.; Vaughn, C.A.; Magee, C.A.; Kantsevoy, S.V. Flexible Transgastric Peritoneoscopy: A Novel Approach to Diagnostic and Therapeutic Interventions in the Peritoneal Cavity. *Gastrointest. Endosc.* **2004**, *60*, 114–117. [CrossRef]
110. Yi, S.W.; Chung, M.J.; Jo, J.H.; Lee, K.J.; Park, J.Y.; Bang, S.; Park, S.W.; Song, S.Y. Gastrojejunostomy by Pure Natural Orifice Transluminal Endoscopic Surgery Using A Newly Designed Anastomosing Metal Stent in A Porcine Model. *Surg. Endosc.* **2014**, *28*, 1439–1446. [CrossRef]

111. Vanbiervliet, G.; Gonzalez, J.M.; Bonin, E.A.; Garnier, E.; Giusiano, S.; Saint Paul, M.C.; Berdah, S.; Barthet, M. Gastrojejunal Anastomosis Exclusively Using the "Notes" Technique in Live Pigs: A Feasibility and Reliability Study. *Surg. Innov.* **2014**, *21*, 409–418. [CrossRef]
112. Barthet, M.; Binmoeller, K.F.; Vanbiervliet, G.; Gonzalez, J.M.; Baron, T.H.; Berdah, S. Natural Orifice Transluminal Endoscopic Surgery Gastroenterostomy with A Biflanged Lumen-Apposing Stent: First Clinical Experience (with Videos). *Gastrointest. Endosc.* **2015**, *81*, 215–218. [CrossRef]
113. Tyberg, A.; Perez-Miranda, M.; Sanchez-Ocana, R.; Penas, I.; De La Serna, C.; Shah, J.; Binmoeller, K.; Gaidhane, M.; Grimm, I.; Baron, T.; et al. Endoscopic Ultrasound-Guided Gastrojejunostomy with A Lumen-Apposing Metal Stent: A Multicenter, International Experience. *Endosc. Int. Open* **2016**, *4*, E276–E281. [CrossRef] [PubMed]
114. Liu, B.R.; Liu, D.; Zhao, L.X.; Ullah, S.; Zhang, J.P.; Zhang, J.Y. Pure Natural Orifice Transluminal Endoscopic Surgery (Notes) Nonstenting Endoscopic Gastroenterostomy: First Human Clinical Experience. *VideoGIE* **2019**, *4*, 206–208. [CrossRef] [PubMed]
115. Caraceni, A.; Portenoy, R.K. Pain Management in Patients with Pancreatic Carcinoma. *Cancer* **1996**, *78*, 639–653. [CrossRef]
116. Anekar, A.A.; Hendrix, J.M.; Cascella, M. *WHO Analgesic Ladder*; Statpearls: Treasure Island, FL, USA, 2023.
117. Koulouris, A.I.; Alexandre, L.; Hart, A.R.; Clark, A. Endoscopic Ultrasound-Guided Celiac Plexus Neurolysis (Eus-Cpn) Technique and Analgesic Efficacy in Patients with Pancreatic Cancer: A Systematic Review and Meta-Analysis. *Pancreatology* **2021**, *21*, 434–442. [CrossRef]
118. Noble, M.; Gress, F.G. Techniques and Results of Neurolysis for Chronic Pancreatitis and Pancreatic Cancer Pain. *Curr. Gastroenterol. Rep.* **2006**, *8*, 99–103. [CrossRef]
119. Lillemoe, K.D.; Cameron, J.L.; Kaufman, H.S.; Yeo, C.J.; Pitt, H.A.; Sauter, P.K. Chemical Splanchnicectomy in Patients with Unresectable Pancreatic Cancer. A Prospective Randomized Trial. *Ann. Surg.* **1993**, *217*, 447–455; discussion 456–447. [CrossRef]
120. Michaels, A.J.; Draganov, P.V. Endoscopic Ultrasonography Guided Celiac Plexus Neurolysis and Celiac Plexus Block in the Management of Pain due to Pancreatic Cancer and Chronic Pancreatitis. *World J. Gastroenterol.* **2007**, *13*, 3575–3580. [CrossRef]
121. Erdine, S. Celiac Ganglion Block. *Agri* **2005**, *17*, 14–22.
122. Dolly, A.; Singh, S.; Prakash, R.; Bogra, J.; Malik, A.; Singh, V. Comparative Evaluation of Different Volumes of 70% Alcohol in Celiac Plexus Block for Upper Abdominal Malignsancies. *S. Asian J. Cancer* **2016**, *5*, 204–209. [CrossRef]
123. Wiersema, M.J.; Wiersema, L.M. Endosonography-Guided Celiac Plexus Neurolysis. *Gastrointest. Endosc.* **1996**, *44*, 656–662. [CrossRef]
124. Eisenberg, E.; Carr, D.B.; Chalmers, T.C. Neurolytic Celiac Plexus Block for Treatment of Cancer Pain: A Meta-Analysis. *Anesth. Analg.* **1995**, *80*, 290–295. [CrossRef] [PubMed]
125. Leblanc, J.K.; Rawl, S.; Juan, M.; Johnson, C.; Kroenke, K.; Mchenry, L.; Sherman, S.; Mcgreevy, K.; Al-Haddad, M.; Dewitt, J. Endoscopic Ultrasound-Guided Celiac Plexus Neurolysis in Pancreatic Cancer: A Prospective Pilot Study of Safety Using 10 Ml Versus 20 Ml Alcohol. *Diagn. Ther. Endosc.* **2013**, *2013*, 327036. [CrossRef] [PubMed]
126. Abedi, M.; Zfass, A.M. Endoscopic Ultrasound-Guided (Neurolytic) Celiac Plexus Block. *J. Clin. Gastroenterol.* **2001**, *32*, 390–393. [CrossRef] [PubMed]
127. Gunaratnam, N.T.; Sarma, A.V.; Norton, I.D.; Wiersema, M.J. A Prospective Study of Eus-Guided Celiac Plexus Neurolysis for Pancreatic Cancer Pain. *Gastrointest. Endosc.* **2001**, *54*, 316–324. [CrossRef]
128. Iwata, K.; Yasuda, I.; Enya, M.; Mukai, T.; Nakashima, M.; Doi, S.; Iwashita, T.; Tomita, E.; Moriwaki, H. Predictive Factors for Pain Relief after Endoscopic Ultrasound-Guided Celiac Plexus Neurolysis. *Dig. Endosc.* **2011**, *23*, 140–145. [CrossRef] [PubMed]
129. Han, C.Q.; Tang, X.L.; Zhang, Q.; Nie, C.; Liu, J.; Ding, Z. Predictors of Pain Response after Endoscopic Ultrasound-Guided Celiac Plexus Neurolysis for Abdominal Pain Caused by Pancreatic Malignancy. *World J. Gastroenterol.* **2021**, *27*, 69–79. [CrossRef]
130. Puli, S.R.; Reddy, J.B.; Bechtold, M.L.; Antillon, M.R.; Brugge, W.R. Eus-Guided Celiac Plexus Neurolysis for Pain due to Chronic Pancreatitis or Pancreatic Cancer Pain: A Meta-Analysis and Systematic Review. *Dig. Dis. Sci.* **2009**, *54*, 2330–2337. [CrossRef]
131. Lu, F.; Dong, J.; Tang, Y.; Huang, H.; Liu, H.; Song, L.; Zhang, K. Bilateral vs. Unilateral Endoscopic Ultrasound-Guided Celiac Plexus Neurolysis for Abdominal Pain Management in Patients with Pancreatic Malignancy: A Systematic Review and Meta-Analysis. *Support. Care Cancer* **2018**, *26*, 353–359. [CrossRef]
132. Levy, M.J.; Topazian, M.D.; Wiersema, M.J.; Clain, J.E.; Rajan, E.; Wang, K.K.; De La Mora, J.G.; Gleeson, F.C.; Pearson, R.K.; Pelaez, M.C.; et al. Initial Evaluation of the Efficacy and Safety of Endoscopic Ultrasound-Guided Direct Ganglia Neurolysis and Block. *Am. J. Gastroenterol.* **2008**, *103*, 98–103. [CrossRef]
133. Doi, S.; Yasuda, I.; Kawakami, H.; Hayashi, T.; Hisai, H.; Irisawa, A.; Mukai, T.; Katanuma, A.; Kubota, K.; Ohnishi, T.; et al. Endoscopic Ultrasound-Guided Celiac Ganglia Neurolysis vs. Celiac Plexus Neurolysis: A Randomized Multicenter Trial. *Endoscopy* **2013**, *45*, 362–369. [CrossRef]
134. Kamata, K.; Kinoshita, M.; Kinoshita, I.; Imai, H.; Ogura, T.; Matsumoto, H.; Minaga, K.; Chiba, Y.; Takenaka, M.; Kudo, M.; et al. Efficacy of Eus-Guided Celiac Plexus Neurolysis in Combination with Eus-Guided Celiac Ganglia Neurolysis for Pancreatic Cancer-Associated Pain: A Multicenter Prospective Trial. *Int. J. Clin. Oncol.* **2022**, *27*, 1196–1201. [CrossRef] [PubMed]

Disclaimer/Publisher's Note: The statements, opinions and data contained in all publications are solely those of the individual author(s) and contributor(s) and not of MDPI and/or the editor(s). MDPI and/or the editor(s) disclaim responsibility for any injury to people or property resulting from any ideas, methods, instructions or products referred to in the content.

Review

Endoscopic Management of Gastro-Entero-Pancreatic Neuroendocrine Tumours: An Overview of Proposed Resection and Ablation Techniques

Rocio Chacchi-Cahuin [1], Edward J. Despott [1], Nikolaos Lazaridis [1], Alessandro Rimondi [1], Giuseppe Kito Fusai [2], Dalvinder Mandair [3], Andrea Anderloni [4], Valentina Sciola [5], Martyn Caplin [3], Christos Toumpanakis [3] and Alberto Murino [1],*

[1] Royal Free Unit for Endoscopy, The Royal Free Hospital and University College London (UCL) Institute for Liver and Digestive Health, London NW3 2QG, UK; rochio.chacchi@nhs.net (R.C.-C.)
[2] Department of HPB Surgery and Liver Transplant, Royal Free Hospital NHS Foundation Trust, London NW3 2QG, UK
[3] Neuroendocrine Tumour Unit, The Royal Free Hospital, Pond Street, London NW3 2QG, UK
[4] Gastroenterology and Endoscopy Unit, Fondazione IRCCS Policlinico San Matteo, 27100 Pavia, Italy; a.anderloni@smatteo.pv.it
[5] Gastroenterology and Endoscopy Unit, Fondazione IRCCS Ca' Granda Ospedale Maggiore Policlinico Milano, 20122 Milan, Italy; valentinasciola@libero.it
* Correspondence: a.murino@ucl.ac.uk; Tel.: +44-(0)-207-794-0500 (ext. 36312)

Simple Summary: Neuroendocrine tumours (NETs) are relatively rare gastrointestinal neoplasms. Many NETs have a favourable prognosis, but some show aggressive features and poor long-term survival. A relatively higher incidence of small lesions amenable to endoscopic resection has been noted. The aim of this review is to present a thorough review of the literature to assist clinicians in the endoscopic management of neuroendocrine tumours (rectal, gastric, duodenal, pancreatic and oesophageal NETs), to highlight novel endoscopic therapeutic techniques of resection.

Abstract: A literature search of MEDLINE/PUBMED was conducted with the aim to highlight current endoscopic management of localised gastro-entero-pancreatic NETs. Relevant articles were identified through a manual search, and reference lists were reviewed for additional articles. The results of the research have been displayed in a narrative fashion to illustrate the actual state-of-the-art of endoscopic techniques in the treatment of NETs. Localised NETs of the stomach, duodenum and rectum can benefit from advanced endoscopic resection techniques (e.g., modified endoscopic mucosal resection, endoscopic full thickness resection, endoscopic submucosal dissection) according to centre expertise. Radiofrequency thermal ablation can be proposed as an alternative to surgery in selected patients with localised pancreatic NETs.

Keywords: NET; endoscopy; ESD; EMR; ablation

1. Introduction

Neuroendocrine tumours (NETs) are neoplasms derived from the diffuse neuroendocrine cell system. NETs can occur in different organs, including the pancreas, the duodenum, the stomach and the rectum, and they show a general relatively indolent growth rate together with the peculiar capacity to secrete a discrete range of active peptides and biogenic amines [1]. Gastro-entero-pancreatic neuroendocrine tumours (GEP-NETs) are still considered rare entities, although their incidence has significantly increased over the last 40 years. The age-adjusted incidence of GEP-NETs has gradually increased up to 3.65-fold in the United States and 3.8- to 4.8-fold in the United Kingdom. The largest increase occurred in gastric and rectal NETs, while the smallest increase occurred for small bowel NETs [2–6]. This increase in incidence is particularly evident for localized, low-grade tumours. Many studies found that even with a marked overall increase in incidence, the number of patients with distant metastases remained stable over 15 years. These data strongly suggest that increased NET incidence may be associated with the enhanced identification of small and asymptomatic lesions [5–7].

Regarding staging and biopsies, both conventional imaging and advanced imaging techniques, such as computed tomography (CT) scans, magnetic resonance imaging (MRI), and somatostatin receptor-based imaging (specifically, positron emission tomography (PET)/CT with 68 Ga-DOTA-peptides), are employed to accurately assess the disease and detect any possible distant metastases. If there is any indication of bone metastases on traditional imaging, it is recommended to conduct magnetic resonance imaging (MRI) of the spine and a 68 Gallium-positron emission tomography (PET) scan [8–11]. Considering the increased identification of small treatable lesions, our aim was to focus on the endoscopic treatment of oesophageal neuroendocrine tumours (O-NETs) gastric neuroendocrine tumours (G-NETs), duodenal neuroendocrine tumours (D-NETs), rectal neuroendocrine tumours (R-NETs) and pancreatic neuroendocrine tumours (P-NETs) that can be treated with endoscopic treatment. The focus on endoscopic treatment lies in its decreased invasiveness in comparison to conventional surgical methods while simultaneously guaranteeing similar effectiveness, as substantiated by recent medical research described hereafter.

2. Materials and Methods

We performed a literature search of EMBASE and MEDLINE databases, using the following keywords: rectal, rectum, gastric, duodenal, duodenum, oesophagus, oesophageal, pancreas, pancreatic, carcinoid, NET, therapy, endoscopy, mucosal resection, and submucosal dissection, to answer the following question "What endoscopic treatments are available for neuroendocrine tumours?". An author reviewed the literature and identified the most relevant articles on this topic. Controversies related to case selection were discussed with two other reviewers, experts in advanced endoscopic resection techniques (EJD and AM).

3. Results

3.1. Oesophageal NETs

Oesophageal NETs represent only 0.2% of GEP-NETs. They are usually diagnosed incidentally as discrete polypoid lesions or in association with adenocarcinoma in Barrett's oesophagus [12–14]. Endoscopic and histological features of O-NETs are not specific. In addition, due to the scarcity of evidence, guidelines for the treatment of O-NETs are missing, and physicians are mostly guided by local expertise and patient preference.

In a publication by Schizas et al. [13], endoscopic resection has been proposed in O-NETs measuring less than 10 mm in size with the absence of regional lymph node metastases. This was also supported by Yazici et al. [14], who proposed a threshold of 10 mm as the maximum size recommended for the endoscopic resection of O-NETs. However, this indication is not supported by a large body of evidence but rather by the extrapolation of data from gastric and rectal NETs, which have shown higher rates of lymph node metastases for lesions measuring over 10 mm in size.

Although EMR allows for the en-bloc excision of small O-NETs, ESD appeared to be more accurate, allowing for a more accurate pathological examination of the respected specimen, which may not be obtained by EMR because of mucosal damage occurring during the resection [14] (Table 1).

Table 1. Endoscopic management of well-differentiated oesophageal neuroendocrine tumours.

Size	Proposed Treatments	Body of Evidence	Pros and Cons
Oesophageal NET < 10 mm	ESD	Case report	Little evidence to support the data.

3.2. Gastric NETs

G-NETs are classified into three categories according to the background gastric pathology: type 1 (prevalence 75%), type 2 (prevalence 5–10%), and type 3 (prevalence 15–25%) (Figure 1). Type 1 G-NETs are typically small, multiple, and associated with chronic atrophic gastritis with hypergastrinaemia and enterochromaffin-like cell hyperplasia. Type 2 G-NETs share the same pathological pathway due to excessive production of gastrin, although they are produced by a gastrinoma in the context of Zollinger–Ellison syndrome, usually in the setting of a multiple endocrine neoplasia type 1 (MEN-1). Type 3 G-NETs are sporadic lesions, typically solitary and undifferentiated; in addition, they are often larger in size when compared to type 1 and 2 G-NETs, and they occur in the setting of normal gastrin levels [9,10]. The tumour cell proliferation index allows for a further grading of G-NETs from G1 to G3 according to the World Health Organization (WHO) [15]. G1 has a mitotic count of <2 per 10 high-power fields (HPF) and/or Ki-67 $\leq$ 2%; G2 has a mitotic count of 2–20 per 10 HPF and/or Ki-67 3–20%; and G3 has a mitotic count and a Ki-67 > 20 [15,16].

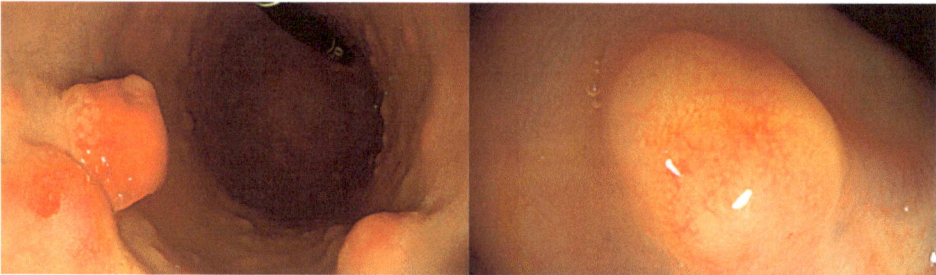

Figure 1. Example of type 1 gastric NET in the context of autoimmune atrophic gastritis on the left; example of rectal NET on the right.

3.2.1. Type 1 Gastric NETs

Type 1 G-NETs occur more frequently in females because of their increased incidence of autoimmune chronic atrophic gastritis. Type 1 lesions are usually G1 tumours; thus, their metastatic risk is extremely low, and the prognosis is excellent. They are generally asymptomatic and usually incidentally detected during screening upper GI endoscopy [9,16–20]. The recent 2023 ENETS guidance paper listed ESD, EMR (standard and modified-EMR—m-EMR, with utilization of cap aspiration or with a ligation device or grasping forceps) and EFTR as possible treatments for localised, low-grade (G1 and G2), type 1 G-NETs [10]. Nevertheless, there are still no significant data documenting the superiority of any method [21]. Prospective studies comparing endoscopic resection methods are scarce, and the majority of data are provided mainly by retrospective studies [17].

A study with 62 patients and 87 type 1 G-NETs sized $\leq$10 mm, treated with ESD or EMR, compared their efficacy. The complete resection rate was higher when ESD was performed (94.9% vs. 83.3%, p-value = 0.174), although this was not statistically significant. However, a statistically significant difference was noted in the vertical margin involvement rate, which was lower in the ESD group (2.6% vs. 16.7%, p = 0.038). No difference was noted in the complication rate between the two groups [22].

A smaller study that included 13 type 1 G-NETs compared ESD and EMR as endoscopic resection techniques and measured effectiveness according to complete resection [23]. Seven ESDs and six EMRs were performed. The horizontal margins of excision were negative for all lesions, but the vertical margins were positive in four lesions (66.7%), all of them in the EMR group.

Recently, a large Korean study evaluated 103 patients with 114 tumours managed with EMR and ESD. En-bloc resection rates were similar, but complete resection was significantly higher in the ESD group. In addition, adverse event rates were similar among the two groups due to the need for surgical management and a disease-free survival rate [23].

A recent systematic review regarding the optimal endoscopic resection technique analysed 6 studies with 112 gastric type 1 NETs removed by EMR and 77 by ESD. Both methods appeared to have similar complete en-bloc resections, complications, and adverse event rates [21].

Surgical treatment is recommended only for patients with type 1 tumours that are predicted as T2 or lesions post-resection with positive margins; local wedge excision or partial gastrectomy should be considered (Table 2) [9,10].

Table 2. Endoscopic management of gastric neuroendocrine tumours.

Size	Proposed Treatments	Body of Evidence	Pros and Cons
Gastric NET Type 1 < 10 mm	ESD m-EMR (cap- or ligation-assisted) Follow up	Multicentre retrospective study Retrospective studies	Slight advantage of ESD over m-EMR in complete resection rates and free vertical margins. No differences in terms of rates of complication. Some evidence for conservative treatment "watch and wait" in G1 NETs.
Gastric NET Type 2 < 10 mm	ESD m-EMR (cap- or ligation-assisted)	Consensus	Same as type 1 gastric NETs.
Gastric NET Type 3	Surgery	Consensus	

It is noteworthy that, in patients with small (<10 mm) type 1 G-NETs who remained under endoscopic surveillance without resection, studies have shown no tumour-related deaths, even after a significant follow-up period (54–68 months). This conservative management seems to be a rational approach in selected patients with type 1 G-NETs, such as elderly patients with small tumours [24,25].

3.2.2. Type 2 Gastric NETs

Type 2 G-NETs are linked with multiple endocrine neoplasia type 1 (MEN1) and Zollinger–Ellison syndrome (ZES). Approximately 13–37% of all patients with MEN1-ZES have been diagnosed with type 2 G-NET, while type 2 NETs are detected in only 0–2% of patients with sporadic ZES (without MEN1). Type 2 G-NETs are equally found in both male and female patients; around 30% of these lesions are metastatic at presentation. Additionally, patients with type 2 G-NETs have a lower survival rate than patients with type 1. Furthermore, although the majority of type 2 lesions are asymptomatic, the most common presenting symptom is related to peptic ulcers. This is due to the increased gastric acid secretion that can subsequently cause the development of peptic ulcers [26–28].

According to the ENETS guidelines, for type 2 G-NETs, treatment is usually dictated by the presence or not of additional duodenal and/or pancreatic lesions as part of MEN-1. Local or selective excision may be recommended, but this should be decided in multidisciplinary NETs centres of excellence [10]. The NCCN and the French Intergroup guidelines state that endoscopic resection may be indicated for lesions measuring up to 2 cm, and multiple biopsies of the surrounding mucosa are recommended for hypergastrinaemic patients [8,9]. However, because of their rarity, data on type 2 G-NETs are scarce; therefore, MDT discussion should always be encouraged.

3.2.3. Type 3 Gastric NETs

Type 3 G-NETs are usually high-grade lesions (G3) with a tendency to infiltrate the muscularis propria, with a higher rate of lymphovascular involvement; this type of G-NETs is usually metastatic at presentation, with spreading to regional lymph nodes or liver. It often appears as a single lesion, usually greater than 10 mm. An atypical presentation (not serotonin related) of "carcinoid syndrome," including flushing, tachycardia, and diarrhoea, occurs rarely in patients with gastric NETs (<1%) and is almost exclusively associated with type 3 tumours with liver metastasis [29–31].

ENETS guidelines recommend that type 3 G-NETs should be managed in the same way as gastric adenocarcinomas [10]. NCCN guidelines recommend radical gastric resection with perigastric lymph node dissection for localized type 3 G-NETs [8]. Interestingly, recent evidence has shown that carefully evaluated patients with small-size (<10 mm) and low-grade type 3 G-NETs could benefit from endoscopic resection [29,32,33]. A multicentre study gathering evidence from six tertiary referral centres found that endoscopic resection or limited surgical resection is feasible and safe in type 3 G-NETs under 10 mm that demonstrate a favourable grade (G1 or low G2) [34]. Nevertheless, MDT discussion in a tertiary referral centre is recommended before considering the endoscopic management of small and G1 type 3 G-NET [35].

3.3. Duodenal NETs

Duodenal NETs are rare, representing 2–4% of gastrointestinal NETs; they present as solitary lesions, confined to the duodenal submucosa [11], measuring different sizes. Endoscopic resection is recommended for D-NETs under 10 mm, although rare cases of local and distant metastases have been reported even for such small lesions [36]. In fact, D-NETs < 10 mm have a 14% rate of nodal metastasis, which increased to 47% for D-NETs measuring 21–50 mm in size. The invasion of the muscularis propria usually involves a size greater than 2 cm, and the presence of mitotic figures is an independent risk factor for metastasis [8–10].

D-NETs smaller than ≤10 mm confined to the submucosal layer and without lymph nodes or distant metastasis [8–10] are treated endoscopically when resection is required. Endoscopic resection in this setting is considered safe and effective. Conversely, surgical treatment is recommended for large (>20 mm) and/or metastatic D-NETs (Table 3).

Table 3. Endoscopic management of well-differentiated duodenal neuroendocrine tumours.

Size	Proposed Treatments	Body of Evidence	Pros and Cons
Duodenal NET < 10 mm	m-EMR EMR ESD Band-and-Slough Laparoscopic endoscopic cooperative surgery	Multicentre retrospective study Small-sample-size retrospective studies Case series	Acceptable rates of en-bloc resection and endoscopic full resection for m-EMR. Role of ESD is still to be decided (little supporting evidence). 50% of R0 in the larger cohorts of patients (90% in surgery patients). Only initial data for laparoscopic endoscopic cooperative surgery.

Therapy is a subject of debate for non-functional, localised, well-differentiated (G1) D-NETs. Both endoscopic therapy and surgical interventions are seen as viable options in this particular scenario. However, there is a dearth of controlled studies evaluating the various techniques. Conversely, surgical intervention is advised for D-NETs that are bigger than 20 mm [8–10,27].

There are only a few retrospective studies that have addressed the endoscopic resection of D-NETs. One compared the outcomes of ligation-assisted endoscopic resection (EMR-L) and conventional EMR of 15 D-NETs with a mean tumour size of 6.6 ± 3.9 mm and mean procedure time of 11.0 ± 11.2 min. En-bloc resection and complete resection rates were higher in the ligation group (100% vs. 87.5%, and 85.7% vs. 62.5%, respectively), although this was not statistically significant. There was no evidence of local or distant metastasis at follow-up (26.1 ± 20.7 months) [37].

In another study by Fujimoto et al. [38], which included 10 patients with D-NETs treated with EMR-L, the en-bloc resection rate and endoscopic complete resection rates were 100%. Nevertheless, complete histopathological resection with clear margins was observed in only 70% of the specimens and vertical margins were negative in all 10. Three patients required additional surgical treatment because of lymphatic vessel invasion. No recurrence was identified at follow-up (mean period: 18.6 months). One major adverse event (perforation) was reported in one patient and was treated conservatively.

A recent retrospective study [36] evaluated the short- and long-term outcomes of ESD for non-ampullary D-NETs. Eight patients with G1 D-NETs with a diameter of less than 10 mm, restricted to the submucosal layer, and no lymph node involvement or distant metastases were included in the study. The majority of these lesions were in the duodenal bulb, with a median size of 6.4 mm. En-bloc resection, R0 resection, and curative resection had respectively 100%, 88%, and 88% success rates. Perforation occurred in one patient and was treated conservatively. Non-recurrencies were reported after a median follow-up period of 34 months.

In a multicentre retrospective study from 2019 [39], 60 non-ampullary D-NETs that underwent either endoscopic mucosal resection with a dual channel endoscope (EMR-D), EMR-L, EMR with a transparent cap (EMR-C), EMR with circumferential mucosal pre-cutting (EMR-P), or ESD were analysed and compared with surgical resection. For the group that received endoscopic treatment (EMR-D, EMR-L, EMR-C, EMR-P and ESD), en bloc resection, endoscopic full resection, and R0 rates were 88%, 92% and 50%, respectively. When the lesion size was more than 11 mm, the R0 rate was lower (50%, $p = 0.003$), and lymphovascular invasion was more common (33.3%, $p = 0.043$). The complete endoscopic resection rate was even lower (50%) in the NETs group with lesions $\geq$11 mm. After the histopathological analysis, patients who were treated surgically had a higher (90.9%) full resection rate than those who were treated endoscopically (50%), showing the advantage of surgical management for lesions larger than 10 mm.

Another new endoscopic technique is the band and slough technique (BAS), described by Hawa et al. [40], which is a minimally invasive endoscopic procedure for the management of small gastric and duodenal NETs (G-NETs and D-NETs). This is a variation of the band ligation procedure without the resection of the lesion. The BAS technique was used to treat three duodenal NETs and one type 1 G-NET, all of which were 10 mm in diameter. After the initial session of banding, both patients reached full recovery with no recurrence at the 3-month follow-up. Furthermore, 12-month monitoring of the site with biopsies revealed no tumour recurrence. The technique is relatively easy to perform; however, it does not provide a histological specimen and therefore its clear advantage in tumour eradication when compared with traditional resection techniques (i.e., m-EMR, ESD) is yet to be proven, and further numbers are therefore required [40].

Bourke et al. [41] conducted a retrospective study to assess patients with D-NETs who underwent ESD at three tertiary referral centres in Australia, France, and Belgium between 2012 and 2022. The results of the study indicated that en-bloc resection rates for D-NETs can reach 100% when performed by experienced endoscopists. There were no instances of distant metastatic spread or local recurrence in this study, suggesting that ESD could be a feasible alternative for patients with D-NETs measuring 10–15 mm who are not suitable candidates for surgery. To the best of our knowledge, only a single case series of three patients diagnosed with D-NETs and treated with EFTR can be found in the literature. The decision of EFTR was based on poor fitness for surgery. To note, disease-free survival at 1-year follow-up was documented [42].

Laparoscopic endoscopic cooperative surgery is an alternative way to achieve the full-thickness resection of D-NETs. The principle of this technique lies in associating the advantages of endoscopic resection (the ability to clearly demarcate the lesion and to avoid unnecessary bowel wall resection) with the suturing effectiveness of surgery. This technique has been adopted mainly in Eastern countries on single cases, with one case series reporting a curative resection in 85% of the cases [43].

3.4. Rectal NETs

The neuroendocrine tumours of the rectum represent 34% of all diagnosed GEP-NETs, and the rectum is the second site of localization by frequency after midgut (Figure 1). The incidence of R-NETs based on the Surveillance, Epidemiology, and End Results United States database (SEER) is approximately 1 per 100,000 population per year, accounting for 17.7% of all NETs [44–46].

As previously mentioned, a significant increase in the incidence in the R-NETs population has been observed; however, this was not followed by a higher incidence of distal metastasis, which remained stable. This scenario is likely to be explained by an improved and earlier diagnosis [7,44–51].

Grade is important when it comes to appointing a prognosis to a patient. Lower grades (G1–G2) are associated with better outcomes in terms of overall survival and risk of metastasis, often taking advantage of localised and endoscopic treatments [52,53]. In contrast, G3 grade is associated with the need for surgery, chemotherapy, and overall poor survival rates (10% at 5 years post-diagnosis) [54].

A systematic review by Mc Dermott et al. included 14 studies with 4575 patients; this showed that 80% of the R-NETs were <10 mm in size, 15% were between 10 and 20 mm, and 5% were >20 mm. Regional lymph nodal metastases were present in 8% of cases, and 4% of all patients had distant metastases. Tumour size greater than 10 mm and muscular and lymphovascular invasion were independently associated with increased risk of metastases. The 5-year survival rate was 93% in patients presenting with localised disease [50].

In their systematic review and meta-analysis, Xin Zhou et al. examined the differences between ESD, EMR, and modified ERM (m-EMR, an EMR performed with additional assistant devices like a ligation band or suction cap). The investigation spanned 650 patients and 10 retrospective studies. The results indicated that the ESD group had a higher rate of complete resection than the EMR group (RR 0.89 95% CI [0.79, 0.99]). In contrast, the rates of complete resection in both the ESD and m-EMR groups were similar (RR 1.03, 95% CI [0.95, 1.11]). Although the procedure duration in the ESD group was considerably longer than in the EMR group, the difference was not statistically significant (STD.50, 95% CI [−3.14, 0.14]). In the EMR group, local recurrence was observed in five cases, whereas no ESD patient experienced it [51].

These results were confirmed in a more recent meta-analysis by Zhang et al., which evaluated the treatment outcomes after ESD, m-EMR, and EMR for R-NETs < 16 mm. Compared with EMR, ESD achieved higher complete resection rates without increasing the overall complication rate (OR = 4.38, 95%CI: 2.43–7.91). Nevertheless, ESD was more time-consuming than EMR and m-EMR (respectively, MD = 6.72, 95%CI: 5.84–7.60 and MD = 12.21, 95%CI: 7.78–16.64). The m-EMR shared comparable outcomes with ESD for R-NETs < 16 mm. Both ESD and m-EMR were superior to conventional EMR in terms of complete resection rate without increasing safety concerns [52].

More recently, multiple retrospective studies comparing different endoscopic resection techniques for R-NETs < 10 mm, in the absence of deep invasion or lymphadenopathy, have been published [53–56]. A study evaluated 77 small rectal R-NET (≤10 mm) patients treated by m-EMR with endoscopic submucosal resection with band ligation (ESMR-L) (n = 53) or ESD (n = 24). En-bloc resection was achieved in all patients. A significantly higher histopathological complete resection rate was observed in the ESMR-L group (100%) than in the ESD group (54.2%) (p = 0.001). The procedure time of ESD was significantly longer than that of ESMR-L [55]. Another similar retrospective study from Japan, including 96 patients treated with EMR (n = 60), m-EMR (n = 21) and ESD (n = 21), showed similar results between the various resection techniques [57].

When endoscopy was compared with surgery in a large monocentric retrospective propensity-matched study of 104 patients, who were equally distributed between ESD and transanal endoscopic microsurgery (TEM) for R-NETs under 20 mm, a similar R0 resection rate was observed in the subgroup analysis divided by tumour size (<10 mm and 10 to 20 mm). However, a shorter procedure time and hospital stay for ESD patients (ESD 22 [range, 11–65] vs. TEM 35 [range, 17–160] minutes; ESD 2.5 [range, 1–5] vs. TEM 4 [range, 3–8] days) was noted [58].

Another technique that can be considered for the removal of R-NETs under 10 mm is endoscopic full-thickness resection. A multicentre retrospective study of 31 German centres reported 501 endoscopic procedures, of which 40 cases were R-NETs, using a full-thickness resection device (FTRD). The median lesion size was 8 mm and resection was endoscopically and histologically complete in all cases whereas full-thickness resection was achieved in 95% of cases. There were no major adverse events, and after follow-up endoscopy, no evidence of residual or recurrent tumour was reported [59].

In another recent study from Korea, 115 patients with R-NETs (<10 mm) were included. Rectal NETs were either removed by ESD ($n = 79$) or underwater endoscopic mucosal resection (UEMR) ($n = 36$). There was no difference in terms of R0 resection rate between the UEMR and ESD groups (86.1% vs. 86.1%, $p = 0.996$), whereas the procedure time was significantly shorter with UEMR (5.8 ± 2.9 vs. 26.6 ± 13.4 min, $p < 0.001$) [55].

According to clinical management guidelines, surgical resection with the removal of associated lymphatic tissue is the preferred treatment for R-NETs greater than 20 mm because of the high risk of lymphatic invasion and metastasis [9,46]. The existence of predictors of nodal involvement, including a tumour size ≥ 15 mm, the atypical endoscopic aspect, muscular layer invasion, a tumour grade of G2-G3, and lymphovascular invasion should guide the management of R-NETs. A low anterior rectal resection with complete mesorectal excision should be performed when one or more of these characteristics are present [9,60,61].

To summarize, in R-NETs < 15 mm, advanced resection techniques (i.e., m-EMR, ESD, EFTR) must be preferred to standard polypectomy or simple endoscopic mucosal resection (EMR), which should be avoided because of the low rate of complete resection. In particular, ESD could be considered as a first-line treatment for fibrotic lesions and should be selected over m-EMR for lesions over 10 mm in diameter as it appears to achieve better complete resection rate outcomes (Table 4) [9,60,61].

Table 4. Endoscopic management of well-differentiated rectal neuroendocrine tumours.

Size	Proposed Treatments	Body of Evidence	Pros and Cons
Rectal NET < 20 mm	ESD m-EMR (ligation-assisted) EMR FTRD	Systematic review and meta-analyses Retrospective multicentre studies	ESD and m-EMR are similar in terms of complete resection rates and complications for NETs under 10 mm. ESD has a longer procedure time compared to m-EMR. m-EMR with ligation devices showed a little benefit over ESD on vertical margin positivity. ESD and m-EMR are both superior to standard EMR in terms of complete resection. FTRD is feasible for R-NETs under 10 mm with a high complete resection rate. For NETs with a size from 10 mm to 20 mm, ESD should be proposed, with a careful evaluation of benefits over surgery techniques.

3.5. Pancreatic NETs

Pancreatic neuroendocrine tumour (P-NET) prevalence is 10% of all pancreatic neoplasms. Their prognosis is good, even in an advanced disease setting [9,62–64].

P-NETs are classified as either sporadic or genetically determined when they occur in the context of inherited syndromes. Additionally, they are categorized according to the manifestation or non-appearance of symptoms caused by hormone secretion, such as insulin, gastrin, and glucagon, which are secreted by functional P-NETs [61,62]. Functional pancreatic nanotubes (P-NETs) are typically detected during the early stages, when the lesions are still minor [42]. In the majority of cases, surgery is the preferred treatment option for these tumours. Conversely, the detection of non-functional P-NETs typically occurs at a later stage; however, the widespread application of CT and MRI imaging has substantially augmented the detection rate of minor incidental lesions [63–65].

Surgery is the most common treatment for P-NETs, but it is associated with potential perioperative risks. Specifically, postoperative complications can be more frequent than those after surgery for ductal adenocarcinoma [66]. For non-functioning P-NETs, the therapeutic approach depends on tumour localizations and size. The ENETs guidelines recommend observation and surveillance for well-differentiated grade 1 tumours <2 cm [61]. However, the interim analysis of the ASPEN trial (a prospective international multicentre study longitudinally following patients with P-NETs <2 cm, either monitored or resected) suggested a more personalized management for non-functional P-NETs between 1 and 2 cm and a mandatory surgical resection for all lesions with a dilated main pancreatic duct, as most of them exhibit a more aggressive biological behaviour [67].

Parenchyma-preserving surgery is preferable for small lesions, mainly those in the head of the pancreas, with a limiting factor being the proximity of the tumour to the pancreatic duct. Alternatively, pancreatoduodenectomy with or without pylorus preservation (Whipples vs. PPPD) and distal pancreatectomy are the standard surgical procedures for P-NETs localized in the head and body/tail of the pancreas, respectively [64].

As an alternative, endoscopic ultrasound-guided radiofrequency ablation (EUS-RFA) has recently been described for functional smaller P-NETs [68,69]. Radiofrequency ablation (RFA) causes irreversible cellular damage, cellular apoptosis, and tissue coagulative necrosis by conveying high temperatures within the tumour mass [70,71]. Compared with surgical methods, the technical advantages of locoregional thermoablative techniques include the preservation of healthy surrounding tissues, lower rates of morbidity, shorter hospital stays, and reduced overall expenses. Furthermore, research suggests that immunomodulation may have a secondary anticancer effect [72].

Two devices are available for performing pancreatic RFA: a cooled needle connected to a dedicated energy source and a 1Fr probe that can be inserted into a 19G needle while being attached to a conventional energy source. Utilizing a high-frequency alternating current and EUS guidance, the needle is introduced into the designated lesion while maintaining a minimum distance of 2 mm from the pancreatic and biliary ducts to prevent injury or duct strictures. Doppler evaluation is employed to ensure that no harm is done to the vasculature [73].

In their recent study, Imperatore et al. [68] undertook a comprehensive review of the literature in order to determine whether EUS-guided RFA treatment is feasible, effective, and safe, and to identify P-NET characteristics that could serve as predictors of response to EUS-RFA. Sixty-nine patients were identified by the authors from twelve investigations. Males comprised thirty (49.2%) of the sixty-one patients, who were aged 65.4 years on average. The investigators identified 73 P-NETs, with a range of 4.5 to 40.0 mm in length, with the following locations: head (35.3%), body (39.7%), uncinate (8.8%), and tail (16.2%). The average measure of each was 16 mm. Out of the total, 30.1% (21 insulinomas and 1 VIPoma) were functional. P-NETs were administered an average of 1.3 RFA sessions over a follow-up period of 11 months (range: 1–34 months), with an overall effectiveness of 96% (75–100%). The response rate was found to be influenced by the size of the tumour. Specifically, larger tumours exhibited a higher frequency of failing to respond to treatment (mean size of 21.8 mm ± 4.71 in the non-response group vs. 15.07 mm ± 7.34 in the response group, $p = 0.048$). A P-NET size of less than 18 mm at EUS was associated with a positive response to EUS-RFA, as determined by the ROC curve, which had the following values: sensitivity (80%), specificity (78.6%), PPV (97.1%), and NPV (30.8%) [68].

A large series comparing EUS-RFA and surgery with propensity-score matching (1:1) in 89 patients affected by insulinoma has been recently published. Notably, clinical efficacy was comparable between RFA and surgery (95.5% vs. 100%, $p = 0.16$), but RFA outperformed surgery in terms of adverse event rate (18.0% vs. 61.8%, $p < 0.001$), severe adverse event rate (0.0% vs. 15.7%) and hospital stay (3.0 ± 2.5 vs. 11.1 ± 9.7, $p < 0.001$). However, recurrence was observed in EUS-RFA-treated patients [74].

A method based on EUS-guided ethanol injection has been described as an alternative to RFA for p-NET under 20 mm and G1-G2 grade without metastasis. A recent pilot study with five patients from Matsumoto et al. showed how a successful ablation could be achieved in four cases, with no recurrence and no adverse events reported [75]. A previous experience on 11 patients and 14 tumours by Park et al. reported a lower response rate (61.5%) with 3 cases of mild pancreatitis (30%) [76] (Table 5).

Table 5. Endoscopic management of well-differentiated pancreatic neuroendocrine tumours.

Size	Proposed Treatments	Body of Evidence	Pros and Cons
Pancreatic NET < 20 mm	Surveillance EUS-RFA EUS-alcohol injection Parenchyma-preserving surgery	Multicentre retrospective study Systematic review with meta-analysis Case reports	Surveillance can be proposed for patients not fit for surgery with non-functioning small P-NETs (<10 mm) EUS-RFA can be considered for P-NETs < 20 mm Parenchyma-preserving surgery is considered the standard of care for fit-for-surgery patients with P-NETs Paucity of data to support EUS-alcohol injection

EUS-guided P–NET localization is another task that could be accomplished by gastrointestinal endoscopy. The localization consists of focal needle tattooing (EUS-FNT) or placing fiducial markers close to or inside lesions to make them easier to identify and resect during surgery. EUS-FNT was performed on a case series of 13 patients, 6 of whom had P-NETs, by labelling the pancreatic parenchyma within 3–5 mm of the lesion with sterile carbon-based ink to permit laparoscopic distal pancreatectomy (LDP). The surgeon's ability to clearly view and palpate the lesion is restricted, if not non-existent, in LDP. Despite a mean of 20.3 days (range, 3–69) between EUS-FNT and surgery, all tattooed tissues were clearly visible at surgery [77].

Finally, when it comes to the diagnosis of P-NETs, fine-needle biopsy (FNB) is the technique recommended to obtain a specimen from pancreatic masses, although a meta-analysis encompassing 11 RCT failed to prove the statistical superiority of FNB over FNA (fine-needle aspiration) [78]. However, a recent network meta-analysis, including 16 studies for a total of 1934 patients, demonstrated how Franseen and Fork-tip needles (FNB), specifically those of the 22-gauge size, were the best performers in terms of obtaining tissue samples from pancreatic masses. At the same time, the level of confidence in the estimates was poor due to relatively few head-to-head trials supporting the comparisons and differences while carrying out the procedure that may have changed the final results [79].

4. Conclusions

Gastrointestinal neuroendocrine tumours are clinically diverse and rare neoplasms that can pose a treatment challenge for endoscopists. The most prevalent detection method for luminal NETs is endoscopy, and histopathology is the gold standard in their diagnosis. EUS has become an essential procedure for the staging and management of pancreatic NETs.

The best treatment depends on the location of the NETs, also requiring an accurate assessment of their size, local lymphadenopathy, and depth of invasion. Endoscopic resection techniques are evolving, with endoscopic submucosal dissection appearing to be an effective and safe method allowing for the en-bloc removal of lesions up to 2 cm in size, possessing an acceptable complete resection and a recurrence rate similar to surgery. The evaluation of the best technique to apply to an NET should be discussed in a multidisciplinary team meeting setting, and the discussion should consider tumour size and location as well as the patient's preferences and local expertise.

Future evidence from a prospective randomised controlled trial is needed to better define the role of therapeutic endoscopy in the resection of luminal NETs measuring up to

2 cm and to investigate whether this could eventually replace surgical intervention when the risk of lymphovascular invasion is low.

Author Contributions: Conceptualization, A.M. and R.C.-C.; methodology, A.M. and E.J.D.; investigation, A.M., E.J.D., R.C.-C., N.L. and A.R.; writing—original draft preparation, A.M., R.C.-C., N.L. and A.R.; writing—review and editing, A.M., E.J.D., R.C.-C., N.L., A.R., G.K.F., V.S., A.A., M.C., D.M. and C.T.; supervision, A.M. and E.J.D.; project administration, A.M. All authors have read and agreed to the published version of the manuscript.

Funding: This research received no external funding.

Conflicts of Interest: R.C.-C.: no disclosures; E.J.D.: educational grants in support of conference organization, and honoraria, from Fujifilm, Pentax, and Olympus, and from Ambu; N.L.: no disclosures; A.R.: no disclosures; G.K.F.: no disclosures; D.M.: no disclosures; A.A.: Consultant for Boston Scientific and Olympus; V.S.: no disclosures; M.C.: no disclosures; C.T.: no disclosures; A.M.: personal payments/honoraria/fees: Olympus, GI supply, Boston Scientific, Fujifilm.

References

1. Cives, M.; Strosberg, J. Gastroenteropancreatic Neuroendocrine Tumors. *CA Cancer J. Clin.* **2018**, *68*, 471–487. [CrossRef]
2. Wyld, D.; Wan, M.H.; Moore, J.; Dunn, N.; Youl, P. Epidemiological trends of neuroendocrine tumours over three decades in Queensland, Australia. *Cancer Epidemiol.* **2019**, *63*, 101598. [CrossRef] [PubMed]
3. Huguet, I.; Grossman, A.B.; O'Toole, D. Changes in the Epidemiology of Neuroendocrine Tumours. *Neuroendocrinology* **2017**, *104*, 105–111. [CrossRef] [PubMed]
4. Fraenkel, M.; Kim, M.; Faggiano, A.; De Herder, W.W.; Valk, G.D. Incidence of gastroenteropancreatic neuroendocrine tumours: A systematic review of the literature. *Endocr. Relat. Cancer.* **2014**, *21*, R153–R163. [CrossRef] [PubMed]
5. Dasari, A.; Shen, C.; Halperin, D.; Zhao, B.; Zhou, S.; Xu, Y.; Shih, T.; Yao, J.C. Trends in the Incidence, Prevalence, and Survival Outcomes in Patients With Neuroendocrine Tumors in the United States. *JAMA Oncol.* **2017**, *3*, 1335–1342. [CrossRef]
6. Alwan, H.; La Rosa, S.; Andreas Kopp, P.; Germann, S.; Maspoli-Conconi, M.; Sempoux, C.; Bulliard, J.L. Incidence trends of lung and gastroenteropancreatic neuroendocrine neoplasms in Switzerland. *Cancer Med.* **2020**, *9*, 9454. [CrossRef]
7. Hallet, J.; Law, C.H.L.; Cukier, M.; Saskin, R.; Liu, N.; Singh, S. Exploring the rising incidence of neuroendocrine tumors: A population-based analysis of epidemiology, metastatic presentation, and outcomes. *Cancer* **2015**, *121*, 589–597. [CrossRef]
8. Kulke, M.H.; Shah, M.H.; Benson, A.B., 3rd; Bergsland, E.; Berlin, J.D.; Blaszkowsky, L.S.; Emerson, L.; Engstrom, P.F.; Fanta, P.; Giordano, T.; et al. Neuroendocrine tumors, version 1.2015. *J. Natl. Compr. Cancer Netw.* **2015**, *13*, 78–108. [CrossRef]
9. De Mestier, L.; Lepage, C.; Baudin, E.; Coriat, R.; Courbon, F.; Couvelard, A.; Do Cao, C.; Frampas, E.; Gaujoux, S.; Gincul, R.; et al. Digestive Neuroendocrine Neoplasms (NEN): French Intergroup clinical practice guidelines for diagnosis, treatment and follow-up (SNFGE, GTE, RENATEN, TENPATH, FFCD, GERCOR, UNICANCER, SFCD, SFED, SFRO, SFR). *Dig. Liver Dis.* **2020**, *52*, 473–492. [CrossRef]
10. Panzuto, F.; Ramage, J.; Pritchard, D.M.; van Velthuysen, M.F.; Schrader, J.; Begum, N.; Sundin, A.; Falconi, M.; O'Toole, D. European Neuroendocrine Tumor Society (ENETS) 2023 guidance paper for gastroduodenal neuroendocrine tumours (NETs) G1-G3. *J. Neuroendocrinol.* **2023**, *35*, e13306. [CrossRef]
11. Rossi, R.E.; Rausa, E.; Cavalcoli, F.; Conte, D.; Massironi, S. Duodenal neuroendocrine neoplasms: A still poorly recognized clinical entity. *Scand. J. Gastroenterol.* **2018**, *53*, 835–842. [CrossRef] [PubMed]
12. Ye, L.; Lu, H.; Wu, L.; Zhang, L.; Shi, H.; Wu, H.M.; Tu, P.; Li, M.; Wang, F.Y. The clinicopathologic features and prognosis of esophageal neuroendocrine carcinomas: A single-center study of 53 resection cases. *BMC Cancer* **2019**, *19*, 1234. [CrossRef] [PubMed]
13. Schizas, D.; Mastoraki, A.; Kirkilesis, G.I.; Sioulas, A.D.; Papanikolaou, I.S.; Misiakos, E.P.; Arkadopoulos, N.; Liakakos, T. Neuroendocrine Tumors of the Esophagus: State of the Art in Diagnostic and Therapeutic Management. *J. Gastrointest. Cancer* **2017**, *48*, 299–304. [CrossRef] [PubMed]
14. Yazici, C.; Boulay, B.R. Evolving role of the endoscopist in management of gastrointestinal neuroendocrine tumors. *World J. Gastroenterol.* **2017**, *23*, 4847–4855. [CrossRef] [PubMed]
15. Rindi, G.; Klimstra, D.S.; Abedi-Ardekani, B.; Asa, S.L.; Bosman, F.T.; Brambilla, E.; Busam, K.J.; de Krijger, R.R.; Dietel, M.; El-Naggar, A.K.; et al. A common classification framework for neuroendocrine neoplasms: An International Agency for Research on Cancer (IARC) and World Health Organization (WHO) expert consensus proposal. *Mod. Pathol.* **2018**, *31*, 1770–1786. [CrossRef] [PubMed]
16. Delle Fave, G.; O'Toole, D.; Sundin, A.; Taal, B.; Ferolla, P.; Ramage, J.K.; Ferone, D.; Ito, T.; Weber, W.; Zheng-Pei, Z.; et al. ENETS Consensus Guidelines Update for Gastroduodenal Neuroendocrine Neoplasms. *Neuroendocrinology* **2016**, *103*, 119–124. [CrossRef]
17. Sato, Y. Endoscopic diagnosis and management of type I neuroendocrine tumors. *World J. Gastrointest. Endosc.* **2015**, *7*, 346. [CrossRef]
18. Ahmed, M. Gastrointestinal neuroendocrine tumors in 2020. *World J. Gastrointest. Oncol.* **2020**, *12*, 791–807. [CrossRef]

19. Attili, F.; Capurso, G.; Vanella, G.; Fuccio, L.; Delle Fave, G.; Costamagna, G.; Larghi, A. Diagnostic and therapeutic role of endoscopy in gastroenteropancreatic neuroendocrine neoplasms. *Dig. Liver Dis.* **2014**, *46*, 9–17. [CrossRef]
20. Shah, S.C.; Piazuelo, M.B.; Kuipers, E.J.; Li, D. AGA Clinical Practice Update on the Diagnosis and Management of Atrophic Gastritis: Expert Review. *Gastroenterology* **2021**, *161*, 1325–1332.e7. [CrossRef]
21. Panzuto, F.; Magi, L.; Esposito, G.; Rinzivillo, M.; Annibale, B. Comparison of Endoscopic Techniques in the Management of Type I Gastric Neuroendocrine Neoplasia: A Systematic Review. *Gastroenterol. Res. Pract.* **2021**, *2021*, 6679397. [CrossRef] [PubMed]
22. Kim, H.H.; Kim, G.H.; Kim, J.H.; Choi, M.G.; Song, G.A.; Kim, S.E. The efficacy of endoscopic submucosal dissection of type I gastric carcinoid tumors compared with conventional endoscopic mucosal resection. *Gastroenterol. Res. Pract.* **2014**, *2014*, 253860. [CrossRef]
23. Noh, J.H.; Kim, D.H.; Yoon, H.; Hsing, L.C.; Na, H.K.; Ahn, J.Y.; Lee, J.H.; Jung, K.W.; Choi, K.D.; Song, H.J.; et al. Clinical Outcomes of Endoscopic Treatment for Type 1 Gastric Neuroendocrine Tumor. *J. Gastrointest. Surg.* **2021**, *25*, 2495–2502. [CrossRef] [PubMed]
24. Ravizza, D.; Fiori, G.; Trovato, C.; Fazio, N.; Bonomo, G.; Luca, F.; Bodei, L.; Pelosi, G.; Tamayo, D.; Crosta, C. Long-term endoscopic and clinical follow-up of untreated type 1 gastric neuroendocrine tumours. *Dig. Liver Dis.* **2007**, *39*, 537–543. [CrossRef] [PubMed]
25. Esposito, G.; Cazzato, M.; Rinzivillo, M.; Pilozzi, E.; Lahner, E.; Annibale, B.; Panzuto, F. Management of type-I gastric neuroendocrine neoplasms: A 10-years prospective single centre study. *Dig. Liver. Dis.* **2021**, *54*, 890–895. [CrossRef] [PubMed]
26. Yang, Z.; Wang, W.; Lu, J.; Pan, G.; Pan, Z.; Chen, Q.; Liu, W.; Zhao, Y. Gastric Neuroendocrine Tumors (G-Nets): Incidence, Prognosis and Recent Trend Toward Improved Survival. *Cell. Physiol. Biochem.* **2018**, *45*, 389–396. [CrossRef]
27. Panzuto, F.; Campana, D.; Massironi, S.; Faggiano, A.; Rinzivillo, M.; Lamberti, G.; Sciola, V.; Lahner, E.; Manuzzi, L.; Colao, A.; et al. Tumour type and size are prognostic factors in gastric neuroendocrine neoplasia: A multicentre retrospective study. *Dig. Liver Dis.* **2019**, *51*, 1456–1460. [CrossRef]
28. Niederle, B.; Selberherr, A.; Bartsch, D.K.; Brandi, M.L.; Doherty, G.M.; Falconi, M.; Goudet, P.; Halfdanarson, T.R.; Ito, T.; Jensen, R.T.; et al. Multiple Endocrine Neoplasia Type 1 and the Pancreas: Diagnosis and Treatment of Functioning and Non-Functioning Pancreatic and Duodenal Neuroendocrine Neoplasia within the MEN1 Syndrome—An International Consensus Statement. *Neuroendocrinology* **2021**, *111*, 609–630. [CrossRef]
29. Min, B.H.; Hong, M.; Lee, J.H.; Rhee, P.L.; Sohn, T.S.; Kim, S.; Kim, K.M.; Kim, J.J. Clinicopathological features and outcome of type 3 gastric neuroendocrine tumours. *Br. J. Surg.* **2018**, *105*, 1480–1486. [CrossRef]
30. Lee, S.H.; Moon, D.; Lee, H.S.; Lee, C.K.; Jeon, Y.D.; Park, J.H.; Kim, H.; Lee, S.K. Multicentric Type 3 Gastric Neuroendocrine Tumors. *Clin. Endosc.* **2015**, *48*, 431–435. [CrossRef]
31. Garcia-Carbonero, R.; Sorbye, H.; Baudin, E.; Raymond, E.; Wiedenmann, B.; Niederle, B.; Sedlackova, E.; Toumpanakis, C.; Anlauf, M.; Cwikla, J.B.; et al. ENETS Consensus Guidelines for High-Grade Gastroenteropancreatic Neuroendocrine Tumors and Neuroendocrine Carcinomas. *Neuroendocrinology* **2016**, *103*, 186–194. [CrossRef]
32. Hirasawa, T.; Yamamoto, N.; Sano, T. Is endoscopic resection appropriate for type 3 gastric neuroendocrine tumors? Retrospective multicenter study. *Dig. Endosc.* **2021**, *33*, 408–417. [CrossRef] [PubMed]
33. Kwon, Y.H.; Jeon, S.W.; Kim, G.H.; Kim, J.I.; Chung, I.K.; Jee, S.R.; Kim, H.U.; Seo, G.S.; Baik, G.H.; Choi, K.D.; et al. Long-term follow up of endoscopic resection for type 3 gastric NET. *World J. Gastroenterol.* **2013**, *19*, 8703–8708. [CrossRef] [PubMed]
34. Exarchou, K.; Kamieniarz, L.; Tsoli, M.; Victor, A.; Oleinikov, K.; Khan, M.S.; Srirajaskanthan, R.; Mandair, D.; Grozinsky-Glasberg, S.; Kaltsas, G.; et al. Is local excision sufficient in selected grade 1 or 2 type III gastric neuroendocrine neoplasms? *Endocrine* **2021**, *74*, 421–429. [CrossRef] [PubMed]
35. Exarchou, K.; Howes, N.; Pritchard, D.M. Systematic review: Management of localised low-grade upper gastrointestinal neuroendocrine tumours. *Aliment. Pharmacol. Ther.* **2020**, *51*, 1247–1267. [CrossRef] [PubMed]
36. Nishio, M.; Hirasawa, K.; Ozeki, Y.; Sawada, A.; Ikeda, R.; Fukuchi, T.; Kobayashi, R.; Makazu, M.; Sato, C.; Maeda, S. Short- and long-term outcomes of endoscopic submucosal dissection for non-ampullary duodenal neuroendocrine tumors. *Ann. Gastroenterol.* **2020**, *33*, 265–271. [CrossRef] [PubMed]
37. Park, S.B.; Kang, D.H.; Choi, C.W.; Kim, H.W.; Kim, S.J. Clinical outcomes of ligation-assisted endoscopic resection for duodenal neuroendocrine tumors. *Medicine* **2018**, *97*, e0533. [CrossRef]
38. Fujimoto, A.; Sasaki, M.; Goto, O.; Maehata, T.; Ochiai, Y.; Kato, M.; Nakayama, A.; Akimoto, T.; Kuramoto, J.; Hayashi, Y.; et al. Treatment Results of Endoscopic Mucosal Resection with a Ligation Device for Duodenal Neuroendocrine Tumors. *Intern. Med.* **2019**, *58*, 773–777. [CrossRef]
39. Lee, S.W.; Sung, J.K.; Cho, Y.S.; Bang, K.B.; Kang, S.H.; Kim, K.B.; Kim, S.H.; Moon, H.S.; Song, K.H.; Kim, S.M.; et al. Comparisons of therapeutic outcomes in patients with nonampullary duodenal neuroendocrine tumors (NADNETs): A multicenter retrospective study. *Medicine* **2019**, *98*, e16154. [CrossRef]
40. Hawa, F.; Sako, Z.; Nguyen, T.; Catanzaro, A.T.; Zolotarevsky, E.; Bartley, A.N.; Gunaratnam, N.T. The band and slough technique is effective for management of diminutive type 1 gastric and duodenal neuroendocrine tumors. *Endosc. Int. Open* **2020**, *8*, E717–E721. [CrossRef]
41. Dwyer, S.; Mok, S. Endoscopic full-thickness resection of well-differentiated T2 neuroendocrine tumors in the duodenal bulb: A case series. *VideoGIE* **2022**, *7*, 196–199. [CrossRef] [PubMed]

42. Gupta, S.; Kumar, P.; Chacchi, R.; Murino, A.; Despott, E.J.; Lemmers, A.; Pioche, M.; Bourke, M. Duodenal neuroendocrine tumors: Short-term outcomes of endoscopic submucosal dissection performed in the Western setting. *Endosc. Int. Open* **2023**, *11*, E1099–E1107; Erratum in *Endosc. Int. Open* **2023**, *30*, C7. [PubMed]
43. Guo, C.G.; Ng, H.I.; Liu, Y.; Sun, C.Y.; Zhang, X.J.; Zhao, D.B.; Wang, G.Q. Laparoscopic endoscopic cooperative surgery for the duodenal neuroendocrine tumor: A single-center case series (How I Do It). *Int. J. Surg.* **2023**, *109*, 1835–1841. [CrossRef] [PubMed]
44. Wu, J.; Man, D.; Wu, J.; Shen, Z.; Zhu, X. Prognosis of patients with neuroendocrine tumor: A SEER database analysis. *Cancer Manag. Res.* **2018**, *10*, 5629–5638.
45. Bertani, E.; Ravizza, D.; Milione, M.; Massironi, S.; Grana, C.M.; Zerini, D.; Piccioli, A.N.; Spinoglio, G.; Fazio, N. Neuroendocrine neoplasms of rectum: A management update. *Cancer Treat. Rev.* **2018**, *66*, 45–55. [CrossRef]
46. Basuroy, R.; Haji, A.; Ramage, J.K.; Quaglia, A.; Srirajaskanthan, R. Review article: The investigation and management of rectal neuroendocrine tumours. *Aliment. Pharmacol. Ther.* **2016**, *44*, 332–345. [CrossRef]
47. Kobori, I.; Katayama, Y.; Kitagawa, T.; Fujimoto, Y.; Oura, R.; Toyoda, K.; Kusano, Y.; Ban, S.; Tamano, M. Pocket Creation Method of Endoscopic Submucosal Dissection to Ensure Curative Resection of Rectal Neuroendocrine Tumors. *GE Port. J. Gastroenterol.* **2019**, *26*, 207–211. [CrossRef]
48. Fine, C.; Roquin, G.; Terrebonne, E.; Lecomte, T.; Coriat, R.; Do Cao, C.; de Mestier, L.; Coffin, E.; Cadiot, G.; Nicolli, P.; et al. Endoscopic management of 345 small rectal neuroendocrine tumours: A national study from the French group of endocrine tumours (GTE). *United Eur. Gastroenterol. J.* **2019**, *7*, 1102–1112. [CrossRef]
49. Huang, J.; Lu, Z.S.; Yang, Y.S.; Yuan, J.; Wang, X.D.; Meng, J.Y.; Du, H.; Wang, H.B. Endoscopic mucosal resection with circumferential incision for treatment of rectal carcinoid tumours. *World J. Surg. Oncol.* **2014**, *12*, 23. [CrossRef]
50. McDermott, F.D.; Heeney, A.; Courtney, D.; Mohan, H.; Winter, D. Rectal carcinoids: A systematic review. *Surg. Endosc.* **2014**, *28*, 2020–2026. [CrossRef]
51. Zhou, X.; Xie, H.; Xie, L.; Li, J.; Cao, W.; Fu, W. Endoscopic resection therapies for rectal neuroendocrine tumors: A systematic review and meta-analysis. *J. Gastroenterol. Hepatol.* **2014**, *29*, 259–268. [CrossRef]
52. Zhang, H.P.; Wu, W.; Yang, S.; Lin, J. Endoscopic treatments for rectal neuroendocrine tumors smaller than 16 mm: A meta-analysis. *Scand. J. Gastroenterol.* **2016**, *51*, 1345–1353. [CrossRef]
53. Kuiper, T.; van Oijen, M.G.H.; van Velthuysen, M.F.; van Lelyveld, N.; van Leerdam, M.E.; Vleggaar, F.D.; Klümpen, H.J. Endoscopically removed rectal NETs: A nationwide cohort study. *Int. J. Colorectal. Dis.* **2021**, *36*, 535–541. [CrossRef] [PubMed]
54. Erstad, D.J.; Dasari, A.; Taggart, M.W.; Kaur, H.; Konishi, T.; Bednarski, B.K.; Chang, G.J. Prognosis for Poorly Differentiated, High-Grade Rectal Neuroendocrine Carcinomas. *Ann. Surg. Oncol.* **2022**, *29*, 2539–2548. [CrossRef]
55. Park, S.S.; Han, K.S.; Kim, B.; Chang Kim, B.; Hong, C.W.; Sohn, D.K.; Chang, H.J. Comparison of underwater endoscopic mucosal resection and endoscopic submucosal dissection of rectal neuroendocrine tumors (with videos). *Gastrointest. Endosc.* **2020**, *91*, 1164–1171.e2. [CrossRef] [PubMed]
56. Bang, B.W.; Park, J.S.; Kim, H.K.; Shin, Y.W.; Kwon, K.S.; Kim, J.M. Endoscopic Resection for Small Rectal Neuroendocrine Tumors: Comparison of Endoscopic Submucosal Resection with Band Ligation and Endoscopic Submucosal Dissection. *Gastroenterol. Res. Pract.* **2016**, *2016*, 6198927. [CrossRef] [PubMed]
57. Kamigaichi, Y.; Yamashita, K.; Oka, S.; Tamari, H.; Shimohara, Y.; Nishimura, T.; Inagaki, K.; Okamoto, Y.; Tanaka, H.; Yuge, R.; et al. Clinical outcomes of endoscopic resection for rectal neuroendocrine tumors: Advantages of endoscopic submucosal resection with a ligation device compared to conventional EMR and ESD. *DEN Open* **2022**, *2*, e35. [CrossRef]
58. Park, S.S.; Kim, B.C.; Lee, D.E.; Han, K.S.; Kim, B.; Hong, C.W.; Sohn, D.K. Comparison of endoscopic submucosal dissection and transanal endoscopic microsurgery for T1 rectal neuroendocrine tumors: A propensity score-matched study. *Gastrointest. Endosc.* **2021**, *94*, 408–415.e2. [CrossRef]
59. Meier, B.; Albrecht, H.; Wiedbrauck, T.; Schmidt, A.; Caca, K. Full-thickness resection of neuroendocrine tumors in the rectum. *Endoscopy* **2020**, *52*, 68–72. [CrossRef]
60. Folkert, I.W.; Sinnamon, A.J.; Concors, S.J.; Bennett, B.J.; Fraker, D.L.; Mahmoud, N.N.; Metz, D.C.; Stashek, K.M.; Roses, R.E. Grade is a Dominant Risk Factor for Metastasis in Patients with Rectal Neuroendocrine Tumors. *Ann. Surg. Oncol.* **2020**, *27*, 855–863. [CrossRef]
61. Gamboa, A.C.; Liu, Y.; Lee, R.M.; Zaidi, M.Y.; Staley, C.A.; Russell, M.C.; Cardona, K.; Sullivan, P.S.; Maithel, S.K. A novel preoperative risk score to predict lymph node positivity for rectal neuroendocrine tumors: An NCDB analysis to guide operative technique. *J. Surg. Oncol.* **2019**, *120*, 932–939. [CrossRef] [PubMed]
62. Fitzgerald, T.L.; Hickner, Z.J.; Schmitz, M.; Kort, E.J. Changing incidence of pancreatic neoplasms: A 16-year review of statewide tumor registry. *Pancreas* **2008**, *37*, 134–138. [CrossRef] [PubMed]
63. Halfdanarson, T.R.; Rubin, J.; Farnell, M.B.; Grant, C.S.; Petersen, G.M. Pancreatic endocrine neoplasms: Epidemiology and prognosis of pancreatic endocrine tumors. *Endocr. Relat. Cancer* **2008**, *15*, 409–427. [CrossRef] [PubMed]
64. Falconi, M.; Eriksson, B.; Kaltsas, G.; Bartsch, D.K.; Capdevila, J.; Caplin, M.; Kos-Kudla, B.; Kwekkeboom, D.; Rindi, G.; Klöppel, G.; et al. ENETS Consensus Guidelines Update for the Management of Patients with Functional Pancreatic Neuroendocrine Tumors and Non-Functional Pancreatic Neuroendocrine Tumors. *Neuroendocrinology* **2016**, *103*, 153–171. [CrossRef] [PubMed]
65. Crippa, S.; Partelli, S.; Zamboni, G.; Scarpa, A.; Tamburrino, D.; Bassi, C.; Pederzoli, P.; Falconi, M. Incidental diagnosis as prognostic factor in different tumor stages of nonfunctioning pancreatic endocrine tumors. *Surgery* **2014**, *155*, 145–153. [CrossRef]

66. Partelli, S.; Tamburrino, D.; Cherif, R.; Muffatti, F.; Moggia, E.; Gaujoux, S.; Sauvanet, A.; Falconi, M.; Fusai, G. Risk and Predictors of Postoperative Morbidity and Mortality After Pancreaticoduodenectomy for Pancreatic Neuroendocrine Neoplasms: A Comparative Study With Pancreatic Ductal Adenocarcinoma. *Pancreas* **2019**, *48*, 504–509. [CrossRef]
67. Partelli, S.; Massironi, S.; Zerbi, A.; Niccoli, P.; Kwon, W.; Landoni, L.; Panzuto, F.; Tomazic, A.; Bongiovanni, A.; Kaltsas, G.; et al. Management of asymptomatic sporadic non-functioning pancreatic neuroendocrine neoplasms no larger than 2 cm: Interim analysis of prospective ASPEN trial. *Br. J. Surg.* **2022**, *109*, 1186–1190. [CrossRef]
68. Imperatore, N.; de Nucci, G.; Mandelli, E.D.; de Leone, A.; Zito, F.P.; Lombardi, G.; Manes, G. Endoscopic ultrasound-guided radiofrequency ablation of pancreatic neuroendocrine tumors: A systematic review of the literature. *Endosc. Int Open* **2020**, *8*, E1759–E1764. [CrossRef]
69. Signoretti, M.; Valente, R.; Repici, A.; Delle Fave, G.; Capurso, G.; Carrara, S. Endoscopy-guided ablation of pancreatic lesions: Technical possibilities and clinical outlook. *World J. Gastrointest. Endosc.* **2017**, *9*, 41. [CrossRef]
70. Paiella, S.; Salvia, R.; Ramera, M.; Girelli, R.; Frigerio, I.; Giardino, A.; Allegrini, V.; Bassi, C. Local Ablative Strategies for Ductal Pancreatic Cancer (Radiofrequency Ablation, Irreversible Electroporation): A Review. *Gastroenterol. Res. Pract.* **2016**, *2016*, 4508376. [CrossRef]
71. Wright, A.S.; Sampson, L.A.; Warner, T.F.; Mahvi, D.M.; Lee, F.T. Radiofrequency versus microwave ablation in a hepatic porcine model. *Radiology* **2005**, *236*, 132–139. [CrossRef] [PubMed]
72. Haen, S.P.; Pereira, P.L.; Salih, H.R.; Rammensee, H.G.; Gouttefangeas, C. More than just tumor destruction: Immunomodulation by thermal ablation of cancer. *Clin. Dev. Immunol.* **2011**, *2011*, 160250. [CrossRef] [PubMed]
73. Rimbaş, M.; Horumbă, M.; Rizzatti, G.; Crinò, S.F.; Gasbarrini, A.; Costamagna, G.; Larghi, A. Interventional endoscopic ultrasound for pancreatic neuroendocrine neoplasms. *Dig. Endosc.* **2020**, *32*, 1031–1041. [CrossRef] [PubMed]
74. Crinò, S.F.; Napoleon, B.; Facciorusso, A.; Lakhtakia, S.; Borbath, I.; Caillol, F.; Do-Cong Pham, K.; Rizzatti, G.; Forti, E.; Palazzo, L.; et al. Endoscopic Ultrasound-guided Radiofrequency Ablation Versus Surgical Resection for Treatment of Pancreatic Insulinoma. *Clin. Gastroenterol. Hepatol.* **2023**, *21*, 2834–2843.e2. [CrossRef]
75. Matsumoto, K.; Kato, H.; Kawano, S.; Fujiwara, H.; Nishida, K.; Harada, R.; Fujii, M.; Yoshida, R.; Umeda, Y.; Hinotsu, S.; et al. Efficacy and safety of scheduled early endoscopic ultrasonography-guided ethanol reinjection for patients with pancreatic neuroendocrine tumors: Prospective pilot study. *Dig. Endosc.* **2020**, *32*, 425–430. [CrossRef]
76. Park, D.H.; Choi, J.H.; Oh, D.; Lee, S.S.; Seo, D.W.; Lee, S.K.; Kim, M.H. Endoscopic ultrasonography-guided ethanol ablation for small pancreatic neuroendocrine tumors: Results of a pilot study. *Clin. Endosc.* **2015**, *48*, 158–164. [CrossRef]
77. Lennon, A.M.; Newman, N.; Makary, M.A.; Edil, B.H.; Shin, E.J.; Khashab, M.A.; Hruban, R.H.; Wolfgang, C.L.; Schulick, R.D.; Giday, S.; et al. EUS-guided tattooing before laparoscopic distal pancreatic resection (with video). *Gastrointest. Endosc.* **2010**, *72*, 1089–1094. [CrossRef]
78. Facciorusso, A.; Bajwa, H.S.; Menon, K.; Buccino, V.R.; Muscatiello, N. Comparison between 22G aspiration and 22G biopsy needles for EUS-guided sampling of pancreatic lesions: A meta-analysis. *Endosc. Ultrasound.* **2020**, *9*, 167–174. [CrossRef]
79. Gkolfakis, P.; Crinò, S.F.; Tziatzios, G.; Ramai, D.; Papaefthymiou, A.; Papanikolaou, I.S.; Triantafyllou, K.; Arvanitakis, M.; Lisotti, A.; Fusaroli, P.; et al. Comparative diagnostic performance of end-cutting fine-needle biopsy needles for EUS tissue sampling of solid pancreatic masses: A network meta-analysis. *Gastrointest. Endosc.* **2022**, *95*, 1067–1077.e15. [CrossRef]

Disclaimer/Publisher's Note: The statements, opinions and data contained in all publications are solely those of the individual author(s) and contributor(s) and not of MDPI and/or the editor(s). MDPI and/or the editor(s) disclaim responsibility for any injury to people or property resulting from any ideas, methods, instructions or products referred to in the content.

Review

Malignant Acute Colonic Obstruction: Multidisciplinary Approach for Endoscopic Management

Aurelio Mauro [1], Davide Scalvini [1,2,*], Sabrina Borgetto [3], Paola Fugazzola [4], Stefano Mazza [1], Ilaria Perretti [5], Anna Gallotti [5], Anna Pagani [3], Luca Ansaloni [4] and Andrea Anderloni [1]

1. Gastroenterology and Endoscopy Unit, Fondazione IRCCS Policlinico San Matteo, Viale Camillo Golgi 19, 27100 Pavia, Italy; a.mauro@smatteo.pv.it (A.M.); a.anderloni@smatteo.pv.it (A.A.)
2. Department of Internal Medicine, PhD in Experimental Medicine Italy, University of Pavia, 27100 Pavia, Italy
3. Medical Oncology Unit, Fondazione IRCCS Policlinico San Matteo, 27100 Pavia, Italy; sabrina.borgetto01@universitadipavia.it (S.B.); a.pagani@smatteo.pv.it (A.P.)
4. Department of General Surgery, Fondazione IRCCS Policlinico San Matteo, 27100 Pavia, Italy; p.fugazzola@smatteo.pv.it (P.F.); l.ansaloni@smatteo.pv.it (L.A.)
5. Institute of Radiology, Fondazione IRCCS Policlinico San Matteo, 27100 Pavia, Italy; ilaria.perretti01@universitadipavia.it (I.P.); a.gallotti@smatteo.pv.it (A.G.)
* Correspondence: davide.scalvini01@universitadipavia.it; Tel.: +39-03-8250-2064

Simple Summary: Acute colonic obstruction is one of the most common manifestations of locally advanced colorectal cancer. Endoscopic stenting has become by far the minimally invasive treatment of choice for malignant colonic obstruction especially in the palliative setting. However, there are still controversies in the literature about the usefulness and safety of endoscopic stenting as a bridge-to-surgery approach or in patients on antiangiogenic therapy. Moreover, endoscopic colonic stenting is an operative procedure that requires adequate pre-interventional management and specific endoscopic knowledge. The present review aimed to summarize the optimization of endoscopic management of patients with malignant acute colonic obstruction based on a multimodal connection with the various medical specialties involved in managing this urgent clinical scenario.

Abstract: Patients presenting with acute colonic obstruction are usually evaluated in the emergency department and multiple specialties are involved in the patients' management. Pre-treatment evaluation is essential in order to establish the correct endoscopic indication for stent implantation. Contrast-enhanced imaging could allow the exclusion of benign causes of colonic obstruction and evaluation of the length of malignant stricture. Endoscopic stenting is the gold standard of treatment for palliative indications whereas there are still concerns about its use as a bridge to surgery. Different meta-analyses showed that stenting as a bridge to surgery improves short-term surgical outcomes but has no role in improving long-term outcomes. Multidisciplinary evaluation is also essential in patients that may be started on or are currently receiving antiangiogenic agents because endoscopic stenting may increase the risk of perforation. Evidence in the literature is weak and based on retrospective data. Here we report on how to correctly evaluate a patient with acute colonic malignant obstruction in collaboration with other essential specialists including a radiologist, surgeon and oncologist, and how to optimize the technique of endoscopic stenting.

Keywords: acute colonic obstruction; colorectal cancer; endoscopic stent; bridge to surgery; antiangiogenic agents; self-expandable metal stent; CT scan

1. Introduction

Colorectal cancer (CRC) is the third most frequently diagnosed malignancy in the world and the second leading cause of cancer-related mortality [1]. About 10–40% of colorectal cancer patients have bowel obstruction at the time of diagnosis, particularly on the left side [2], and large bowel obstruction (LBO) is a common condition that accounts for about 24% of admissions for acute mechanical bowel obstruction [3]. Acute colonic obstruction requires urgent management in order to avoid further complications such as perforation or ischemia. In the past, urgent surgery with colonic resection and/or stoma formation was the only available treatment. Endoscopic decompression with the application of self-expandable metal stents (SEMS) within the stricture has been proposed in the last decades as a less invasive option and has become the treatment of choice for patients who need palliative treatment [4].

The surgeon is usually the first specialist to evaluate patients with acute colonic obstruction in the emergency department; the consultation with the endoscopist aims to share the indication of endoscopic stenting, especially for resectable patients for whom there are still no clear indications for the bridge-to-surgery stenting [5]. When evaluating the patient for colonic stenting, the on-call endoscopist usually performs a virtual urgent multidisciplinary consultation with the main figures involved in the management of the patient. Radiological consultation is then essential before endoscopic stenting in order to confirm the malignant etiology of the acute obstruction, to better define the colonic anatomy and to identify urgent criteria for the timing of endoscopic decompression. Lastly, acute colonic obstruction may develop in patients with a known colon cancer under active chemotherapy treatment. Antiangiogenic therapy (e.g., bevacizumab) has been associated with an increased risk of perforation in patients treated with colonic stenting [6]. However, evidence is lacking and based on retrospective studies. Thus, multidisciplinary evaluation with an oncologist and surgeon is crucial in order to identify the correct management strategy and to discuss the risks and benefits of the endoscopic procedure.

Endoscopic colonic stenting is an interventional procedure that requires specific endoscopic skills and experience in the interpretation of intraprocedural radiological frames [7]. The placement of a colonic stent is usually easy in the case of short and linear strictures, whereas it could become particularly challenging in the case of difficult anatomic locations, including lesions close to the anal verge, in the right colon or colonic flexures.

The present review intends to assess the appropriate clinical indications for colonic stenting in a multidisciplinary context and to provide practical tips and tricks for the endoscopic procedure.

2. Surgeon–Endoscopist Collaboration for Indication of Endoscopic Stenting

Acute LBO remains a surgical urgency and initial evaluation should be performed by the general surgeon. Preliminary surgical evaluation is aimed at confirming the diagnosis of malignant etiology, excluding other benign causes of LBO that usually do not require endoscopic management except for specific cases (e.g., endoscopic decompression for sigmoid volvulus). Surgical consultation is also essential for the interpretation of CT scan images in order to exclude the presence of abdominal complications that require urgent surgical therapy such as bowel perforation and/or ischemia.

Once a diagnosis of malignant LBO is confirmed, it is essential for the surgeon and endoscopist to collaboratively determine the indication for endoscopic stenting. Endoscopic treatment is sometimes contraindicated when there are signs of colonic ischemia or perforation and therefore emergency surgery (ES) is the only possible treatment [8]. ES contemplates the emergency resection of the primary lesion with an immediate colorectal anastomosis (possibly associated with a diverting loop ileostomy) or without a prompt recanalization and the creation of a colostomy ("Hartmann's procedure"). When contraindications to stenting are excluded, it is important to evaluate the clinical context of the patient. Patients unsuitable for surgery due to advanced disease (e.g., metastatic CRC) or for the presence of multiple comorbidities should be referred for endoscopic palliative stenting. When the patient has a resectable CRC, it is possible to perform a two-step approach consisting of the endoscopic placement of a SEMS to resolve the obstruction and in the elective surgical resection a few weeks after (stent as a bridge to surgery).

Acute colonic obstruction may also be caused by a non-primary colonic tumor such as pelvic tumors, advanced gastric or other metastatic cancers that cause extrinsic compression.

Usually, patients with extrinsic obstruction have milder symptoms than patients with strictures caused by primary colonic cancer [5]. In this subgroup of patients, the aim of the treatment is usually palliative and endoscopic stenting, which demonstrates feasibility but with lower rate of clinical success [9]. Surgical treatment should be evaluated according to the performance status of the patients and to the resectability of the primary neoplasia.

2.1. Palliative Colonic Stenting

In patients with LBO due to colorectal cancer and advanced/metastatic disease that are not eligible for curative treatment, surgical or non-surgical palliation should be considered. Resection, bypass, and colostomy are the available surgical options, but the European Society of Gastrointestinal Endoscopy (ESGE) Guideline [10] strongly recommends colonic stenting as the preferred treatment for the palliation of malignant colonic obstruction. Some studies comparing colonic stenting and ES for palliation [11–18] showed a that technical success of stent placement ranged from 88% to 100%, while the initial clinical relief of obstruction was significantly higher after palliative surgery compared to colonic stenting [12]. Conflicting results have been reported regarding short-term mortality and overall morbidity [11–15], but colonic stenting was associated with a shorter length of stay, lower costs, a lower intensive care unit admission rate and a shorter time to the initiation of chemotherapy [11–15,17,18]. Furthermore, patients treated with endoscopic stenting had better quality of life if compared to patients treated with palliative surgery until 12 months after the procedure [18].

However, some observational studies [9,19,20] showed lower technical success and an increased complication rate for colonic stenting in patients with peritoneal metastases, because the main limitations to the success of bowel stenting are the presence of multiple sites of obstruction. In this situation, a surgical approach could be considered.

2.2. SEMS Role as Bridge to Surgery

Placement of an SEMS before elective surgery has the rationale to allow resolution of the obstruction and consequently to obtain patients' stabilization, improvement of general conditions and nutritional status, accurate staging and definition of a tailored treatment for the patients [21]. As a result, higher quality oncologic resections could be performed using minimally invasive approaches and without the need for permanent stoma [22]. However, some controversies have emerged regarding the oncological safety of SEMS. It has been speculated that the increased interstitial pressure in the neoplastic mass can cause cell dissemination, cell shedding and tumor embolization into lymphatic vessels, as a consequence of a higher rate of recurrence observed in patients with SEMS [23,24]. For these reasons, choosing the most appropriate decompression method can be challenging given the need to balance short- and long-term outcomes. Moreover, guidelines on this topic are inconsistent about the optimal treatment to choose [10,25,26].

Several randomized trials investigated this issue and were summarized in several meta-analyses [21,27–35], which compared short- and long-term outcomes of ES and stenting as a bridge to surgery in malignant LBO (Table 1). Considering short-term outcomes, multiple studies demonstrated that post-operative morbidity, such as the rate of anastomotic leak and wound infection, was significantly lower in patients who underwent stenting as a bridge to surgery [27,29,31–33,35]; among these, only one study also showed a significantly lower post-operative mortality [33] in this group of patients. Other short-term surgical outcomes that have a significant impact on patients' quality of life, such as the rate of temporary [27,29,32,33] or permanent [27,29,35] stoma, were significantly lower in the group of stenting as a bridge to surgery with an odds ratio (OR) of 0.39 [33]. Finally, two meta-analyses [32,33] compared the rate of laparoscopic versus open resection in the two groups, finding a significantly higher rate of laparoscopic resection in the endoscopic stenting group. On the other hand, the implantation of an endoscopic stent may increase the length of stay in order to wait for the time for recanalization and optimal timing for surgery [29,32]; however, this hospitalization time is usually exploited for the correct clinical staging, for patients' stabilization and restarting of enteral nutrition.

Table 1. Short- and long-term outcomes of Emergency Surgery and Stent as Bridge to Surgery (SBTS) in malignant bowel obstruction reported in the most recent meta-analyses. Green color: outcomes in favor of SBTS; red color: outcomes in favor of ES; yellow color: outcomes in which ES and SBTS are not significantly different.

Meta-Analyses	Included Studies	Short-Term Outcomes							Long-Term Outcomes			
		Morbidity	Mortality	Stoma Rate	Permanent Stoma	VLS	LOS		OS	DFS	Local Recurrence
Allievi 2017, [27]	7 RCTs (448 pts)	ES 54.8% vs. SBTS 37.8% ($p = 0.02$)	ns	ES 46.0% vs. SBTS 28.9% ($p < 0.0001$)	ES 35.2% vs. SBTS 24.5% ($p < 0.02$)	/	/		/	/	/
Amelung 2018, [28]	5 RCTs, 4 prospective, 12 retrospective (1919 pts)	/	/	/	/	/	/		3 years and 5 years: ns	3 years and 5 years: ns	ns
Arezzo 2017, [29]	8 RCTs (497 pts)	ES 51.2% vs. SBTS 33.9% ($p = 0.023$)	ns	ES 51.4% vs. SBTS 33.9% ($p < 0.001$)	ES 35.2% vs. SBTS 22.2% ($p = 0.003$)	/	ES 14.5 d vs. SBTS 15.5 ($p = 0.039$)		/	/	ES 26.6% vs. SBTS 40.5% ($p = 0.09$)
Cao 2019, [30]	5 RCTs, 3 prospective, 16 retrospective (2508 pts)	/	/	/	/	/	/		3 years and 5 years: ns	3 years and 5 years: ns	ns
Ceresoli 2017, [21]	5 RCTs, 3 prospective, 9 retrospective	/	/	/	/	/	/		ns	Ns	ns
Foo 2019, [31]	7 RCTs (448 pts)	Reduced in SBTS (RR 0.6, $p = 0.032$)	ns	/	/	/	/		3 years: ns	3 years: ns	ns
Jain 2020, [32]	8 RCTs, 25 observational (15,224 pts)	SBTS vs. ES: anastomotic leak OR 0.59 ($p = 0.006$), wound infection OR 0.64 ($p = 0.004$)	SBTS vs. ES: OR 0.69 ($p = 0.010$)	SBTS vs. ES: OR 0.39 ($p < 0.001$)	ns	SBTS vs. ES: OR 5.9 ($p < 0.001$)	SBTS vs. ES: Reduction in ES (<0.001)		ns	Ns	ns
Kanaka 2022, [33]	Right-sided MLBO:#break#7 observational studies (5136 pts)	SBTS vs. ES: OR 0.78 ($p = 0.003$)	SBTS vs. ES: OR 0.51 ($p = 0.03$)	SBTS 2.0% vs. ES 11.0% ($p < 0.01$)		SBTS 48.5% vs. ES 15.7% ($p < 0.01$)	/		/	/	/
Matsuda 2015, [34]	2 RCTs, 9 observational studies (1136 pts)	/	/	/	/	/	/		3 y and 5 y: ns	3 y and 5 y: ns	/
McKechnie 2023, [35]	Network meta-analysis: 53 studies	ES vs. SBTS: OR 2.14 ($p < 0.0019$)	ns	/	ES vs. SBTS: OR 2.91 ($p < 0.001$)	/	/		3 y and 5 y: ns	/	/

ES, emergency surgery; SBTS, stent as a bridge to surgery; VLS, videolaparoscopy; LOS, length of stay; OS, overall survival; DFS, disease-free survival; RCT, randomized controlled trials; pts, patients; OR, odd ratio.

While short-term outcomes are globally in favor of the bridge-to-surgery approach, less evidence supports this kind of approach for long-term outcomes. Different meta-analyses failed to find any differences between ES and stenting as a bridge to surgery in terms of overall survival [21,28,30–32,34,35] or disease-free survival [21,28,30–32,34] at three or five years. Moreover, the meta-analysis by Arezzo et al. found a higher local recurrence rate in the stent as a bridge to surgery group that was not even statistically significant (40.5% vs. 26.6%, $p = 0.09$) [29]. However, several other meta-analyses did not confirm this finding [21,28,30–32].

In conclusion, the current available evidence on the role of endoscopic stenting as a bridge to surgery are globally weak. However, short-term outcomes such as the rate of permanent or temporary stoma, which significantly affect patients' quality of life, are significantly improved with the endoscopic stenting. It is therefore recommended to propose the endoscopic stenting as a first line of treatment in this situation only if the global organization of the Institution (i.e., on-call endoscopist with expertise in radiological procedure availability 24 h a day) and an agreed pathway with surgeons allow this kind of approach. As an alternative, ES remains a valid option for the treatment of resectable patients with acute malignant colonic obstruction.

3. Pre-Operative Evaluation: How the Radiologist Can Help the Endoscopist

3.1. Diagnosis of Large Bowel Obstruction

Clinical suspicion of bowel obstruction can be confirmed by various imaging methods that assist radiologists in addressing critical issues, such as identifying the specific location and underlying reason for the blockage, as well as determining the presence of any associated complications.

Plain abdominal radiography is usually first prescribed in the emergency department in patients with a suspicion of acute bowel obstruction because of its speed of acquisition, low cost, wide availability and low radiation exposure. Plain abdominal radiography, in the dependent and nondependent position, can provide diagnostic confirmation in approximately 50–70% of cases [36]. Typical radiographic indicators of LBO include enlargement of the colon and cecum, with diameters greater than 6 cm and 9 cm, respectively. Significant gas depletion in the rectum and accumulation of fecal material in the proximal colon is often observed [37]. In cases of LBO, the dilation of the small intestine may be variable, which depends on factors such as the duration of obstruction, the presence of a closed loop or a functional ileocecal valve [38]. In particular, when the ileocecal valve is incompetent in the context of a distal LBO, diffuse distention of the small and large intestines may mimic the appearance of pseudo-obstruction upon radiography. In this situation, distention of the large bowel is not marked and endoscopic stenting may not be organized as an emergent procedure. Moreover, temporary placement of a nasogastric tube could reduce bowel distension.

On the contrary, where a closed loop develops, the absence of proximal decompression increases the risk of progressive and localized colonic dilatation, ischemia and perforation. In this situation, the addition of computed tomography (CT) is necessary in order to obtain further information on the possible development of complications. A meta-analysis reported a CT sensitivity of 92% (range 81–100%) and specificity of 93% (range 68–100%) in detecting complete obstruction [39]. An abdominal CT scan is also necessary in order to investigate the etiology of acute obstruction.

3.2. Differential Diagnosis and Characteristics of Malignant Acute Colonic Obstruction from CT Scan

Multiplanar thin-layer acquisitions could delineate precisely the morphology of the colon, allow for the diagnosis of intraluminal, mural, and extramural causes of LBO and detect any potential complications. Performing a CT scan also in the urgent setting of malignant acute colonic obstruction also offers the advantage of detecting local and distant metastases. Intravenous contrast medium is recommended to help in identifying the presence of a mass and signs of inflammation and/or ischemia of the bowel wall.

Intravenous iodinated contrast agent is essential to evaluate the enhancement of the colonic wall and can be administered with a weight-based protocol and a rate of 3 mL/s [40]. Indeed, an unenhanced CT scan for the evaluation of abdominal pain in the emergency department has an accuracy 30% lower than an enhanced CT scan [41] (Figure 1).

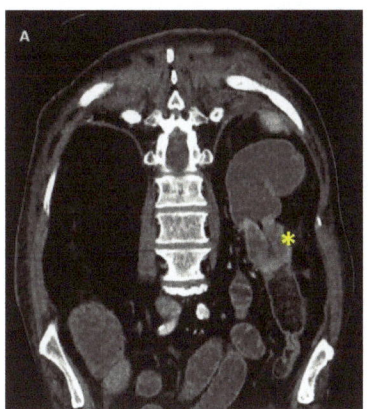
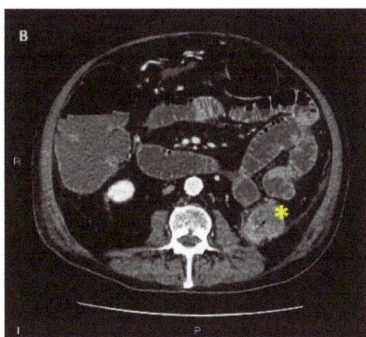

Figure 1. Enhanced CT scan of acute colonic obstruction with presence of neoplastic stricture at the level of the descending colon. In the coronal section: (**A**) stenosis of the descending with marked thickening of its walls, non-homogeneous contrast enhancement, possible extra-visceral extension; in the ax, (**B**) overdistention of large and small bowel with air fluids levels. Yellow stars at the level of the neoplastic stricture.

Coronal and multiplanar reformulations allow for the identification of the course of the distended bowel and the exact location of the obstruction, particularly in the case of rectosigmoid carcinoma where the distance of the stricture from the anal sphincter can be determined with good accuracy.

The diagnosis of LBO is primarily based on the presence of a dilated large bowel proximal to the colonic wall thickening with luminal narrowing and decompressed bowel distal to the stricture. The identification of a transition point is considered a clear sign for the diagnosis of colonic obstruction. CT-scan signs of malignant colonic obstruction are a localized thickening of the colonic wall with an uneven pattern, resembling an "applecore", or the presence of a contrast-enhanced soft tissue mass situated at the center of the colon, causing a narrowing of the colonic passage [42]. In rare instances, a central necrosis or the presence of air within the lesion resembling an abscess-like lesion may also be observed.

It is important to consider the possibility of diverticulitis as a differential diagnosis, especially when there is evidence of colon cancer extending into the surrounding adipose tissue. Diverticulitis is typically represented by segmental, symmetric bowel wall thickening with hyperemia, the presence of fluid in the root of the mesentery and vascular engorgement [43]. Additional indications of active disease include inflammation of the mesenteric fat, the presence of abscesses and the formation of phlegmon. In a retrospective study conducted by Chintapalli et al., they examined all potential CT signs that could aid in distinguishing diverticulitis from CRC. They concluded that the most specific findings for diverticulitis were pericolonic stranding and a length of the involved segment of more than 10 cm, whereas the presence of pericolic lymphadenopathies and presence of a luminal mass were more commonly found in CRC [42].

To summarize, to enable endoscopists and surgeons to choose the best therapeutic strategy, the radiologist should make a diagnosis of a likely malignant colonic lesion, indicate the site and length of stenosis and its distance from the anal sphincter (in the case of rectal stenosis), and identify the presence of any complications (especially perforation and ischemia). Post-stent radiological control is not necessary except for cases where stent migration is suspected and plain radiography is sufficient.

4. Influence of Chemotherapy on Endoscopic Colonic Stenting: The Oncologist Point of View

Patients treated with stenting, especially in the palliative setting, may also experience adverse events related to chemotherapy or tumor progression. Delayed stent migration may occur under chemotherapy due to tumor regression, whereas re-obstruction or perforation could happen in relation to tumor growth [44]. However, the risk of long-term complications must be weighed against the lower mortality and earlier start of chemotherapy compared to surgery. Several small retrospective studies have shown that colonic stenting, compared to ES, reduces the duration of a hospital stay and allows the earlier initiation of chemotherapy, thus showing an advantage in quality of life and overall survival [14,17,45].

4.1. Timing and Safety of Chemotherapy Initiation in Patients with Acute Colonic Malignant Obstruction

Karoui et al. conducted a retrospective study with 58 stage IV obstructive CRCs comparing stent insertion or surgery as a palliative treatment. They demonstrated that the median time to chemotherapy beginning was shorter after stent insertion than after ES (14 vs. 28.5 days) with a potential benefit in quality of life and survival [17]. Lee et colleagues analyzed 88 patients with occlusive stage IV CRC who underwent surgery or stent insertion. In this study also, the median chemotherapy initiation was shorter in the stent group (8.1 days vs. 21.7 days) than in the surgery group [45]. In a meta-analysis, Zhao et al. considered 837 patients, of which 404 were treated with colonic stent and 433 with surgery. They concluded that the median time to start chemotherapy was significantly lower in the stent group than in the surgery group (15.5 days vs. 33.4 days) [14].

When determining the appropriateness of stent placement, the clinician should consider the risk of long-term stent-related complications in relation to the lower short-term mortality and faster onset of systemic chemotherapy treatment, considering that survival advantage with chemotherapy could expose patients with a colonic stent to an increased risk of long-term complications [14]. However, a multicenter retrospective study evaluating the outcomes of palliative chemotherapy without target therapy after colonic stent insertion for an obstructive primary tumor demonstrated that this is feasible and safe. The response rate (38%) and the disease control rate (62%) with Folfox or Folfiri were similar to other previous clinical trials using similar regimes, and the perforation rate after 2–15 months of placement was acceptable [46].

4.2. Effects of Antiangiogenic Agents following Colonic Stenting

Bevacizumab is a recombinant humanized monoclonal antibody that blocks the activity of the vascular endothelial growth factor, which is well-known to be a risk factor of intestinal perforation, which occurs in 1–2% of the patients [6]. Chemotherapy with antiangiogenic agents showed a higher response rate and extended overall survival, thereby it is established as the first-line treatment in stage IV CRC. However, safety issues were raised in patients undergoing colonic stent placement, especially due to the risk of perforation. Several mechanisms have been proposed to explain the increased perforation rate in patients that received bevacizumab, such as tumor regression, necrosis, and a weakened serosa, combined with the pressure of the radial force of the stent on the colonic tumor [47].

Two retrospective analyses showed that bevacizumab increased the risk of perforation in patients with previous placement of a colonic stent. Small et al. collected data of stage IV CRC patients who had a stent placed for palliation or as a bridge to surgery. Eight patients of twenty-six (34.8%) reported a stent–related complication and a median of 44 days after bevacizumab started vs. a 22.8% complication rate in untreated patients. Four patients treated with antiangiogenic agents developed colonic perforation after 21 days of bevacizumab starting (15.4%) compared to a lower incidence in untreated patients (6.8%) [48]. The same result was obtained by Imbulgoda et al. who found a 20% rate of perforations in patients treated with chemotherapy and bevacizumab after stent placement [49]. A meta-analysis of 86 studies including 4086 patients with colonic stenting in malignant LBO showed that bevacizumab-based therapy was associated with a perforation rate of 12.5%. However, in this study only 2.1% of the included patients were treated with bevacizumab and the timing between the placement of the colonic stent and the start of antiangiogenic therapy was not specified [50].

Conversely, Lee and colleagues recently conducted a study of 104 patients who had received a colonic stent and were subsequently treated with bevacizumab. In the group treated with bevacizumab, only one patient had perforation as opposed to the three patients in the non-bevacizumab group [51]. Similarly, a relatively large retrospective study of 353 patients conducted by Park et al. demonstrated that the rate of perforation in patients with bevacizumab and without bevacizumab were equivalent (7.5% and 7%, respectively). Furthermore, the majority of the patients receiving bevacizumab who experienced perforation had started antiangiogenic treatment 1 month after stent insertion [19]. Fuccio et colleagues reported a retrospective case series of 91 patients who underwent palliative stent placement and they found that the complication rate was higher, though not statistically different, in the group treated with bevacizumab alone (29.4%) compared to those treated with chemotherapy alone (19.6%) or chemotherapy plus antiangiogenic agents (18%). It is also interesting to highlight that patients receiving chemotherapy and molecular-targeted drugs had longer overall survival compared to those receiving only chemotherapy, without showing an increased risk of stent-related complications [52]. Additionally, the study conducted by Pacheco-Barcia et al. on 78 patients demonstrated that chemotherapy plus antiangiogenetic agents improved overall survival compared to chemotherapy alone (43 months vs. 20 months) without an increase in risk perforation. The authors concluded that the increased risk of long-term stent complications in patients treated with chemotherapy and bevacizumab should not preclude the administration of these therapies for patients with metastatic colon cancer, considering the substantial response rate and its correlation with OS [53].

Nonetheless, due to the limited incidence, the lack of evidence, the retrospective nature of performed studies and their contradictory results, it is not possible to determine a definitive risk of stent-related perforation following bevacizumab treatment beginning [19,48,50,51]. Consequently, the most recent European guidelines suggest considering the use of bevacizumab following colonic stenting [10]. In clinical practice, colonic stenting serves as a life-saving procedure in acute LBO situations and it should be considered as a first option where local expertise is available, in elderly patients or in patients unfit for surgery or emergency surgery, regardless of the risk occurring with a potential therapy with bevacizumab [54]. Multidisciplinary discussion, weighing the potential benefits and harm for each patient, is therefore the decision-making moment in this acute situation.

No data are available about the risk of perforation for the new antiangiogenics recently prescribed in colorectal cancer (regorafenib, aflibercept). It could be speculated that considering the comparable mechanism of action with bevacizumab, the risk of gastrointestinal perforation induced by these new drugs is similar to that reported for bevacizumab.

4.3. Colonic Stenting during or after Antiangiogenic Treatment

Approximately 20% of patients with previous systemic chemotherapy develop primary tumor obstruction. In these patients, the benefits of minimally invasive SEMS over surgery would be greater since a high mortality rate and major complications following ES have been reported in this patient setting. Regarding the perforation risk of SEMS placement in patients with previous bevacizumab use, there are very limited evidence and clinical data.

Imbulgoda et al. collected the data of 87 stage IV CRC patients treated with or without chemotherapy and bevacizumab before or after stent placement. Perforation occurred globally in four patients (13%). Two out of the ten patients (20%) that received chemotherapy plus bevacizumab (3 before stenting, 6 after and 1 unknown) had a perforation. This result was not statistically different than the rate of perforation that occurred in patients treated with chemotherapy alone. However, the low number of events could have affected the statistical result. Patients who received chemotherapy with or without bevacizumab before stenting were treated for at least 50 days before the procedure and in this category of patients, the rate of perforation was the lowest reported (6%). It could be argued that a major determinant for perforation is the interval between stenting and chemotherapy infusion [49].

Bong et al. retrospectively collected the data of 1008 patients with metastatic colorectal cancer treated with bevacizumab and the incidence rate of complications requiring surgery was approximately 5.9% (60/1008). In this study, twenty-three patients during bevacizumab exposure required SEMS insertion and seven of these patients (7/23, 30.4%) experienced perforation requiring surgery. The authors concluded that SEMS was a significant risk factor for complication requiring surgery in patients already receiving bevacizumab [55]. In addition, in Park's previously mentioned study, bevacizumab was administered both before and after colonic stenting. Only 2 out of 96 (2%) patients who developed perforation received stent implantation within 30 days before or after the start of bevacizumab treatment [19].

Based on these very low-quality data, ESGE guidelines suggest against colonic stenting while patients are receiving antiangiogenic therapy such as bevacizumab [10]. However, in this situation is fundamental to discuss with the oncologist whether the patient could benefit from a minimally invasive treatment (namely colonic stenting) and eventually try to extend the time between the last infusion of bevacizumab and colonic stenting.

5. Endoscopic Stenting

5.1. Preparation to Endoscopy

Colonic stenting, similar to other interventional endoscopic procedures, necessitates an adequate endoscopic setting in order to facilitate the procedure and minimize the adverse events (Figure 2). A room equipped with fluoroscopy is mandatory, whereas a radiology technician is not necessary if the endoscopic team is able to understand radiological images and to maneuver the C-arm.

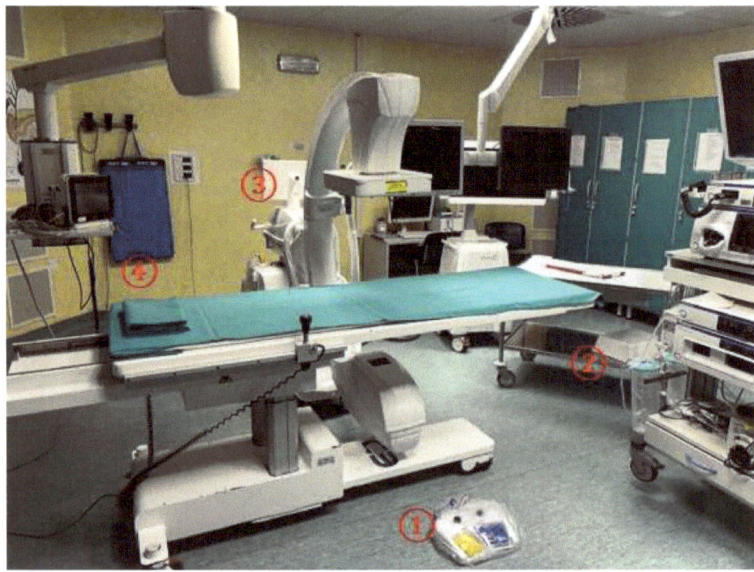

Figure 2. Example of an endoscopic setting for the placement of a colonic stent. ① Endoscopist position with radiological and endoscopic monitor placed frontally and the endoscopic processor in the right position. ② Nurse position at the end of the radiological bed with an instrument shelf on his side. ③ Position of the assistant at the C-arm. ④ Anesthesiological position at the patient's head with vital monitoring on his side.

A reasonable bowel preparation is suggested before the procedure in order to reach the stenosis quicker, to identify the residual lumen and to expedite the entire procedure. A standard bowel preparation is contraindicated in consideration of the occlusive status; thus, cleansing enemas are suggested before the procedure [56].

Considering the type of sedation to perform and the anesthesiologic support, it is important to evaluate patients' comorbidities and the degree of respiratory distress related to the abdominal distension. In the case of compromised patients, especially those requiring non-invasive ventilation support, it is suggested to perform the endoscopic procedure in an equipped room with anesthesiologic support. If the patient is not in critical condition, colonic stenting is not a painful procedure and can be performed under conscious sedation with a combination of benzodiazepines and opioids.

Lastly, regarding which endoscopist could perform colonic stenting, it is necessary to consider that there is a learning curve for the endoscopist performing the stenting procedure. The level of expertise has a direct correlation to the rates of successful stenting and complications [57]. To increase the chances for success, the endoscopist is expected to have attempted more than 20 procedures and be familiar with other fluoroscopic endoscopic procedures like the endoscopic retrograde cholangiopancreatography (ERCP) [58].

5.2. Technique

The procedure requires an endoscope with a large operative channel, with a diameter of at least 3.7 mm, such as a standard colonoscope or an operative gastroscope. The presence of the carbon dioxide pump is also suggested in order to minimize abdominal distension and to prevent air dissemination in case of perforation [7]. Prophylactic antibiotics are not routinely suggested because the risk of bacteremia after stent insertion is very low [10].

The patient can be indistinctly supine or on the left side: the former facilitates the interpretation of radiological frames while the latter, which is the standard colonoscopy position, allows for an easier passage through the sigmoid colon. We usually start the procedure on the left side and once the stricture is reached, we evaluate radiologically the extent and location of the stricture with the contrastography obtained with air insufflation (Figure 3). If the radiological orientation of the stricture is not clear, we rotate the patient in the supine position in order to obtain a clearer radiological view of the stricture.

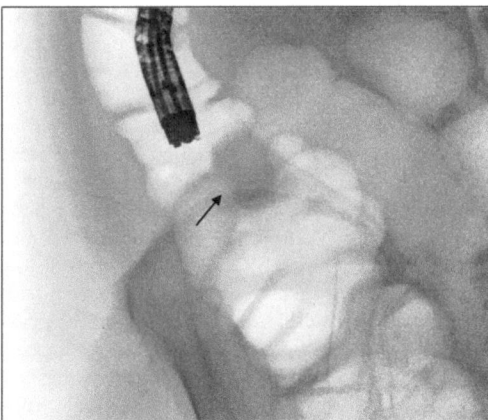

Figure 3. Contrastography obtained with air insufflation showing colonic stricture (black arrow) at the level of ascending colon.

A long 0.035 inch guidewire (450 cm) with a hydrophilic tip is inserted through a cannula in order to pass the stricture. In case of angulated strictures, the use of a sphincterotome or an angle tip guidewire could help the passage of the stricture. Changing the patient's position could sometimes help to better visualize the stricture. After the evaluation of the correct position and evaluation of the length of the stricture with the use of a contrast agent, the guidewire is pulled over to create safe loops over the tumor and the cannula is then retracted.

Histological confirmation of the malignant etiology of the stricture is mandatory [10]; however, the urgent setting and the location of the stricture could increase the difficulty of obtaining biopsies. Moreover, the post-biopsy bleeding could obscure the stricture visualization and consequently increases the difficulty of guidewire passage. What we usually suggest is to first put the guidewire over the stricture and then take the biopsies before stent placement in order to guarantee access at the stricture.

Colonic stent placement could be performed with the through-the-scope (TTS) technique or the over-the-wire (OTW) technique. Conflicting results have been produced about technical and clinical benefits between the two procedures [59]. However, with the widespread availability of TTS stents it has become a common practice to proceed with the TTS technique, which allows an endoscopic control during stent release, and to reserve the OTW technique when larger stents are needed or when an endoscope with a smaller operative channel is used instead of standard scopes for anatomical reasons.

The SEMS is then advanced over the guidewire under radiologic and endoscopic visualization. The cranial flange is the first opened and then retracted close to the cranial extreme of the stricture when a little feeling of resistance at the catheter is appreciated. Then, the entire stent is released. In the case of angulated strictures, it is suggested to maintain an adequate distance with the endoscope from the stricture and to pay attention during the release maneuver. In this situation, it is suggested to pull the SEMS catheter vigorously when the stent is released in order to avoid misdeployment of the stent in the proximal colon. Dilation of the stricture before or after the stent placement is contraindicated because a retrospective analysis evidenced a higher rate of colonic perforation [60].

Different stents are available for colonic stenting and all of them are composed by nitinol, a nonferromagnetic blend. Considering the type of the stent chosen, some meta-analyses evidenced the superiority of uncovered stents compared to covered ones in terms of global complications, tumor overgrowth and stent migration resulting in longer stent patency. Only the risk of tumor ingrowth is reduced with covered SEMS. Partially covered stents potentially could reduce the risk of migration; however, no studies found a superiority of this type of stents compared to uncovered or fully covered stents [61]. The most recent guideline suggests with a weak grade of recommendation to use an uncovered stent as a first line approach [62], however the stent choice should also be made on a clinical basis.

The length of the stent should be tailored to the length of the stricture and its position. It is generally suggested to extend the length of the stent at least 2 cm before and after the stenosis, in particular for tumors at flexures, for which the risk of migration is higher [10]. No definitive conclusions have been reached about the ideal stent diameter, but few studies reported a higher rate of stent migration with a diameter less than 24 mm [63,64].

Tumor localization requires careful consideration. Some authors have expressed reservations about the use of colonic stents for proximal colonic obstruction or rectal malignant stricture near to the anal verge. Studies comparing the clinical success rates between right-sided and left-sided stenting showed conflicting results. However, several retrospective studies evidenced the feasibility of the stenting of the proximal colon (Figure 4) [65–68].

In summary, in case of right-sided LBO it is important to evaluate the patency of the ileocecal valve. In the case of involvement of the ileocecal valve by the tumor, it is difficult to proceed with the stenting and have a clinical benefit. On the other hand, in the case of a patency of the ileocecal valve and identification of the stricture distally to the ileocecal valve, it is possible to attempt the placement of an SEMS after a multidisciplinary consultation.

Tumor location in the distal rectum, defined as a mass within 8–10 cm of the anal verge, raises concerns due to the potential post-procedural pain and increased risk of migration, tenesmus and bleeding. Data on rectal stenting are scarce, with only one systematic review and meta-analysis evaluating short-term outcomes. This study found that rectal and rectosigmoid stenting had a high technical success rate, but this did not always translate into clinical success (97% and 69%, respectively). Additionally, the overall complication rate was 32%, with stent re-obstruction and migration being the most common issues (10.5% and 9.3%, respectively) [69]. Consequently, no definitive conclusions can be drawn from these data and in the case of acute colonic obstruction related to a tumor close to the anal verge, it is fundamental to establish the clinical indication case by case.

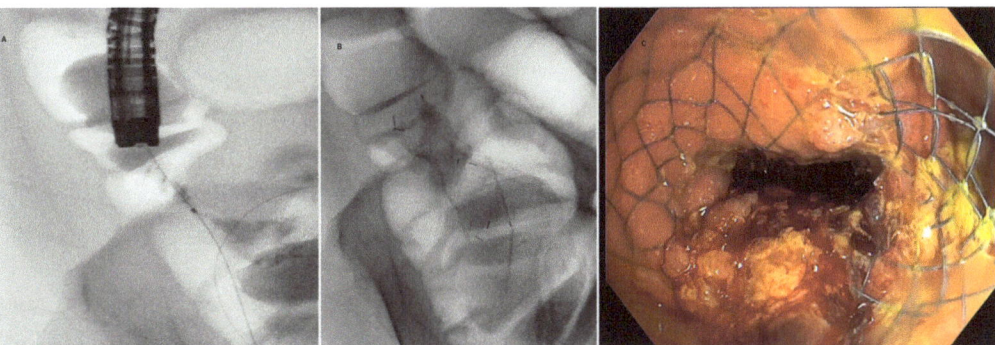

Figure 4. Colonic stenting of a malignant stricture at the level of the ascending colon. (**A**): Contrast injection at the level of the stricture with loop of the guidewire in the cecum; (**B**) radiological evidence of the self-expandable metal stent (SEMS) within the malignant stricture; (**C**) endoscopic view of the stricture with the metallic mesh of the distal part of the uncovered SEMS.

5.3. Procedural Adverse Events and Post Endoscopic Management

Complications arising from the insertion of SEMS can be divided into minor and major complications according to the degree of severity [70]. Perforation is the most common (4–7%) and most feared major complication that can occur during endoscopic stent placement [71,72]. Immediate causes of perforation include wire or catheter misplacement beyond the colonic wall. Delayed perforation usually occurs in the distended right colon when there is a thin walled caecum; this type of perforation usually occurs concomitantly to the placement of SEMS in a difficult endoscopic position that leads to an excessive amount of air insufflation in a distended large bowel [73]. Conservative treatment of perforation is possible when the SEMS is correctly placed and there is only a modest quantity of air in the abdomen without the presence of fluid and without clinical worsening. Otherwise, surgical treatment is the only therapeutic choice for colonic perforation. Lastly, late perforation can be caused by concomitant therapy with antiangiogenic agents, as mentioned before.

Stent migration is the other major complication that may occur during endoscopic positioning. In the case of distal migration, the stent could be retrieved endoscopically with caution in order to avoid mucosal damage and a new stent could be repositioned. Stent migration proximally to the stricture makes impossible the endoscopic retrieval of the stent. In the case of palliative indications, the migrated stent could be left in the proximal colon but clinical monitoring in search of signs of decubitus of the stent is needed [5]. Bleeding is a minor complication that may occur after stent insertion but it is usually self-limiting.

The clinical success of endoscopic stenting is obtained when there is an adequate passage of air and feces from the anus after the endoscopic procedure. Studies do not specify which is the ideal timing to evaluate an effective colonic recanalization. Usually, in the case of a correct stent deployment, the passage of air and of liquid feces is almost immediate and patients' relief is evident. In the case of angulated (>165°) and long strictures, colonic canalization could require more hours. These types of strictures are also those that, despite a complete technical success (usually obtained in 95% of patients), have a short term clinical failure and therefore require surgery resolution of the obstruction [73].

The optimal interval between SEMS placement and subsequent surgery in patients with MBO is still unclear and few studies evaluated this question. The most recent guidelines, based on low-quality studies, and randomized trial suggest surgery after 5 to 14 days to the colonic stenting. In our Institution, in cooperation with surgeons we usually candidate the patient to surgery after a bridge with endoscopic stenting when there are no more signs of obstruction and it is possible to perform a complete bowel preparation [74,75].

Following colonic stenting for malignant LBO, it is necessary to perform a complete colonoscopy to exclude the presence of a synchronous lesion, which occurs in 3–13% of patients with CRC [76,77]. Through-the-stent colonoscopy is feasible after an adequate PEG-based bowel preparation, with a success rate ranging from 63% to 100%; notably, the lowest rate increased to 87.5% when a gastroscope was utilized instead of a colonoscope [77,78]. It is important to state that through-the-stent colonoscopy is a safe procedure with no instances of migration or bleeding related to this implementation [77,78]. There is a complete lack of research concerning the optimal timing for proximal colonic evaluation after the revealing of a CRC. For these reasons, the most recent ESGE guidelines suggest conducting a complete colonoscopy no more than 6 months after colonic stenting and surgery or even before the surgery in the curative setting [17].

6. Conclusions

Acute malignant colonic obstruction is an urgent clinical condition that still has some grey areas regarding endoscopic management. Robust data support colonic stenting in the palliative setting [11–18] and demonstrate improved quality of life and better clinical outcomes compared to ES. However, such robust scientific support is lacking for other more challenging situations such as resectable patients who may benefit from the bridge-to-surgery approach and those experiencing acute obstruction under antiangiogenic treatment. The available scientific data show that colonic stenting certainly improves the short-term outcomes of patients who undergo the bridge-to-surgery approach and that endoscopic stenting close to bevacizumab administration may increase the risk of perforation [19,27,29,31–33,35,49,55]. The location of malignant stricture in debated positions, such as in the right colon or close to the anal verge, may limit the clinical success of colonic stenting. The literature suggests that there are no absolute contraindications to colonic stenting even in these technical and clinically challenging situations and a minimally invasive treatment compared to ES could improve patients' quality of life.

Acute malignant colonic obstruction is therefore a critical situation with some challenging clinical and technical issues that frequently necessitates real-time consultation among different specialists to establish the most suitable therapeutic strategy. In our opinion, we suggest organizing a local pathway shared with the radiologist, surgeon, oncologist and endoscopist to create a predefined algorithm based on the available local resources. For instance, as highlighted in the present review, it is fundamental to perform an enhanced CT scan in the diagnostic process of LBO to establish its etiology and to study the extension of the stricture in preparation for endoscopic stenting. However, many non-tertiary centers do not have the 24 h availability of enhanced radiology and this aspect could restrict the endoscopic stenting. The absence of specialized figures that could approach patients with acute colonic obstruction could limit endoscopic treatment in favor of surgery. Moreover, the local diagnostic-treatment algorithm should consider the number of available on-call endoscopists to perform colonic stenting. Lastly, it is reasonable that in more structured institutions with a high availability of human resources and expertise, the clinical indications for endoscopic stenting may be expanded to include borderline situations (e.g., bridge-to-surgery indication, challenging endoscopic positions).

Author Contributions: Conceptualization, A.M. and A.A.; writing—original draft preparation, A.M.; D.S.; S.B.; P.F.; A.G.; S.M.; I.P. and A.P.; writing—review and editing: L.A. and A.A. All authors have read and agreed to the published version of the manuscript.

Funding: This research received no external funding.

Institutional Review Board Statement: Not applicable.

Data Availability Statement: No new data were created or analyzed in this study. Data sharing is not applicable to this article.

Conflicts of Interest: The authors declare no conflicts of interest.

References

1. Sung, H.; Ferlay, J.; Siegel, R.L.; Laversanne, M.; Soerjomataram, I.; Jemal, A.; Bray, F. Global Cancer Statistics 2020: GLOBOCAN Estimates of Incidence and Mortality Worldwide for 36 Cancers in 185 Countries. *CA Cancer J. Clin.* **2021**, *71*, 209–249. [CrossRef]
2. Seoane Urgorri, A.; Saperas, E.; O'Callaghan Castella, E.; Pera Román, M.; Raga Gil, A.; Riu Pons, F.; Barranco Priego, L.; Dedeu Cusco, J.M.; Pantaleón Sánchez, M.; Alvarez, M.A.; et al. Colonic stent vs surgical resection of the primary tumor. Effect on survival from stage-IV obstructive colorectal cancer. *Rev. Esp. Enferm. Dig.* **2020**, *112*, 694–700. [CrossRef] [PubMed]
3. Sarani, B.; Paspulati, R.M.; Hambley, J.; Efron, D.; Martinez, J.; Perez, A.; Bowles-Cintron, R.; Yi, F.; Hill, S.; Meyer, D.; et al. A multidisciplinary approach to diagnosis and management of bowel obstruction. *Curr. Probl. Surg.* **2018**, *55*, 394–438. [CrossRef]
4. Davids, P.H.; Groen, A.K.; Rauws, E.A.; Tytgat, G.N.; Huibregtse, K. Randomised trial of self-expanding metal stents versus polyethylene stents for distal malignant biliary obstruction. *Lancet* **1992**, *340*, 1488–1492. [CrossRef] [PubMed]
5. Ormando, V.M.; Palma, R.; Fugazza, A.; Repici, A. Colonic stents for malignant bowel obstruction: Current status and future prospects. *Expert. Rev. Med. Devices* **2019**, *16*, 1053–1061. [CrossRef]
6. Botrel, T.E.A.; Clark, L.G.O.; Paladini, L.; Clark, O.A.C. Efficacy and safety of bevacizumab plus chemotherapy compared to chemotherapy alone in previously untreated advanced or metastatic colorectal cancer: A systematic review and meta-analysis. *BMC Cancer* **2016**, *16*, 677. [CrossRef]
7. Baron, T.H. Technique of colonic stenting. *Tech. Gastrointest. Endosc.* **2014**, *16*, 108–111. [CrossRef]
8. Trompetas, V.; Saunders, M.; Gossage, J.; Anderson, H. Shortcomings in colonic stenting to palliate large bowel obstruction from extracolonic malignancies. *Int. J. Color. Dis.* **2010**, *25*, 851–854. [CrossRef]
9. Faraz, S.; Salem, S.B.; Schattner, M.; Mendelsohn, R.; Markowitz, A.; Ludwig, E.; Zheng, J.; Gerdes, H.; Shah, P.M. Predictors of clinical outcome of colonic stents in patients with malignant large-bowel obstruction because of extracolonic malignancy. *Gastrointest. Endosc.* **2018**, *87*, 1310–1317. [CrossRef] [PubMed]
10. van Hooft, J.E.; Bemelman, W.A.; Oldenburg, B.; Marinelli, A.W.; Lutke Holzik, M.F.; Grubben, M.J.; Sprangers, M.A.; Dijkgraaf, M.G.; Fockens, P. Self-expandable metal stents for obstructing colonic and extracolonic cancer: European Society of Gastrointestinal Endoscopy (ESGE) Guideline—Update 2020. *Endoscopy* **2020**, *52*, 389–407. [CrossRef]
11. Liang, T.W.; Sun, Y.; Wei, Y.C.; Yang, D.X. Palliative treatment of malignant colorectal obstruction caused by advanced malignancy: A system review and meta-analysis. *Surg. Today* **2014**, *44*, 22–33. [CrossRef] [PubMed]
12. Ribeiro, I.B.; Bernardo, W.M.; Martins, B.D.C.; de Moura, D.T.H.; Baba, E.R.; Josino, I.R.; Miyahima, N.T.; Coronel Cordero, M.A.; Visconti, T.A.C.; Ide, E.; et al. Colonic stent versus emergency surgery as treatment of malignant colonic obstruction in the palliative setting: A systematic review and meta-analysis. *Endosc. Int. Open* **2018**, *6*, E558–E567. [PubMed]
13. Takahashi, H.; Okabayashi, K.; Tsuruta, M.; Hasegawa, H.; Yahagi, M.; Kitagawa, Y. Self-Expanding Metallic Stents Versus Surgical Intervention as Palliative Therapy for Obstructive Colorectal Cancer: A Meta-analysis. *World J. Surg.* **2015**, *39*, 2037–2044. [CrossRef]
14. Zhao, X.D.; Cai, B.B.; Cao, R.S.; Shi, R.H. Palliative treatment for incurable malignant colorectal obstructions: A meta-analysis. *World J. Gastroenterol.* **2013**, *19*, 5565–5574. [CrossRef]
15. Abelson, J.S.; Yeo, H.L.; Mao, J.; Milsom, J.W.; Sedrakyan, A. Long-term Postprocedural Outcomes of Palliative Emergency Stenting vs Stoma in Malignant Large-Bowel Obstruction. *JAMA Surg.* **2017**, *152*, 429–435. [CrossRef] [PubMed]
16. Young, C.J.; Zahid, A. Randomized controlled trial of colonic stent insertion in non-curable large bowel obstruction: A post hoc cost analysis. *Color. Dis.* **2018**, *20*, 288–295. [CrossRef] [PubMed]
17. Karoui, M.; Charachon, A.; Delbaldo, C.; Loriau, J.; Laurent, A.; Sobhani, I.; Tran Van Nhieu, J.; Delchier, J.C.; Fagniez, P.L.; Piedbois, P.; et al. Stents for palliation of obstructive metastatic colon cancer: Impact on management and chemotherapy administration. *Arch. Surg.* **2007**, *142*, 619–623. [CrossRef] [PubMed]
18. Young, C.J.; De-Loyde, K.J.; Young, J.M.; Solomon, M.J.; Chew, E.H.; Byrne, C.M.; Salkeld, G. and Faragher, I.G. Improving Quality of Life for People with Incurable Large-Bowel Obstruction: Randomized Control Trial of Colonic Stent Insertion. *Dis. Colon. Rectum.* **2015**, *58*, 838–849. [CrossRef]
19. Park, Y.E.; Park, Y.; Park, S.J.; Cheon, J.H.; Kim, W.H.; Kim, T.I. Outcomes of stent insertion and mortality in obstructive stage IV colorectal cancer patients through 10 year duration. *Surg. Endosc.* **2019**, *33*, 1225–1234. [CrossRef]
20. Park, J.J.; Rhee, K.; Yoon, J.Y.; Park, S.J.; Kim, J.H.; Youn, Y.H.; Kim, T.I.; Park, H.; Kim, W.H.; Cheon, J.H. Impact of peritoneal carcinomatosis on clinical outcomes of patients receiving self-expandable metal stents for malignant colorectal obstruction. *Endoscopy* **2018**, *50*, 1163–1174. [CrossRef]
21. Ceresoli, M.; Allievi, N.; Coccolini, F.; Montori, G.; Fugazzola, P.; Pisano, M.; Sartelli, M.; Catena, F.; Ansaloni, L. Long-term oncologic outcomes of stent as a bridge to surgery versus emergency surgery in malignant left side colonic obstructions: A meta-analysis. *J. Gastrointest. Oncol.* **2017**, *8*, 867–876. [CrossRef]

22. Shin, S.J.; Kim, T.I.; Kim, B.C.; Lee, Y.C.; Song, S.Y.; Kim, W.H. Clinical application of self-expandable metallic stent for treatment of colorectal obstruction caused by extrinsic invasive tumors. *Dis. Colon. Rectum.* **2008**, *51*, 578–583. [CrossRef]
23. Lemoine, L.; Sugarbaker, P.; Van der Speeten, K. Pathophysiology of colorectal peritoneal carcinomatosis: Role of the peritoneum. *World J. Gastroenterol.* **2016**, *22*, 7692–7707. [CrossRef]
24. Hayashi, K.; Jiang, P.; Yamauchi, K.; Yamamoto, N.; Tsuchiya, H.; Tomita, K.; Moossa, A.R.; Bouvet, M. and Hoffman, R.M. Real-time imaging of tumor-cell shedding and trafficking in lymphatic channels. *Cancer Res.* **2007**, *67*, 8223–8228. [CrossRef]
25. Pisano, M.; Zorcolo, L.; Merli, C.; Cimbanassi, S.; Poiasina, E.; Ceresoli, M.; Agresta, F.; Allievi, N.; Bellanova, G.; Coccolini, F.; et al. 2017 WSES guidelines on colon and rectal cancer emergencies: Obstruction and perforation. *World J. Emerg. Surg.* **2018**, *13*, 36. [CrossRef]
26. Ferrada, P.; Patel, M.B.; Poylin, V.; Bruns, B.R.; Leichtle, S.W.; Wydo, S.; Sultan, S.; Haut, E.R.; Robinson, B. Surgery or stenting for colonic obstruction: A practice management guideline from the Eastern Association for the Surgery of Trauma. *J. Trauma. Acute Care Surg.* **2016**, *80*, 659–664. [CrossRef]
27. Allievi, N.; Ceresoli, M.; Fugazzola, P.; Montori, G.; Coccolini, F.; Ansaloni, L. Endoscopic Stenting as Bridge to Surgery versus Emergency Resection for Left-Sided Malignant Colorectal Obstruction: An Updated Meta-Analysis. *Int. J. Surg. Oncol.* **2017**, *2017*, 2863272. [CrossRef]
28. Amelung, F.J.; Burghgraef, T.A.; Tanis, P.J.; van Hooft, J.E.; Ter Borg, F.; Siersema, P.D.; Bemelman, W.A.; Consten, E.C.J. Critical appraisal of oncological safety of stent as bridge to surgery in left-sided obstructing colon cancer: A systematic review and meta-analysis. *Crit. Rev. Oncol. Hematol.* **2018**, *131*, 66–75. [CrossRef]
29. Arezzo, A.; Passera, R.; Lo Secco, G.; Verra, M.; Bonino, M.A.; Targarona, E.; Morino, M. Stent as bridge to surgery for left-sided malignant colonic obstruction reduces adverse events and stoma rate compared with emergency surgery: Results of a systematic review and meta-analysis of randomized controlled trials. *Gastrointest. Endosc.* **2017**, *86*, 416–426. [CrossRef]
30. Cao, Y.; Gu, J.; Deng, S.; Li, J.; Wu, K.; Cai, K. Long-term tumour outcomes of self-expanding metal stents as 'bridge to surgery' for the treatment of colorectal cancer with malignant obstruction: A systematic review and meta-analysis. *Int. J. Color. Dis.* **2019**, *34*, 1827–1838. [CrossRef]
31. Foo, C.C.; Poon, S.H.T.; Chiu, R.H.Y.; Lam, W.Y.; Cheung, L.C.; Law, W.L. Is bridge to surgery stenting a safe alternative to emergency surgery in malignant colonic obstruction: A meta-analysis of randomized control trials. *Surg. Endosc.* **2019**, *33*, 293–302. [CrossRef] [PubMed]
32. Jain, S.R.; Yaow, C.Y.L.; Ng, C.H.; Neo, V.S.Q.; Lim, F.; Foo, F.J.; Wong, N.W.; Chong, C.S. Comparison of colonic stents, stomas and resection for obstructive left colon cancer: A meta-analysis. *Tech Coloproctol.* **2020**, *24*, 1121–1136. [CrossRef]
33. Kanaka, S.; Matsuda, A.; Yamada, T.; Ohta, R.; Sonoda, H.; Shinji, S.; Takahashi, G.; Iwai, T.; Takeda, K.; Ueda, K.; et al. Colonic stent as a bridge to surgery versus emergency resection for right-sided malignant large bowel obstruction: A meta-analysis. *Surg. Endosc.* **2022**, *36*, 2760–2770. [CrossRef]
34. Matsuda, A.; Miyashita, M.; Matsumoto, S.; Matsutani, T.; Sakurazawa, N.; Takahashi, G.; Kishi, T.; Uchida, E. Comparison of long-term outcomes of colonic stent as "bridge to surgery" and emergency surgery for malignant large-bowel obstruction: A meta-analysis. *Ann. Surg. Oncol.* **2015**, *22*, 497–504. [CrossRef]
35. McKechnie, T.; Springer, J.E.; Cloutier, Z.; Archer, V.; Alavi, K.; Doumouras, A.; Hong, D.; Eskicioglu, C. Management of left-sided malignant colorectal obstructions with curative intent: A network meta-analysis. *Surg. Endosc.* **2023**, *37*, 4159–4178. [CrossRef]
36. Cappell, M.S.; Batke, M. Mechanical obstruction of the small bowel and colon. *Med. Clin. N. Am.* **2008**, *92*, 575–597. [CrossRef]
37. Nelms, D.W.; Kann, B.R. Imaging Modalities for Evaluation of Intestinal Obstruction. *Clin. Colon. Rectal Surg.* **2021**, *34*, 205–218. [CrossRef]
38. Love, L. The role of the ileocecal valve in large bowel obstruction. A preliminary report. *Radiology* **1960**, *75*, 391–398. [CrossRef]
39. Mallo, R.D.; Salem, L.; Lalani, T.; Flum, D.R. Computed tomography diagnosis of ischemia and complete obstruction in small bowel obstruction: A systematic review. *J. Gastrointest. Surg.* **2005**, *9*, 690–694. [CrossRef]
40. Jaffe, T.; Thompson, W.M. Large-Bowel Obstruction in the Adult: Classic Radiographic and CT Findings, Etiology, and Mimics. *Radiology* **2015**, *275*, 651–663. [CrossRef]
41. Shaish, H.; Ream, J.; Huang, C.; Troost, J.; Gaur, S.; Chung, R.; Kim, S.; Patel, H.; Newhouse, J.H.; Khalatbari, S.; et al. Diagnostic Accuracy of Unenhanced Computed Tomography for Evaluation of Acute Abdominal Pain in the Emergency Department. *JAMA Surg.* **2023**, *158*, e231112. [CrossRef]
42. Chintapalli, K.N.; Chopra, S.; Ghiatas, A.A.; Esola, C.C.; Fields, S.F.; Dodd, G.D., 3rd. Diverticulitis versus colon cancer: Differentiation with helical CT findings. *Radiology* **1999**, *210*, 429–435. [CrossRef]
43. Ramanathan, S.; Ojili, V.; Vassa, R.; Nagar, A. Large Bowel Obstruction in the Emergency Department: Imaging Spectrum of Common and Uncommon Causes. *J. Clin. Imaging Sci.* **2017**, *7*, 15. [CrossRef]
44. Lueders, A.; Ong, G.; Davis, P.; Weyerbacher, J.; Saxe, J. Colonic stenting for malignant obstructions—A review of current indications and outcomes. *Am. J. Surg.* **2022**, *224*, 217–227. [CrossRef] [PubMed]
45. Lee, W.S.; Baek, J.H.; Kang, J.M.; Choi, S.; Kwon, K.A. The outcome after stent placement or surgery as the initial treatment for obstructive primary tumor in patients with stage IV colon cancer. *Am. J. Surg.* **2012**, *203*, 715–719. [CrossRef]
46. Cézé, N.; Charachon, A.; Locher, C.; Aparicio, T.; Mitry, E.; Barbieux, J.P.; Landi, B.; Dorval, E.; Moussata, D.; Lecomte, T. Safety and efficacy of palliative systemic chemotherapy combined with colorectal self-expandable metallic stents in advanced colorectal cancer: A multicenter study. *Clin. Res. Hepatol. Gastroenterol.* **2016**, *40*, 230–238. [CrossRef] [PubMed]

47. Cennamo, V.; Fuccio, L.; Mutri, V.; Minardi, M.E.; Eusebi, L.H.; Ceroni, L.; Laterza, L.; Ansaloni, L.; Pinna, A.D.; Salfi, N.; et al. Does stent placement for advanced colon cancer increase the risk of perforation during bevacizumab-based therapy? *Clin. Gastroenterol. Hepatol.* **2009**, *7*, 1174–1176. [CrossRef] [PubMed]
48. Small, A.J.; Coelho-Prabhu, N.; Baron, T.H. Endoscopic placement of self-expandable metal stents for malignant colonic obstruction: Long-term outcomes and complication factors. *Gastrointest. Endosc.* **2010**, *71*, 560–572. [CrossRef]
49. Imbulgoda, A.; MacLean, A.; Heine, J.; Drolet, S.; Vickers, M.M. Colonic perforation with intraluminal stents and bevacizumab in advanced colorectal cancer: Retrospective case series and literature review. *Can. J. Surg.* **2015**, *58*, 167–171. [CrossRef]
50. van Halsema, E.E.; van Hooft, J.E.; Small, A.J.; Baron, T.H.; García-Cano, J.; Cheon, J.H.; Lee, M.S.; Kwon, S.H.; Mucci-Hennekinne, S.; Fockens, P.; et al. Perforation in colorectal stenting: A meta-analysis and a search for risk factors. *Gastrointest. Endosc.* **2014**, *79*, 970–982.e7. [CrossRef]
51. Lee, J.H.; Emelogu, I.; Kukreja, K.; Ali, F.S.; Nogueras-Gonzalez, G.; Lum, P.; Coronel, E.; Ross, W.; Raju, G.S.; Lynch, P.; et al. Safety and efficacy of metal stents for malignant colonic obstruction in patients treated with bevacizumab. *Gastrointest. Endosc.* **2019**, *90*, 116–124. [CrossRef]
52. Fuccio, L.; Correale, L.; Arezzo, A.; Repici, A.; Manes, G.; Trovato, C.; Mangiavillano, B.; Manno, M.; Cortelezzi, C.C.; Dinelli, M.; et al. Influence of K-ras status and anti-tumour treatments on complications due to colorectal self-expandable metallic stents: A retrospective multicentre study. *Dig. Liver Dis.* **2014**, *46*, 561–567. [CrossRef]
53. Pacheco-Barcia, V.; Mondéjar, R.; Martínez-Sáez, O.; Longo, F.; Moreno, J.A.; Rogado, J.; Donnay, O.; Santander, C.; Carrato, A.; Colomer, R. Safety and Oncological Outcomes of Bevacizumab Therapy in Patients with Advanced Colorectal Cancer and Self-expandable Metal Stents. *Clin. Color. Cancer* **2019**, *18*, e287–e293. [CrossRef]
54. van Halsema, E.E.; van Hooft, J.E. Bevacizumab in patients treated with palliative colonic stent placement: Is it safe? *Gastrointest Endosc.* **2019**, *90*, 125–126. [CrossRef] [PubMed]
55. Bong, J.W.; Lee, J.L.; Kim, C.W.; Yoon, Y.S.; Park, I.J.; Lim, S.B.; Yu, C.S.; Kim, T.W.; Kim, J.C. Risk Factors and Adequate Management for Complications of Bevacizumab Treatment Requiring Surgical Intervention in Patients with Metastatic Colorectal Cancer. *Clin. Color. Cancer* **2018**, *17*, e639–e645. [CrossRef]
56. Kuwai, T.; Yamaguchi, T.; Imagawa, H.; Yoshida, S.; Isayama, H.; Matsuzawa, T.; Yamada, T.; Saito, S.; Shimada, M.; Hirata, N.; et al. Factors related to difficult self-expandable metallic stent placement for malignant colonic obstruction: A post-hoc analysis of a multicenter study across Japan. *Dig. Endosc.* **2019**, *31*, 51–58. [CrossRef]
57. Bonin, E.A.; Baron, T.H. Update on the indications and use of colonic stents. *Curr. Gastroenterol. Rep.* **2010**, *12*, 374–382. [CrossRef] [PubMed]
58. Ansaloni, L.; Andersson, R.E.; Bazzoli, F.; Catena, F.; Cennamo, V.; Di Saverio, S.; Fuccio, L.; Jeekel, H.; Leppäniemi, A.; Moore, E.; et al. Guidelenines in the management of obstructing cancer of the left colon: Consensus conference of the world society of emergency surgery (WSES) and peritoneum and surgery (PnS) society. *World J. Emerg. Surg.* **2010**, *5*, 29. [CrossRef]
59. Kim, J.W.; Jeong, J.B.; Lee, K.L.; Kim, B.G.; Jung, Y.J.; Kim, W.; Kim, H.Y.; Ahn, D.W.; Koh, S.J.; Lee, J.K. Comparison of clinical outcomes between endoscopic and radiologic placement of self-expandable metal stent in patients with malignant colorectal obstruction. *Korean J. Gastroenterol.* **2013**, *61*, 22–29. [CrossRef] [PubMed]
60. Tanaka, A.; Sadahiro, S.; Yasuda, M.; Shimizu, S.; Maeda, Y.; Suzuki, T.; Tokunaga, N.; Ogoshi, K. Endoscopic balloon dilation for obstructive colorectal cancer: A basic study on morphologic and pathologic features associated with perforation. *Gastrointest. Endosc.* **2010**, *71*, 799–805. [CrossRef]
61. Kim, E.J.; Kim, Y.J. Stents for colorectal obstruction: Past, present, and future. *World J. Gastroenterol.* **2016**, *22*, 842–852. [CrossRef] [PubMed]
62. Mashar, M.; Mashar, R.; Hajibandeh, S. Uncovered versus covered stent in management of large bowel obstruction due to colorectal malignancy: A systematic review and meta-analysis. *Int. J. Color. Dis.* **2019**, *34*, 773–785. [CrossRef] [PubMed]
63. Manes, G.; de Bellis, M.; Fuccio, L.; Repici, A.; Masci, E.; Ardizzone, S.; Mangiavillano, B.; Carlino, A.; Rossi, G.B.; Occhipinti, P.; et al. Endoscopic palliation in patients with incurable malignant colorectal obstruction by means of self-expanding metal stent: Analysis of results and predictors of outcomes in a large multicenter series. *Arch. Surg.* **2011**, *146*, 1157–1162. [CrossRef] [PubMed]
64. Kim, B.C.; Han, K.S.; Hong, C.W.; Sohn, D.K.; Park, J.W.; Park, S.C.; Kim, S.Y.; Baek, J.Y.; Choi, H.S.; Chang, H.J.; et al. Clinical outcomes of palliative self-expanding metallic stents in patients with malignant colorectal obstruction. *J. Dig. Dis.* **2012**, *13*, 258–266. [CrossRef] [PubMed]
65. Ji, W.B.; Kwak, J.M.; Kang, D.W.; Kwak, H.D.; Um, J.W.; Lee, S.I.; Min, B.W.; Sung, N.S.; Kim, J.; Kim, S.H. Clinical benefits and oncologic equivalence of self-expandable metallic stent insertion for right-sided malignant colonic obstruction. *Surg. Endosc.* **2017**, *31*, 153–158. [CrossRef] [PubMed]
66. Geraghty, J.; Sarkar, S.; Cox, T.; Lal, S.; Willert, R.; Ramesh, J.; Bodger, K.; Carlson, G.L. Management of large bowel obstruction with self-expanding metal stents. A multicentre retrospective study of factors determining outcome. *Color. Dis.* **2014**, *16*, 476–483. [CrossRef]
67. Schoonbeek, P.K.; Genzel, P.; van den Berg, E.H.; van Dobbenburgh, O.A.; Ter Borg, F. Outcomes of Self-Expanding Metal Stents in Malignant Colonic Obstruction are Independent of Location or Length of the Stenosis: Results of a Retrospective, Single-Center Series. *Dig. Surg.* **2018**, *35*, 230–235. [CrossRef]

68. Siddiqui, A.; Cosgrove, N.; Yan, L.H.; Brandt, D.; Janowski, R.; Kalra, A.; Zhan, T.; Baron, T.H.; Repici, A.; Taylor, L.J.; et al. Long-term outcomes of palliative colonic stenting versus emergency surgery for acute proximal malignant colonic obstruction: A multicenter trial. *Endosc. Int. Open* **2017**, *5*, E232–E238. [CrossRef]
69. Trabulsi, N.H.; Halawani, H.M.; Alshahrani, E.A.; Alamoudi, R.M.; Jambi, S.K.; Akeel, N.Y.; Farsi, A.H.; Nassif, M.O.; Samkari, A.A.; Saleem, A.; et al. Short-term outcomes of stents in obstructive rectal cancer: A systematic review and meta-analysis. *Saudi J. Gastroenterol.* **2021**, *27*, 127–135. [CrossRef]
70. Lim, T.Z.; Tan, K.K. Endoscopic stenting in colorectal cancer. *J. Gastrointest. Oncol.* **2019**, *10*, 1171–1182. [CrossRef]
71. Huang, X.; Lv, B.; Zhang, S.; Meng, L. Preoperative colonic stents versus emergency surgery for acute left-sided malignant colonic obstruction: A meta-analysis. *J. Gastrointest. Surg.* **2014**, *18*, 584–591. [CrossRef] [PubMed]
72. Khot, U.P.; Lang, A.W.; Murali, K.; Parker, M.C. Systematic review of the efficacy and safety of colorectal stents. *Br. J. Surg.* **2002**, *89*, 1096–1102. [CrossRef] [PubMed]
73. Han, Y.M.; Lee, J.M.; Lee, T.H. Delayed colon perforation after palliative treatment for rectal carcinoma with bare rectal stent: A case report. *Korean J. Radiol.* **2000**, *1*, 169–171. [CrossRef] [PubMed]
74. van Hooft, J.E.; Bemelman, W.A.; Oldenburg, B.; Marinelli, A.W.; Lutke Holzik, M.F.; Grubben, M.J.; Sprangers, M.A.; Dijkgraaf, M.G.; Fockens, P. Colonic stenting versus emergency surgery for acute left-sided malignant colonic obstruction: A multicentre randomised trial. *Lancet Oncol.* **2011**, *12*, 344–352. [CrossRef] [PubMed]
75. Pirlet, I.A.; Slim, K.; Kwiatkowski, F.; Michot, F.; Millat, B.L. Emergency preoperative stenting versus surgery for acute left-sided malignant colonic obstruction: A multicenter randomized controlled trial. *Surg. Endosc.* **2011**, *25*, 1814–1821. [CrossRef] [PubMed]
76. Kodeda, K.; Nathanaelsson, L.; Jung, B.; Olsson, H.; Jestin, P.; Sjövall, A.; Glimelius, B.; Påhlman, L.; Syk, I. Population-based data from the Swedish Colon Cancer Registry. *Br. J. Surg.* **2013**, *100*, 1100–1107. [CrossRef]
77. Itonaga, S.; Hamada, S.; Ihara, E.; Honma, H.; Fukuya, H.; Ookubo, A.; Sasaki, T.; Yoshimura, D.; Nakamuta, M.; Sumida, Y.; et al. Importance of preoperative total colonoscopy and endoscopic resection after self-expandable metallic stent placement for obstructive colorectal cancer as a bridge-to-surgery. *BMC Gastroenterol.* **2023**, *23*, 251. [CrossRef]
78. Kim, J.S.; Lee, K.M.; Kim, S.W.; Kim, E.J.; Lim, C.H.; Oh, S.T.; Choi, M.G.; Choi, K.Y. Preoperative colonoscopy through the colonic stent in patients with colorectal cancer obstruction. *World J. Gastroenterol.* **2014**, *20*, 10570–10576. [CrossRef]

Disclaimer/Publisher's Note: The statements, opinions and data contained in all publications are solely those of the individual author(s) and contributor(s) and not of MDPI and/or the editor(s). MDPI and/or the editor(s) disclaim responsibility for any injury to people or property resulting from any ideas, methods, instructions or products referred to in the content.

Review

Endoscopic Diagnosis of Small Bowel Tumor

Tomonori Yano and Hironori Yamamoto *

Department of Medicine, Division of Gastroenterology, Jichi Medical University, Shimotsuke 329-0498, Japan; tomonori@jichi.ac.jp
* Correspondence: ireef@jichi.ac.jp

Simple Summary: Recent technological advances, including capsule endoscopy (CE) and balloon-assisted endoscopy (BAE), have revealed that small intestinal disease is more common than previously thought. Early diagnosis of small intestinal tumors is essential for favorable outcomes. For early diagnosis, after examination of the upper and lower gastrointestinal tract, the possibility of small bowel lesions should be considered in patients with unexplained symptoms and signs, including gastrointestinal bleeding, chronic anemia, abdominal pain, obstructive symptoms, body weight loss, palpable abdominal mass, and fever of unknown origin.

Abstract: Recent technological advances, including capsule endoscopy (CE) and balloon-assisted endoscopy (BAE), have revealed that small intestinal disease is more common than previously thought. CE has advantages, including a high diagnostic yield, discomfort-free, outpatient basis, and physiological images. BAE enabled endoscopic diagnosis and treatment in the deep small bowel. Computed tomography (CT) enterography with negative oral contrast can evaluate masses, wall thickening, and narrowing of the small intestine. In addition, enhanced CT can detect abnormalities outside the gastrointestinal tract that endoscopy cannot evaluate. Each modality has its advantages and disadvantages, and a good combination of multiple modalities leads to an accurate diagnosis. As a first-line modality, three-phase enhanced CT is preferred. If CT shows a mass, stenosis, or wall thickening, a BAE should be selected. If there are no abnormal findings on CT and no obstructive symptoms, CE should be selected. If there are significant findings in the CE, determine the indication for BAE and its insertion route based on these findings. Early diagnosis of small intestinal tumors is essential for favorable outcomes. For early diagnosis, the possibility of small bowel lesions should be considered in patients with unexplained symptoms and signs after examination of the upper and lower gastrointestinal tract.

Keywords: balloon-assisted enteroscopy; capsule endoscopy; CT enterography

Citation: Yano, T.; Yamamoto, H. Endoscopic Diagnosis of Small Bowel Tumor. *Cancers* **2024**, *16*, 1704. https://doi.org/10.3390/cancers16091704

Academic Editors: Alain P. Gobert and Takuji Tanaka

Received: 25 January 2024
Revised: 22 April 2024
Accepted: 24 April 2024
Published: 27 April 2024

Copyright: © 2024 by the authors. Licensee MDPI, Basel, Switzerland. This article is an open access article distributed under the terms and conditions of the Creative Commons Attribution (CC BY) license (https:// creativecommons.org/licenses/by/ 4.0/).

1. Introduction

Small bowel tumors (SBTs) are relatively rare in incidence. They account for only approximately 3–6% of all gastrointestinal neoplasms. According to population-based cancer incidence data in the United States, the incidence of SBTs has increased over the past 20 years, from 5260 per year in 2004 to 12,440 in 2024 in the USA, and deaths due to SBTs have increased from 1130 in 2004 to 2090 in 2024 [1,2].

SBTs comprise different histological subtypes, including adenocarcinoma (30–45%), neuroendocrine tumors (20–40%), lymphomas (10–20%), and sarcomas (10–15%) [3]. Distribution varies geographically; in the United States, neuroendocrine tumors are most common (35–42%), followed by adenocarcinoma (30–40%) [4]. In Japan, lymphomas (47%) are the most common, followed by gastrointestinal stromal tumors (25%), and adenocarcinoma (24%) [5].

The term SBT often includes not only malignant neoplasms but also benign neoplasms and non-neoplastic lesions in the small bowel. In this review, the term SBT includes them.

2. Symptoms and Signs of Small Bowel Tumors

Recent technological advances, including capsule endoscopy (CE) and balloon-assisted endoscopy (BAE), have revealed that small intestinal disease is more common than previously thought. Although small bowel tumors are less common than in other gastrointestinal tracts, any disease is difficult to diagnose without "suspecting" it. When the following symptoms and signs are unexplained after examination of the upper and lower gastrointestinal tracts, small bowel disease may be the cause.

- Gastrointestinal bleeding
- Anemia
- Abdominal pain
- Obstructive symptom
- Body weight loss
- Palpable abdominal mass
- Fever of unknown origin

3. Family History

Family history is another important clue to the diagnosis of small bowel tumors because the following hereditary diseases have an increasing risk of small bowel tumors.

Lynch syndrome (HNPCC; hereditary non-polyposis colorectal cancer) is defined by germline mutations in one of the mismatch repair (MMR) genes, mostly *MLH1*, *MSH2*, and *MSH6*. Patients with Lynch syndrome have a 100-fold increased risk of small bowel cancer compared to the general population [6].

Patients affected by neurofibromatosis type 1 (NF-1), also known as von Recklinghausen disease, have an increased risk of developing gastrointestinal stromal tumors (GIST) [7].

Familial adenomatous polyposis of the colon (FAP) is a disease of autosomal dominant inheritance and is caused by a pathogenic germline variant in the adenomatous polyposis coli (APC) gene. In patients with FAP, the cumulative risk of duodenal cancer is estimated at 4% at 70 years of age [8].

Peutz–Jeghers syndrome (PJS) is caused by germline mutations in the serine-threonine kinase 11 (*STK11*) gene (formerly known as *LKB1*) located on chromosome 19p13.3 [9]. The lifetime risk of developing cancers in a PJS patient ranges from 55% to 83% by age 60–70, including colon cancer (39%), pancreatic cancer (11–36%), and small bowel cancer (29%) [10].

4. Characteristics of Each Modality for Small Intestinal Tumors

4.1. Capsule Endoscopy (CE)

CE has advantages, including a high diagnostic yield, being discomfort-free, its outpatient basis, and physiological images.

However, CE has several disadvantages. Because the lumen is not inflated by gas insufflation, diverticula and submucosal tumors (SMT) are often missed. Lack of irrigation and aspiration capabilities makes it difficult to detect lesions if there are a lot of residues. In patients with intestinal stenosis, there is a risk of retention. When the patency of the gastrointestinal tract cannot be confirmed, CE should not be used for patients with definitive obstructive symptoms.

CE depends on peristalsis to move, so it takes at least several hours to complete the test. CE cannot evaluate the bypassed intestinal tracts or the afferent limb after Roux-en-Y reconstruction in patients with surgically altered anatomy.

Because CE passage is too rapid in the duodenum and the proximal jejunum, CE can recognize only 42.7% to 43.6% of the duodenal papillae [11,12]. The lesions in the duodenum and the proximal jejunum can be missed by CE. Han et al. reported that small bowel tumors were not detected by CE but were eventually diagnosed by DBE in nine (16.7%) of 54 patients. Five lesions (55.6%) of the nine missed lesions were located in the proximal jejunum [13].

When there are multiple similar lesions, it is difficult to distinguish each and count the lesions. CE is not suitable for counting lesions in polyposis syndromes.

Even large tumors can be missed by CE. If the CE is caught on the proximal side of a large tumor, the tumor may not be detected, depending on the direction of the CE's camera. After staying for a while, the CE quickly slips through the large tumor area, and the large tumor is not captured by the CE. If the CE remains in the same location for more than 15 min, this is an abnormal finding known as a regional transit abnormality (RTA) [14].

CE can be very useful if it is used with an understanding of the above characteristics.

4.2. Balloon-Assisted Endoscopy (BAE)

The double-balloon endoscopy (DBE) is equipped with two balloons attached at the endoscope's tip and the overtube's tip. It enabled endoscopic diagnosis (Figure 1) and treatment in the deep small bowel. The single-balloon endoscopy (SBE), developed after DBE's launch, omits the balloon at the tip of the endoscope to simplify the preparation. Both are collectively referred to as balloon-assisted endoscopies (BAE). The BAE can be inserted with the aid of a ballooned overtube to suppress unnecessary deflection of the intestine. In addition, because the bowel is folded over the overtube, the BAE can be inserted into intestinal tracts longer than the working length of the scope. BAE can be inserted without relying on intestinal peristalsis, allowing evaluation of the afferent limb and bypassed intestinal tracts.

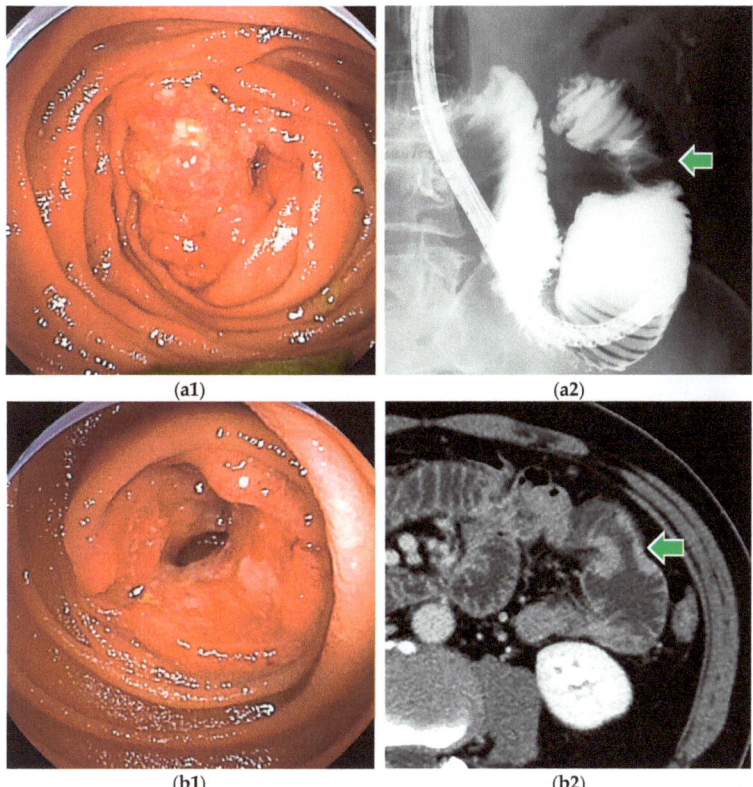

Figure 1. Small bowel adenocarcinomas, which are often advanced at the time of diagnosis, and endoscopic findings often include ulceration and stenosis. (**a1,a2**) DBE revealed adenocarcinoma in the proximal jejunum. Endoscopic enteroclysis showed the stenosis as an apple core sign. (**b1,b2**) DBE revealed adenocarcinoma with ulceration. CT showed mild stenosis.

The BAE is equipped with a working channel that allows for procedures such as tissue biopsy, marking, and treatment. Some kinds of small bower tumors can be treated by endoscopic treatment or chemotherapy. Before starting the chemotherapy, its histological diagnosis should be confirmed. When a small bowel tumor requires surgical treatment, endoscopic tattooing facilitates laparoscopic surgery. Chromoendoscopy with indigo-carmine makes it easy to detect small lesions in patients with FAP [15]. The miniature probe can be inserted into the working channel and enables endoscopic ultrasound evaluation for submucosal tumors [16].

In the setting of X-ray fluoroscopy, endoscopic enteroclysis can be performed by injecting contrast through the working channel. The size and shape of the lesion can be evaluated by fluoroscopy. During endoscopic enteroclysis, the scope balloon of DBE can inflate to reduce the backflow of contrast and evaluate the wide range of the small bowel by fluoroscopy.

BAE has several disadvantages. BAE requires endoscopic skills. Severe adhesions or stenosis make it difficult to achieve total enteroscopy with BAE. Especially near large tumors, maneuverability may be poor due to compression and adhesions caused by the tumor, and it may not be possible to reach the lesion.

4.3. Computed Tomography (CT)

Recent technological advances have increased the usefulness of computed tomography (CT) in the diagnosis of small bowel lesions. Since the introduction of MDCT (Multi-Detector-Row CT) with 4-row detectors in 1998, the number of detectors has increased and evolved to 16-row, 64-row, and 320-row. As a result, images with high spatial resolution can be obtained in a short time over the entire abdomen. In addition to axial section, coronal and sagittal section images can be reconstructed, and multiplanar reformation (MPR) images, virtual enteroscopy, and virtual enteroclysis are also available.

CT can evaluate ascites, misty mesentery, and abnormal blood vessels. In diagnosing small bowel tumors, CT is very useful in evaluating lymphadenopathy associated with malignant lymphoma and small intestinal cancer, as well as extra-luminal GIST, which is difficult to evaluate by endoscopy. However, plain CT provides a very limited amount of information and makes it difficult to detect lesions, so it is preferable to use at least contrast-enhanced CT and, if possible, dynamic CT. Some kinds of small bowel tumors are difficult to detect with conventional contrast-enhanced CT and are easily detected with dynamic CT. Shinya et al. reported that gastrointestinal tumors and neuroendocrine tumors demonstrated a hyper-vascular pattern in the multiphasic dynamic CT. Adenocarcinomas and lymphomas showed a delayed enhancement pattern [17].

One problem with CT is that CT images are momentary images, and depending on the timing of imaging, the shape of the intestinal tract due to peristalsis or spasm may appear as a stenosis or mass. Dynamic CT can solve this problem by comparing the intestinal geometry between images taken at different times (plain, early contrast, and late contrast). Dynamic CT makes it easier to distinguish intestinal stricture from peristalsis and spasms of the intestinal tract.

CT enterography, in which a bowel cleansing medium such as polyethylene glycol is taken as a negative contrast agent before the CT scan, provides detailed imaging of the small bowel by adequate lumen distention and provides information on masses, wall thickening, and narrowing of the small bowel. According to a meta-analysis, the sensitivity and specificity of CT enterography for small bowel tumors were 0.93 and 0.83, respectively [18]. Although there are problems with radiation exposure and side effects from contrast media, it is a minimally invasive test that provides a large amount of information quickly. Magnetic resonance enterography has similar sensitivity and specificity [18] and can be an alternative with no radiation exposure when it is available.

5. Diagnostic Strategy for Small Bowel Tumors

Each modality has its advantages and disadvantages, and a good combination of multiple modalities leads to an accurate diagnosis since a false negative or false positive result is possible with a single modality alone.

Honda et al. reported a comparative study of the diagnostic yields of contrast-enhanced CT, fluoroscopic enteroclysis, CE, and DBE for small bowel tumors [19]. In their comparing study, diagnostic yields for small bowel tumors </=10 mm were significantly low in contrast-enhanced CT and fluoroscopic enteroclysis. However, the diagnostic yields of CE and DBE were high for small bowel tumors, regardless of size. In contrast-enhanced CT, the diagnostic yield of epithelial tumors was significantly lower compared with subepithelial tumors. The diagnostic yields of CE and DBE were significantly higher than those of contrast-enhanced CT, and the diagnostic yield of DBE was significantly higher than that of CE. However, a combination of contrast-enhanced CT and CE had a diagnostic yield similar to that of DBE. Because CE and CT can cover each other's shortcomings in detecting small bowel tumors, the combination use of CE and contrast-enhanced CT is recommended for detecting small bowel tumors. After the screening, DBE is useful for histologic diagnosis and endoscopic treatment.

Based on the results of the above study, we recommend the following strategies (graphic abstract).

As a first-line modality, dynamic CT is preferred because CT findings are informative to select the next test, CE or BAE. In addition to axial images, coronal images should be produced for precise reading. CT enterography makes it easier to detect masses in the small intestine. In clinical practice, CT and colonoscopy can be scheduled on the same day, and CT can be taken before the colonoscopy to obtain CT enterography images.

If CT shows a mass, stenosis, or wall thickening, a BAE should be selected because of its capability for biopsy and marking. The route of insertion of the BAE should be selected based on the information obtained from the CT.

If there are no abnormal findings on CT and no obstructive symptoms, CE should be selected. If there are significant findings in the CE, determine the indication for BAE and its insertion route based on these findings. However, CE often misses diverticula and submucosal tumors due to its inability to insufflate gas. It can also miss lesions in the duodenum and the proximal jejunum due to its rapid movement. CE cannot evaluate the bypassed intestinal tracts or the afferent limb in patients with surgically altered anatomy. After negative CE, the indication of further examinations should be decided with an understanding of the above characteristics of CE.

6. The Role of Enteroscopy in Each Disease

6.1. Small Bowel Adenocarcinoma

The rate of primary small bowel adenocarcinoma is less than 3% of all gastrointestinal cancers. Risk factors for small bowel adenocarcinoma include FAP, HNPCC, PJS, Crohn's disease, and celiac disease.

Most primary small bowel adenocarcinomas arise in the duodenum, the proximal jejunum, or the distal ileum. They are often found with obstructive symptoms or chronic iron deficiency anemia.

Although they can sometimes be reached with a conventional endoscope, the range of routine upper gastrointestinal endoscopy and colonoscopy does not include the deep duodenum, the proximal jejunum, or the distal ileum. To endoscopically diagnose small bowel adenocarcinoma, intentional deep insertion is necessary. As a result, at the time of diagnosis, most of the patients were in an advanced stage with metastasis to other organs or peritoneal dissemination. The multi-center retrospective study, which included 354 patients with primary small bowel adenocarcinoma, reported that the rates for clinical stages 0, I, II, III, and IV at the time of diagnosis were 5.4%, 2.5%, 27.1%, 26.0%, and 35.6%, respectively [5]. The tumor stage is the most important prognostic factor for small bowel adenocarcinoma. Therefore, for early diagnosis, the possibility of small bowel lesions should be considered in patients with unexplained symptoms and signs after routine upper gastrointestinal endoscopy and colonoscopy.

BAE, or push enteroscopy, can reach the lesion (Figure 1), take a biopsy for histopathologic diagnosis, and mark it by tattooing for surgical treatment. Endoscopic findings of small bowel adenocarcinoma often include ulceration and stenosis. Type 2 (54.2%) was the most common among the macroscopic types, followed by Type 3 (18.2%) [5].

6.2. Gastrointestinal Stromal Tumor

Gastrointestinal stromal tumors (GIST) are often caused by gastrointestinal bleeding but can also be found incidentally on contrast-enhanced CT for the evaluation of other diseases. GIST is a mesenchymal malignancy derived from the interstitial cells of Cajal that control intestinal peristalsis. Patients with NF-1 are often associated with GISTs. NF1-associated GISTs occur in younger patients compared with sporadic GISTs and often multiple tumors, mainly incidental, localized at the small bowel and in the absence of KIT/PDGRFα mutations [20,21].

GIST growth patterns include extraluminal, intraluminal, or mixed (dumbbell-shaped) patterns. GIST is a submucosal tumor covered by normal mucosa. Because it is difficult to distinguish from extraluminal compression, GISTs are often missed by CE. Ulcers/erosions or dilated abnormal vessels are important findings for detecting GIST using CE.

Although GIST with intraluminal and mixed patterns can be detected by BAE (Figure 2), GIST with extraluminal patterns is hardly detected by endoscopy, except for abnormal vessels and unnatural traction findings due to lesions.

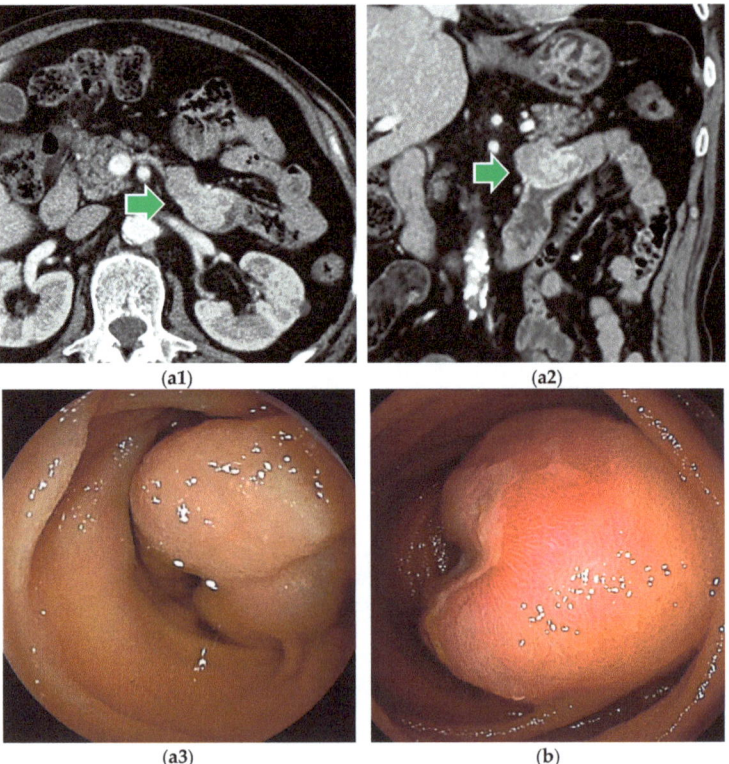

Figure 2. Gastrointestinal stromal tumor (GIST): (**a1–a3**) Contrast-enhanced CT revealed a well-enhanced lesion. DBE showed a submucosal tumor covered by normal mucosa. (**b**) In patients with bleeding symptoms, erosions, ulcers, or dilated blood vessels are seen on the surface.

Tattooing by BAE is helpful for identifying the lesion during laparoscopic-assisted partial resection of the small intestine.

Endoscopic ultrasound fine needle aspiration (EUS-FNA) became a routine examination for evaluating gastric GIST. However, EUS-FNA is not available for small-bowel GIST due to a lack of dedicated equipment. Endoscopic biopsy for GIST has a low diagnostic rate and cannot determine malignancy accurately due to its subepithelial nature [22,23]. Endoscopic biopsy also carries the risk of post-biopsy bleeding. Biopsy is indicated only when contrast-enhanced CT or endoscopic findings are atypical or when histopathology is necessary prior to chemotherapy for unresectable lesions.

Symptomatic GISTs, regardless of size, are indicated for surgical resection if they are resectable. The indication for surgical treatment for asymptomatic GISTs is determined by their size and rate of growth. Small-bowel GISTs have a significantly higher metastatic risk than gastric GISTs [20]. However, for multiple GISTs in NF-1 patients, small lesions may be followed up without surgical resection since they have favorable histologic parameters (relatively low mitotic rates) [24], and it is difficult to resect all multiple lesions.

6.3. Malignant Lymphoma

Malignant lymphomas in the small intestine have been conventionally diagnosed by radiographic examinations, and surgical resections were required for histological diagnosis. BAE enables tissue biopsy and histopathological diagnosis of primary small intestinal lymphoma without surgical resection.

The macroscopic findings of small bowel lymphoma are classified as polypoid, ulcerative (including stricturing, non-stricturing, and aneurysmal forms), polyposis (multiple lymphomatous polyposis), diffusely infiltrating, or mixed type. There is some correlation between the macroscopic and histological types. Many cases of ulcerative type are histologically diffuse large B-cell lymphoma (DLBCL). Most cases of polyposis type (multiple lymphomatous polyposis) are follicular lymphoma or mantle cell lymphoma, while diffusely infiltrating type tends to comprise either T-cell lymphomas or immunoproliferative small intestinal disease [25].

Although endoscopic findings of malignant lymphomas vary by histologic type (Figure 3), a definitive histopathologic diagnosis can be made by biopsy in most cases. Based on the histopathologic diagnosis, lymphomas can be treated with chemotherapy [26,27]. However, in cases of bleeding or obstructive symptoms, chemotherapy is given after surgical treatment.

Endoscopic tattooing is useful for recognizing lesions during surgical treatment and for identifying lesion sites for follow-up BAE after chemotherapy when complete remission is achieved.

Endoscopic balloon dilation for post-chemotherapy stenosis is an alternative option that avoids surgical treatment [28].

6.4. Neuroendocrine Tumor

Gastrointestinal neuroendocrine tumors (GI NET), formerly known as carcinoids. According to the WHO classification, NETs are classified into low-grade neuroendocrine tumors (NETs) and high-grade neuroendocrine carcinomas (NECs). NETs are also classified into G1, less malignant, and G2, more malignant than G1.

The secretion of serotonin and other hormones from NET can cause facial flushing, diarrhea, bronchospasm, and other symptoms known as carcinoid syndrome. The carcinoid syndrome is more frequent in NET, especially those occurring in the jejunum, ileum, and appendix, than in other gastrointestinal sites.

Multifocal NET may occur in 30–50% of patients. Patients with multiple lesions are younger than those with solitary tumors, have a significantly higher risk of developing carcinoid syndrome, and have a poorer prognosis [29].

The endoscopic image of NET is yellowish SMT-like, but it is precisely a tumor of epithelial origin. It is often seen as an elastic, hard, mobile tumor with atrophied surface villi and dilated capillaries (Figure 4a). Depressions, ulcers, or erosions on the surface may indicate a high-grade lesion (Figure 4b).

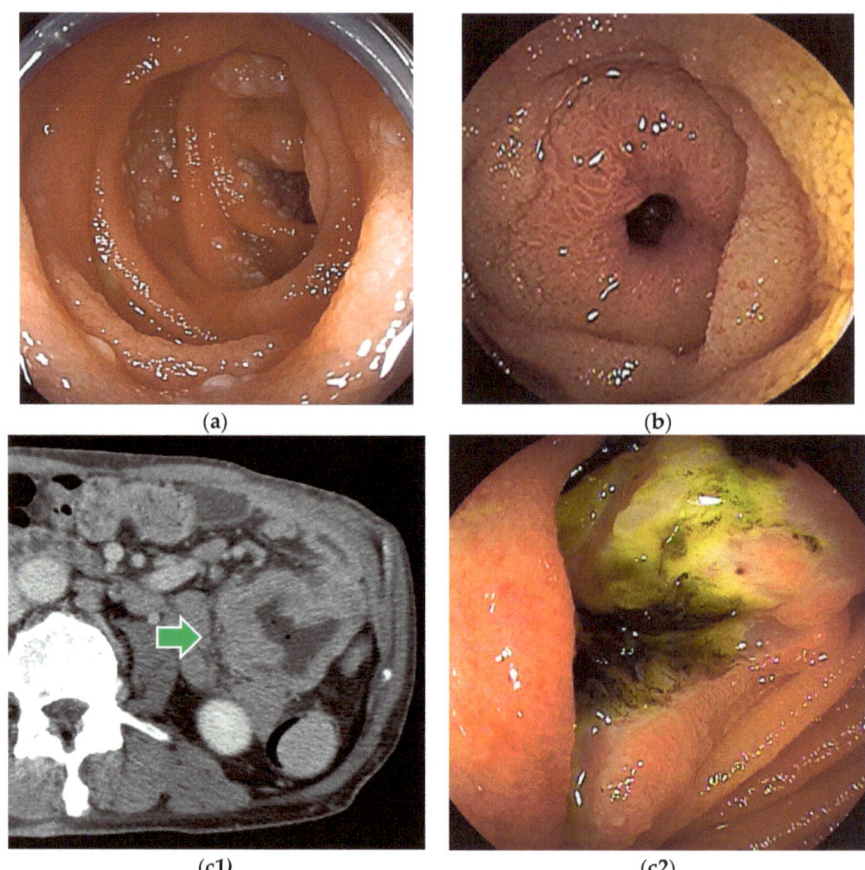

Figure 3. Malignant lymphoma: (**a**) Follicular lymphoma is characterized by aggregations of large and small white granules. The lesions are distributed focally from the duodenum to the jejunum. (**b**) Rarely, it may be a form of concentric stenosis with ulceration. (**c1,c2**) Diffuse large B-cell lymphoma (DLBCL) often shows an ulcerated or polypoid morphology, and the biopsy should be taken from the ulcer bed rather than from the edges. CT revealed a wall-thickened intestine with a dilated lumen. DBE showed an ulcerated lesion.

Although CE may be useful in identifying NETs, it cannot confirm the correct number of multiple lesions or perform marking at the lesions.

The sensitivity of BAE for the primary SB-NET was 88%, compared to 60% for CT, 54% for MRI, and 56% for somatostatin receptor imaging. BAE could also be considered for detecting multifocal NETs before surgery. In patients who underwent small bowel resection, additional lesions were found in 54% of patients at preoperative BAE, but only 18% of patients at preoperative CE [30].

Endoscopic resection is not recommended for jejunal and ileal NETs due to the risk of invasion and lymphatic spread, even with diminutive lesions, which may necessitate a more extensive surgical resection.

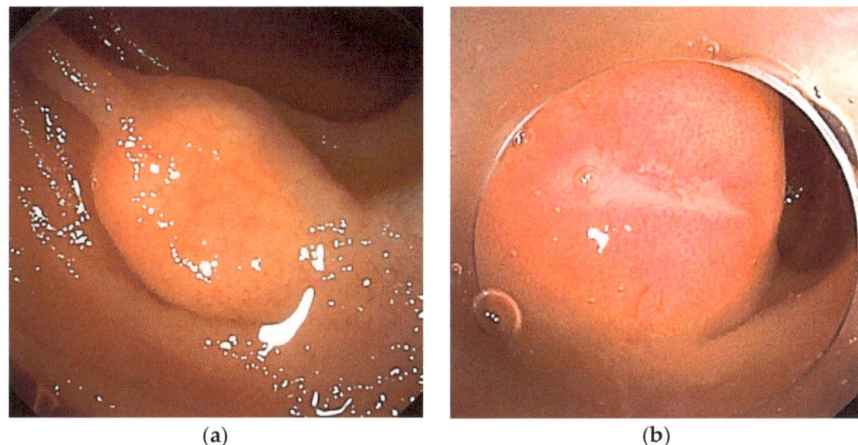

Figure 4. Neuroendocrine tumor: (**a**) SMT-like lesion with atrophied surface villi and dilated capillaries (**b**) A high-grade lesion with ulcers on the surface.

6.5. Metastatic Tumors

Metastatic tumors of the small bowel include direct invasion from other organs, intraperitoneal disseminated tumors, and metastatic tumors due to hematogenous metastasis. Pancreatic cancer frequently directly invades the duodenum. Colon, ovary, uterus, and stomach cancers metastasize to the small intestine through direct invasion or intraperitoneal dissemination. Lung cancer, breast cancer, and malignant melanoma metastasize hematogenously to the small intestine. The most common primary tumor for metastatic tumors in the small intestine is lung cancer.

The endoscopic image of the metastatic lesion is variable (Figure 5). Metastatic lesions can present as single or multiple polypoid lesions, with or without ulcers. Metastatic tumors arise from the submucosa and are sometimes difficult to distinguish from malignant lymphomas on endoscopic images. They may be seen as focal bowel wall thickening and cause luminal narrowing.

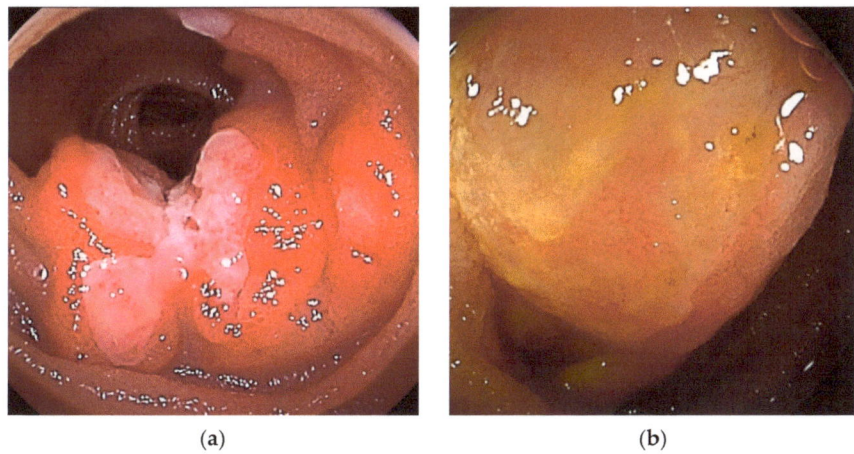

Figure 5. Metastatic tumors: (**a**) metastatic jejunal tumor from angiosarcoma of the breast. (**b**) Metastatic jejunal tumor from lung cancer.

The management of the metastatic lesion depends on the symptoms and stage of the primary tumor. Surgical resection of the affected intestine can be useful to relieve symptoms such as obstruction and bleeding.

6.6. Benign Tumors

Many benign tumors of the small intestine are asymptomatic, making it difficult to calculate their exact prevalence. Benign tumors can cause overt or obscure bleeding with chronic anemia. Larger tumors can cause obstructive symptoms due to a narrowing of the lumen or intussusception. In the past, they were often found by chance during surgery or autopsy of other diseases, but with the widespread use of CE and BAE, there are more and more opportunities for diagnosis by endoscopy.

Benign small bowel tumors include adenomas, hamartomas, lipomas (Figure 6a), hemangiomas (Figure 6b), lymphangiomas (Figure 7a,b), and inflammatory fibroid polyps (Figure 8a). The ectopic pancreas can be found as a submucosal tumor-like lesion (Figure 8b).

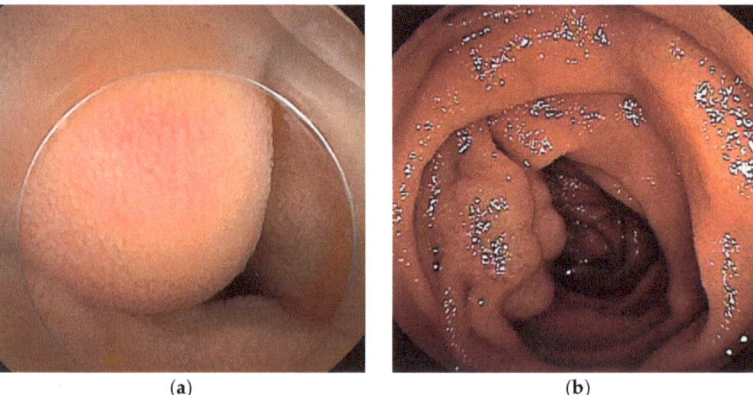

(a) (b)

Figure 6. (**a**) Lipoma: Lipomas are yellowish-white submucosal tumors that are soft and deform when pressed with forceps. It is characterized by low density on CT as well as fatty tissue and high echoic lesions on ultrasound. (**b**) Cavernous hemangioma: Localized cavernous hemangiomas are soft, pale to dark red submucosal tumors with a smooth surface.

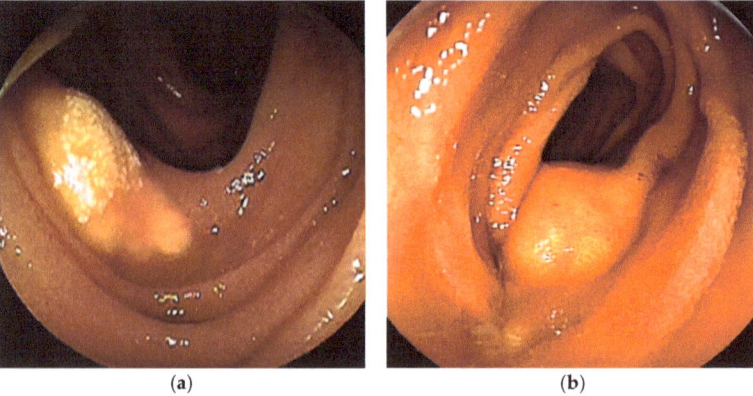

(a) (b)

Figure 7. Lymphangioma: Lymphangiomas have different endoscopic appearances depending on the depth of the dilated lymphatic channels. (**a**) If the lesion has dilated lymphatic channels in the mucosa, it will be an elevation with white dots on the surface. (**b**) If the lesion is primarily in the submucosa, it will have a smooth surface and a yellowish-white to pale blue submucosal tumor without white dots.

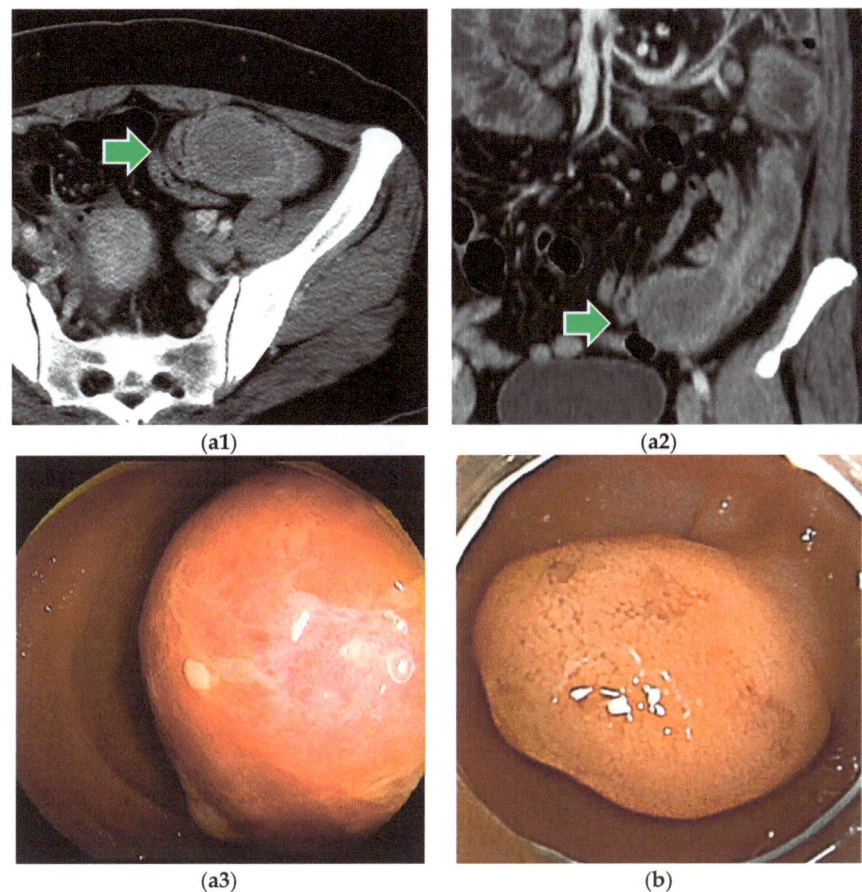

Figure 8. (**a1–a3**) Inflammatory fibroid polyp (IFP): IFP is a pedunculated or sub-pedunculated lesion that may be found as a submucosal tumor, but as it grows, the mucosa is shed by mechanical stimulation and is found as a smooth, protruding lesion. A large IFP can cause intussusception. This lesion was treated by a laparoscopy-assisted partial small bowel resection. (**b**) Ectopic pancreas: Ectopic pancreas presents as a 10~20 mm-sized, hemispherical, or sub-pedunculated SMT-like appearance with multiple nodules covered by thin, normal mucosa reflecting the internal multifocal structure, with a slight depression on the surface.

Adenoma can be treated by various techniques, such as cold snare polypectomy (CSP), endoscopic mucosal resection (EMR), underwater endoscopic mucosal resection (UEMR) [31,32], gel immersion endoscopic mucosal resection (GIEMR) [33–38], and endoscopic submucosal dissection (ESD). Hamartomas [39], lipomas [40,41], lymphangiomas [42], and inflammatory fibroid polyps [38] can also be resected endoscopically. Hemangioma, especially in patients with blue rubber bleb nevus syndrome [43], can be treated by various techniques, such as polidocanol injection therapy [44], electro-coagulation [45], polypectomy [46], band-ligation [47], and loop-ligation [48]. Of course, surgical treatment should be considered for massive lesions, even benign tumors.

6.7. Peutz–Jeghers Syndrome

Peutz–Jeghers syndrome (PJS) is an inherited polyposis syndrome characterized by the presence of multiple hamartomatous polyps (Figure 9) throughout the gastrointestinal tract, excluding the esophagus. It is accompanied by mucocutaneous melanin pigmentation and an elevated lifetime risk of both gastrointestinal and extra-gastrointestinal malignancies [10,49]. PJS is inherited in an autosomal dominant manner, yet around 45% of PJS patients are de novo cases. The estimated incidence has been reported at 1/200,000 [50].

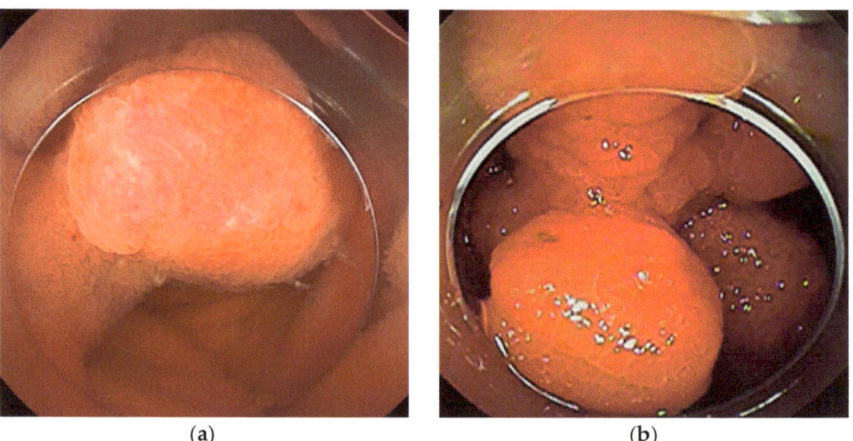

Figure 9. Polyps in patients with Peutz–Jeghers syndrome: (**a**) Most polyps in patients with Peutz–Jeghers syndrome are pedunculated or sub-pedunculated polyps. (**b**) Some of the growing lesions are branched, bifid, or multinodular, reflecting dendritic growth of the muscularis mucosa.

The malignant potential of polyps is low, but the polyps can enlarge, resulting in intussusception and emergency laparotomy. Polyps can develop and grow throughout life, and repeated surgical treatment can lead to short bowel syndrome. Because intra-abdominal adhesions due to surgery can cause difficulty in total enteroscopy with BAE, the digestive tract should be examined, and endoscopic treatment with BAE should be initiated before emergency laparotomy for intussusception. In patients with PJS, gastrointestinal surveillance through upper gastrointestinal endoscopy, colonoscopy, and CE should begin by the age of 8 years old at the latest. European and Japanese guidelines recommend that SB polyps > 15 mm be treated to prevent intussusception [51].

The conventional techniques for polyp removal in PJS are snare polypectomy and endoscopic mucosal resection [52,53]. However, these conventional techniques have a risk of complications such as perforation and bleeding. Recently, endoscopic ischemic polypectomy (EIP) has been described, which involves strangulating the stalk of a polyp using a detachable snare [54] or clips [55] to induce polyp destruction. Performing EIP with clips is technically easier compared to conventional techniques because EIP requires visualization of only the stalk of the polyp, even within a limited working space. The advantages of EIP include the removal of a larger number of polyps and the prevention of complications after polypectomy, such as bleeding, perforation, or post-polypectomy syndrome [56,57]. Considering the low risk of adverse events in EIP, EIP has the potential to change the size threshold for endoscopic treatment. The disadvantage of EIP is a lack of histopathological evaluation for treated polyps.

Endoscopic reduction of intussusception in patients with PJS is a viable alternative to surgery, except for patients with necrosis or perforation. The reported success rate of endoscopic reduction of 22 sites in 19 patients was 95%, with only two mild pancreatitis adverse events [58].

6.8. Familial Adenomatous Polyposis

Familial adenomatous polyposis (FAP) is a rare genetic predisposition primarily to digestive cancers, inherited in a dominant manner (APC gene) or recessive manner (MUTYH gene) for the main types of FAP. The definition of classical FAP is based on the presence of at least 100 colorectal adenomas [59]. Adenomatous polyps in the duodenum are found in nearly 100% of individuals with classical APC-related FAP and in approximately 30% of patients with biallelic MUTYH mutations, and less frequently in the distal small intestine [60]. In patients with FAP, other than colorectal cancer, duodenal cancer and desmoid tumors are significant contributors to mortality [61].

A systematic gradient from the duodenum to the jejunum is observed, as indicated by the low risk of small bowel cancer as far as the distal small bowel is considered [62]. Endoscopic surveillance for duodenal adenomas is recommended to start around 25 years old [6,63]. Chromoendoscopy with indigo carmine is useful in detecting small lesions (Figure 10a). A large adenoma in the proximal jejunum is sometimes detected by double-balloon enteroscopy after a long period of surveillance and treatment using only conventional upper gastrointestinal endoscopy (Figure 10b) [64]. Endoscopic surveillance for jejunal adenomas should be considered at several-year intervals, depending on the severity of the patient. Chromoendoscopy with indigo carmine can increase the detection of adenomas [15,65].

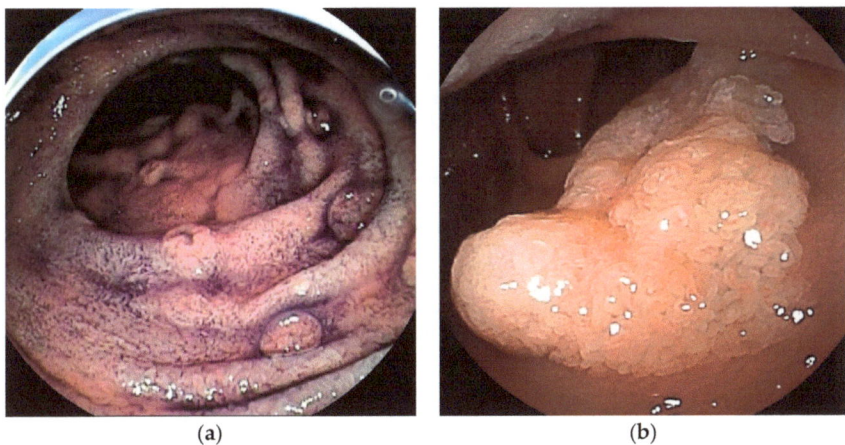

Figure 10. Polyps in patients with familial adenomatous polyposis: (**a**) Small, flat lesions often occur from the duodenum to the jejunum. (**b**) A large adenoma in the proximal jejunum was detected by double-balloon enteroscopy after a long period of surveillance and treatment using only conventional upper gastrointestinal endoscopy. This lesion was treated by endoscopic submucosal dissection (ESD) using the pocket-creation method and balloon-assisted endoscopy [64].

In the recent ESGE guideline, endoscopic treatment of lesions in the duodenum and jejunum has been recommended for lesions larger than 10 mm in size [6] because of endoscopic maneuverability and the risk of complications. However, the size threshold for endoscopic resection should be optimized by weighing the risks and benefits. The feasibility and safety of performing a cold snare polypectomy for duodenal adenomas in patients with FAP were reported [66]. CSP enabled the removal of a greater number of polyps in a shorter duration while maintaining safety [67]. Underwater endoscopic mucosal resection [31] for sporadic nonampullary duodenal adenoma [32] is reported as a safe and effective procedure. These new techniques have the potential to change the size threshold for endoscopic resection. Takeuchi et al. reported that intensive downstaging polypectomy using the new techniques showed significant downstaging with acceptable adverse events for multiple duodenal adenomas in patients with FAP [68].

Each adenoma may be fused in patients with multiple adenomas and become a larger lesion. Endoscopic resection of larger adenomas may increase the risk of adverse events. At least in patients with a large number of adenomas, intensive downstaging polypectomy should be considered, even without adenomas larger than 10 mm.

7. Conclusions

Advances in various medical technologies have greatly advanced the diagnosis and treatment of small intestinal diseases. However, small intestinal cancer is often found at an advanced stage. Early diagnosis of small intestinal tumors is essential for favorable outcomes. For early diagnosis, the possibility of small bowel lesions should be considered in patients with unexplained symptoms and signs after examination of the upper and lower gastrointestinal tract.

Author Contributions: Writing—original draft preparation, T.Y.; writing—review and editing, H.Y. All authors have read and agreed to the published version of the manuscript.

Funding: This review received no external funding.

Conflicts of Interest: Hironori Yamamoto has patents for double-balloon endoscopy and a consultant relationship with Fujifilm. Tomonori Yano has received research funding and honoraria from Fujifilm.

Abbreviations

BAE	balloon-assisted endoscopy
CE	capsule endoscopy
CSP	cold snare polypectomy
CT	computed tomography
DBE	double-balloon endoscopy
DLBCL	diffuse large B-cell lymphoma
EIP	endoscopic ischemic polypectomy
EMR	endoscopic mucosal resection
ESD	endoscopic submucosal dissection
FAP	familial adenomatous polyposis
GIST	gastrointestinal stromal tumor
HNPCC	hereditary non-polyposis colorectal cancer
IFP	inflammatory fibroid polyp
MDCT	multidetector-row computed tomography
NEC	neuroendocrine carcinoma
NET	neuroendocrine tumor
PJS	Peutz–Jeghers syndrome
RTA	regional transit abnormality
SBE	single-balloon endoscopy
SBT	small bowel tumor
SMT	submucosal tumor
UEMR	underwater endoscopic mucosal resection

References

1. Jemal, A.; Tiwari, R.C.; Murray, T.; Ghafoor, A.; Samuels, A.; Ward, E.; Feuer, E.J.; Thun, M.J. Cancer statistics, 2004. *CA Cancer J. Clin.* **2004**, *54*, 8–29. [CrossRef] [PubMed]
2. Siegel, R.L.; Giaquinto, A.N.; Jemal, A. Cancer statistics, 2024. *CA Cancer J. Clin.* **2024**, *74*, 12–49. [CrossRef] [PubMed]
3. Rondonotti, E.; Koulaouzidis, A.; Georgiou, J.; Pennazio, M. Small bowel tumours: Update in diagnosis and management. *Curr. Opin. Gastroenterol.* **2018**, *34*, 159–164. [CrossRef]
4. Haselkorn, T.; Whittemore, A.S.; Lilienfeld, D.E. Incidence of small bowel cancer in the United States and worldwide: Geographic, temporal, and racial differences. *Cancer Causes Control* **2005**, *16*, 781–787. [CrossRef]
5. Yamashita, K.; Oka, S.; Yamada, T.; Mitsui, K.; Yamamoto, H.; Takahashi, K.; Shiomi, A.; Hotta, K.; Takeuchi, Y.; Kuwai, T.; et al. Clinicopathological features and prognosis of primary small bowel adenocarcinoma: A large multicenter analysis of the JSCCR database in Japan. *J. Gastroenterol.* **2024**, *59*, 376–388. [CrossRef] [PubMed]
6. van Leerdam, M.E.; Roos, V.H.; van Hooft, J.E.; Dekker, E.; Jover, R.; Kaminski, M.F.; Latchford, A.; Neumann, H.; Pellise, M.; Saurin, J.C.; et al. Endoscopic management of polyposis syndromes: European Society of Gastrointestinal Endoscopy (ESGE) Guideline. *Endoscopy* **2019**, *51*, 877–895. [CrossRef] [PubMed]
7. Mussi, C.; Schildhaus, H.U.; Gronchi, A.; Wardelmann, E.; Hohenberger, P. Therapeutic consequences from molecular biology for gastrointestinal stromal tumor patients affected by neurofibromatosis type 1. *Clin. Cancer Res.* **2008**, *14*, 4550–4555. [CrossRef] [PubMed]

8. Saurin, J.C.; Gutknecht, C.; Napoleon, B.; Chavaillon, A.; Ecochard, R.; Scoazec, J.Y.; Ponchon, T.; Chayvialle, J.A. Surveillance of duodenal adenomas in familial adenomatous polyposis reveals high cumulative risk of advanced disease. *J. Clin. Oncol.* **2004**, *22*, 493–498. [CrossRef]
9. Jenne, D.E.; Reimann, H.; Nezu, J.; Friedel, W.; Loff, S.; Jeschke, R.; Müller, O.; Back, W.; Zimmer, M. Peutz-Jeghers syndrome is caused by mutations in a novel serine threonine kinase. *Nat. Genet.* **1998**, *18*, 38–43. [CrossRef]
10. Yehia, L.; Heald, B.; Eng, C. Clinical Spectrum and Science Behind the Hamartomatous Polyposis Syndromes. *Gastroenterology* **2023**, *164*, 800–811. [CrossRef]
11. Monteiro, S.; de Castro, F.D.; Carvalho, P.B.; Moreira, M.J.; Rosa, B.; Cotter, J. PillCam(R) SB3 capsule: Does the increased frame rate eliminate the risk of missing lesions? *World J. Gastroenterol.* **2016**, *22*, 3066–3068. [CrossRef] [PubMed]
12. Kong, H.; Kim, Y.S.; Hyun, J.J.; Cho, Y.J.; Keum, B.; Jeen, Y.T.; Lee, H.S.; Chun, H.J.; Um, S.H.; Lee, S.W.; et al. Limited ability of capsule endoscopy to detect normally positioned duodenal papilla. *Gastrointest. Endosc.* **2006**, *64*, 538–541. [CrossRef] [PubMed]
13. Han, J.W.; Hong, S.N.; Jang, H.J.; Jeon, S.R.; Cha, J.M.; Park, S.J.; Byeon, J.S.; Ko, B.M.; Kim, E.R.; Choi, H.; et al. Clinical Efficacy of Various Diagnostic Tests for Small Bowel Tumors and Clinical Features of Tumors Missed by Capsule Endoscopy. *Gastroenterol. Res. Pract.* **2015**, *2015*, 623208. [CrossRef] [PubMed]
14. Tang, S.J.; Zanati, S.; Dubcenco, E.; Christodoulou, D.; Cirocco, M.; Kandel, G.; Kortan, P.; Haber, G.B.; Marcon, N.E. Capsule endoscopy regional transit abnormality: A sign of underlying small bowel pathology. *Gastrointest. Endosc.* **2003**, *58*, 598–602. [PubMed]
15. Monkemuller, K.; Fry, L.C.; Ebert, M.; Bellutti, M.; Venerito, M.; Knippig, C.; Rickes, S.; Muschke, P.; Rocken, C.; Malfertheiner, P. Feasibility of double-balloon enteroscopy-assisted chromoendoscopy of the small bowel in patients with familial adenomatous polyposis. *Endoscopy* **2007**, *39*, 52–57. [CrossRef] [PubMed]
16. Wada, M.; Lefor, A.T.; Mutoh, H.; Yano, T.; Hayashi, Y.; Sunada, K.; Nishimura, N.; Miura, Y.; Sato, H.; Shinhata, H.; et al. Endoscopic ultrasound with double-balloon endoscopy in the evaluation of small-bowel disease. *Surg. Endosc.* **2014**, *28*, 2428–2436. [CrossRef] [PubMed]
17. Shinya, T.; Inai, R.; Tanaka, T.; Akagi, N.; Sato, S.; Yoshino, T.; Kanazawa, S. Small bowel neoplasms: Enhancement patterns and differentiation using post-contrast multiphasic multidetector CT. *Abdom. Radiol.* **2017**, *42*, 794–801. [CrossRef] [PubMed]
18. Li, R.; Ye, S.; Zhou, C.; Liu, F.; Li, X. A systematic review and meta-analysis of magnetic resonance and computed tomography enterography in the diagnosis of small intestinal tumors. *PeerJ* **2023**, *11*, e16687. [CrossRef] [PubMed]
19. Honda, W.; Ohmiya, N.; Hirooka, Y.; Nakamura, M.; Miyahara, R.; Ohno, E.; Kawashima, H.; Itoh, A.; Watanabe, O.; Ando, T.; et al. Enteroscopic and radiologic diagnoses, treatment, and prognoses of small-bowel tumors. *Gastrointest. Endosc.* **2012**, *76*, 344–354. [CrossRef]
20. Miettinen, M.; Lasota, J. Gastrointestinal stromal tumors. *Gastroenterol. Clin. N. Am.* **2013**, *42*, 399–415. [CrossRef]
21. Salvi, P.F.; Lorenzon, L.; Caterino, S.; Antolino, L.; Antonelli, M.S.; Balducci, G. Gastrointestinal stromal tumors associated with neurofibromatosis 1: A single centre experience and systematic review of the literature including 252 cases. *Int. J. Surg. Oncol.* **2013**, *2013*, 398570. [CrossRef] [PubMed]
22. Nakano, A.; Nakamura, M.; Watanabe, O.; Yamamura, T.; Funasaka, K.; Ohno, E.; Kawashima, H.; Miyahara, R.; Goto, H.; Hirooka, Y. Endoscopic Characteristics, Risk Grade, and Prognostic Prediction in Gastrointestinal Stromal Tumors of the Small Bowel. *Digestion* **2017**, *95*, 122–131. [CrossRef] [PubMed]
23. Martinez-Alcalá, A.; Fry, L.C.; Kröner, T.; Peter, S.; Contreras, C.; Mönkemüller, K. Endoscopic spectrum and practical classification of small bowel gastrointestinal stromal tumors (GISTs) detected during double-balloon enteroscopy. *Endosc. Int. Open* **2021**, *9*, E507–E512. [CrossRef] [PubMed]
24. Miettinen, M.; Lasota, J. Histopathology of gastrointestinal stromal tumor. *J. Surg. Oncol.* **2011**, *104*, 865–873. [CrossRef] [PubMed]
25. Nakamura, S.; Matsumoto, T. Gastrointestinal lymphoma: Recent advances in diagnosis and treatment. *Digestion* **2013**, *87*, 182–188. [CrossRef] [PubMed]
26. Kobayashi, H.; Nagai, T.; Omine, K.; Sato, K.; Ozaki, K.; Suzuki, T.; Mori, M.; Muroi, K.; Yano, T.; Yamamoto, H.; et al. Clinical outcome of non-surgical treatment for primary small intestinal lymphoma diagnosed with double-balloon endoscopy. *Leuk. Lymphoma* **2013**, *54*, 731–736. [CrossRef] [PubMed]
27. Tian, F.Y.; Wang, J.X.; Huang, G.; An, W.; Ai, L.S.; Wang, S.; Wang, P.Z.; Yu, Y.B.; Zuo, X.L.; Li, Y.Q. Clinical and endoscopic features of primary small bowel lymphoma: A single-center experience from mainland China. *Front. Oncol.* **2023**, *13*, 1142133. [CrossRef] [PubMed]
28. Magome, S.; Sakamoto, H.; Shinozaki, S.; Okada, M.; Yano, T.; Sunada, K.; Lefor, A.K.; Yamamoto, H. Double-Balloon Endoscopy-Assisted Balloon Dilation of Strictures Secondary to Small-Intestinal Lymphoma. *Clin. Endosc.* **2020**, *53*, 101–105. [CrossRef] [PubMed]
29. Yantiss, R.K.; Odze, R.D.; Farraye, F.A.; Rosenberg, A.E. Solitary versus multiple carcinoid tumors of the ileum: A clinical and pathologic review of 68 cases. *Am. J. Surg. Pathol.* **2003**, *27*, 811–817. [CrossRef]
30. Manguso, N.; Gangi, A.; Johnson, J.; Harit, A.; Nissen, N.; Jamil, L.; Lo, S.; Wachsman, A.; Hendifar, A.; Amersi, F. The role of pre-operative imaging and double balloon enteroscopy in the surgical management of small bowel neuroendocrine tumors: Is it necessary? *J. Surg. Oncol.* **2018**, *117*, 207–212. [CrossRef]
31. Binmoeller, K.F.; Shah, J.N.; Bhat, Y.M.; Kane, S.D. "Underwater" EMR of sporadic laterally spreading nonampullary duodenal adenomas (with video). *Gastrointest. Endosc.* **2013**, *78*, 496–502. [CrossRef] [PubMed]

32. Yamasaki, Y.; Uedo, N.; Takeuchi, Y.; Higashino, K.; Hanaoka, N.; Akasaka, T.; Kato, M.; Hamada, K.; Tonai, Y.; Matsuura, N.; et al. Underwater endoscopic mucosal resection for superficial nonampullary duodenal adenomas. *Endoscopy* **2018**, *50*, 154–158. [CrossRef]
33. Toyonaga, H.; Harada, T.; Katanuma, A. Gel Immersion Endoscopic Mucosal Resection for a Large Non-Ampullary Duodenal Adenoma. *Am. J. Gastroenterol.* **2022**, *117*, 1402. [CrossRef]
34. Miyakawa, A.; Kuwai, T.; Miyauchi, T.; Shimura, H.; Shimura, K. Gel immersion endoscopy-facilitated endoscopic mucosal resection of a superficial nonampullary duodenal epithelial tumor: A novel approach. *VideoGIE* **2021**, *6*, 422–426. [CrossRef]
35. Amino, H.; Yamashina, T.; Marusawa, H. Under-gel Endoscopic Mucosal Resection without Injection: A Novel Endoscopic Treatment Method for Superficial Nonampullary Duodenal Epithelial Tumors. *JMA J.* **2021**, *4*, 415–419. [PubMed]
36. Yamashina, T.; Shimatani, M.; Takahashi, Y.; Takeo, M.; Saito, N.; Matsumoto, H.; Kasai, T.; Kano, M.; Sumimoto, K.; Mitsuyama, T.; et al. Gel Immersion Endoscopic Mucosal Resection (EMR) for Superficial Nonampullary Duodenal Epithelial Tumors May Reduce Procedure Time Compared with Underwater EMR (with Video). *Gastroenterol. Res. Pract.* **2022**, *2022*, 2040792. [CrossRef]
37. Miyakawa, A.; Kuwai, T.; Sakuma, Y.; Kubota, M.; Nakamura, A.; Itobayashi, E.; Shimura, H.; Suzuki, Y.; Shimura, K. A feasibility study comparing gel immersion endoscopic resection and underwater endoscopic mucosal resection for superficial nonampullary duodenal epithelial tumors. *Endoscopy* **2023**, *55*, 261–266. [CrossRef]
38. Matsubara, Y.; Tsuboi, A.; Hirata, I.; Sumioka, A.; Tanaka, H.; Yamashita, K.; Urabe, Y.; Oka, S. Gel immersion EMR of small-bowel inflammatory fibroid polyp using double-balloon endoscopy. *VideoGIE* **2024**, *9*, 92–94. [CrossRef] [PubMed]
39. Kitaoka, F.; Shiogama, T.; Mizutani, A.; Tsurunaga, Y.; Fukui, H.; Higami, Y.; Shimokawa, I.; Taguchi, T.; Kanematsu, T. A solitary Peutz-Jeghers-type hamartomatous polyp in the duodenum. A case report including results of mutation analysis. *Digestion* **2004**, *69*, 79–82. [CrossRef]
40. Hizawa, K.; Kawasaki, M.; Kouzuki, T.; Aoyagi, K.; Fujishima, M. Unroofing technique for the endoscopic resection of a large duodenal lipoma. *Gastrointest. Endosc.* **1999**, *49*, 391–392. [CrossRef]
41. Toya, Y.; Endo, M.; Orikasa, S.; Sugai, T.; Matsumoto, T. Lipoma of the small intestine treated with endoscopic resection. *Clin. J. Gastroenterol.* **2014**, *7*, 502–505. [CrossRef]
42. Wu, T.L.; Hsu, H.T.; Yen, H.H. Jejunum lymphangioma: A rare case of obscure gastrointestinal bleeding with successful endoscopic therapy. *Endoscopy* **2021**, *53*, E307–E308. [CrossRef]
43. Rimondi, A.; Sorge, A.; Murino, A.; Nandi, N.; Scaramella, L.; Vecchi, M.; Tontini, G.E.; Elli, L. Treatment options for gastrointestinal bleeding blue rubber bleb nevus syndrome: Systematic review. *Dig. Endosc.* **2024**, *36*, 162–171. [CrossRef] [PubMed]
44. Igawa, A.; Oka, S.; Tanaka, S.; Kunihara, S.; Nakano, M.; Chayama, K. Polidocanol injection therapy for small-bowel hemangioma by using double-balloon endoscopy. *Gastrointest. Endosc.* **2016**, *84*, 163–167. [CrossRef] [PubMed]
45. Kopacova, M.; Tacheci, I.; Koudelka, J.; Kralova, M.; Rejchrt, S.; Bures, J. A new approach to blue rubber bleb nevus syndrome: The role of capsule endoscopy and intra-operative enteroscopy. *Pediatr. Surg. Int.* **2007**, *23*, 693–697. [CrossRef] [PubMed]
46. Emami, M.H.; Haghdani, S.; Tavakkoli, H.; Mahzouni, P. Endoscopic polypectomy resection of blue rubber bleb nevus lesions in small bowel. *Indian J. Gastroenterol.* **2008**, *27*, 165–166.
47. Grammatopoulos, A.; Petraki, K.; Katsoras, G. Combined use of band ligation and detachable snares (endoloop) in a patient with blue rubber bleb nevus syndrome. *Ann. Gastroenterol.* **2013**, *26*, 264–266.
48. Lazaridis, N.; Murino, A.; Koukias, N.; Kiparissi, F.; Despott, E.J. Blue rubber bleb nevus syndrome in a 10-year-old child treated with loop ligation facilitated by double-balloon enteroscopy. *VideoGIE* **2020**, *5*, 412–414. [CrossRef]
49. Latchford, A.R.; Clark, S.K. Gastrointestinal aspects of Peutz-Jeghers syndrome. *Best Pract. Res. Clin. Gastroenterol.* **2022**, *58–59*, 101789. [CrossRef]
50. Patel, R.; Hyer, W. Practical management of polyposis syndromes. *Frontline Gastroenterol.* **2019**, *10*, 379–387. [CrossRef]
51. Pennazio, M.; Rondonotti, E.; Despott, E.J.; Dray, X.; Keuchel, M.; Moreels, T.; Sanders, D.S.; Spada, C.; Carretero, C.; Cortegoso Valdivia, P.; et al. Small-bowel capsule endoscopy and device-assisted enteroscopy for diagnosis and treatment of small-bowel disorders: European Society of Gastrointestinal Endoscopy (ESGE) Guideline—Update 2022. *Endoscopy* **2023**, *55*, 58–95. [CrossRef] [PubMed]
52. Ohmiya, N.; Taguchi, A.; Shirai, K.; Mabuchi, N.; Arakawa, D.; Kanazawa, H.; Ozeki, M.; Yamada, M.; Nakamura, M.; Itoh, A.; et al. Endoscopic resection of Peutz-Jeghers polyps throughout the small intestine at double-balloon enteroscopy without laparotomy. *Gastrointest. Endosc.* **2005**, *61*, 140–147. [CrossRef] [PubMed]
53. Sakamoto, H.; Yamamoto, H.; Hayashi, Y.; Yano, T.; Miyata, T.; Nishimura, N.; Shinhata, H.; Sato, H.; Sunada, K.; Sugano, K. Nonsurgical management of small-bowel polyps in Peutz-Jeghers syndrome with extensive polypectomy by using double-balloon endoscopy. *Gastrointest. Endosc.* **2011**, *74*, 328–333. [CrossRef] [PubMed]
54. Takakura, K.; Kato, T.; Arihiro, S.; Miyazaki, T.; Arai, Y.; Nakao, Y.; Komoike, N.; Itagaki, M.; Odagi, I.; Hirohama, K.; et al. Selective ligation using a detachable snare for small-intestinal polyps in patients with Peutz-Jeghers syndrome. *Endoscopy* **2011**, *43* (Suppl. S2), E264–E265. [CrossRef]
55. Yano, T.; Shinozaki, S.; Yamamoto, H. Crossed-clip strangulation for the management of small intestinal polyps in patients with Peutz-Jeghers syndrome. *Dig. Endosc.* **2018**, *30*, 677. [CrossRef] [PubMed]
56. Khurelbaatar, T.; Sakamoto, H.; Yano, T.; Sagara, Y.; Dashnyam, U.; Shinozaki, S.; Sunada, K.; Lefor, A.K.; Yamamoto, H. Endoscopic ischemic polypectomy for small-bowel polyps in patients with Peutz-Jeghers syndrome. *Endoscopy* **2021**, *53*, 744–748. [CrossRef] [PubMed]

57. Kamiya, K.J.L.L.; Hosoe, N.; Takabayashi, K.; Okuzawa, A.; Sakurai, H.; Hayashi, Y.; Miyanaga, R.; Sujino, T.; Ogata, H.; Kanai, T. Feasibility and Safety of Endoscopic Ischemic Polypectomy and Clinical Outcomes in Patients with Peutz-Jeghers Syndrome (with Video). *Dig. Dis. Sci.* **2023**, *68*, 252–258. [CrossRef] [PubMed]
58. Oguro, K.; Sakamoto, H.; Yano, T.; Funayama, Y.; Kitamura, M.; Nagayama, M.; Sunada, K.; Lefor, A.K.; Yamamoto, H. Endoscopic treatment of intussusception due to small intestine polyps in patients with Peutz-Jeghers Syndrome. *Endosc. Int. Open* **2022**, *10*, E1583–E1588. [CrossRef] [PubMed]
59. Gardner, E.J. Follow-up study of a family group exhibiting dominant inheritance for a syndrome including intestinal polyps, osteomas, fibromas and epidermal cysts. *Am. J. Hum. Genet.* **1962**, *14*, 376–390.
60. Saurin, J.C.; Ligneau, B.; Ponchon, T.; Leprêtre, J.; Chavaillon, A.; Napoléon, B.; Chayvialle, J.A. The influence of mutation site and age on the severity of duodenal polyposis in patients with familial adenomatous polyposis. *Gastrointest. Endosc.* **2002**, *55*, 342–347. [CrossRef]
61. Iwama, T.; Tamura, K.; Morita, T.; Hirai, T.; Hasegawa, H.; Koizumi, K.; Shirouzu, K.; Sugihara, K.; Yamamura, T.; Muto, T.; et al. A clinical overview of familial adenomatous polyposis derived from the database of the Polyposis Registry of Japan. *Int. J. Clin. Oncol.* **2004**, *9*, 308–316. [CrossRef] [PubMed]
62. Jagelman, D.G.; DeCosse, J.J.; Bussey, H.J. Upper gastrointestinal cancer in familial adenomatous polyposis. *Lancet* **1988**, *1*, 1149–1151. [CrossRef] [PubMed]
63. Yang, J.; Gurudu, S.R.; Koptiuch, C.; Agrawal, D.; Buxbaum, J.L.; Abbas Fehmi, S.M.; Fishman, D.S.; Khashab, M.A.; Jamil, L.H.; Jue, T.L.; et al. American Society for Gastrointestinal Endoscopy guideline on the role of endoscopy in familial adenomatous polyposis syndromes. *Gastrointest. Endosc.* **2020**, *91*, 963–982.e962. [CrossRef] [PubMed]
64. Iida, M.; Sakamoto, H.; Miura, Y.; Yano, T.; Hayashi, Y.; Lefor, A.K.; Yamamoto, H. Jejunal endoscopic submucosal dissection is feasible using the pocket-creation method and balloon-assisted endoscopy. *Endoscopy* **2018**, *50*, 931–932. [CrossRef] [PubMed]
65. Dekker, E.; Boparai, K.S.; Poley, J.W.; Mathus-Vliegen, E.M.; Offerhaus, G.J.; Kuipers, E.J.; Fockens, P.; Dees, J. High resolution endoscopy and the additional value of chromoendoscopy in the evaluation of duodenal adenomatosis in patients with familial adenomatous polyposis. *Endoscopy* **2009**, *41*, 666–669. [CrossRef]
66. Hamada, K.; Takeuchi, Y.; Ishikawa, H.; Ezoe, Y.; Arao, M.; Suzuki, S.; Iwatsubo, T.; Kato, M.; Tonai, Y.; Shichijo, S.; et al. Safety of cold snare polypectomy for duodenal adenomas in familial adenomatous polyposis: A prospective exploratory study. *Endoscopy* **2018**, *50*, 511–517. [CrossRef] [PubMed]
67. Sekiya, M.; Sakamoto, H.; Yano, T.; Miyahara, S.; Nagayama, M.; Kobayashi, Y.; Shinozaki, S.; Sunada, K.; Lefor, A.K.; Yamamoto, H. Double-balloon endoscopy facilitates efficient endoscopic resection of duodenal and jejunal polyps in patients with familial adenomatous polyposis. *Endoscopy* **2021**, *53*, 517–521. [CrossRef]
68. Takeuchi, Y.; Hamada, K.; Nakahira, H.; Shimamoto, Y.; Sakurai, H.; Tani, Y.; Shichijo, S.; Maekawa, A.; Kanesaka, T.; Yamamoto, S.; et al. Efficacy and safety of intensive downstaging polypectomy (IDP) for multiple duodenal adenomas in patients with familial adenomatous polyposis: A prospective cohort study. *Endoscopy* **2023**, *55*, 515–523. [CrossRef]

Disclaimer/Publisher's Note: The statements, opinions and data contained in all publications are solely those of the individual author(s) and contributor(s) and not of MDPI and/or the editor(s). MDPI and/or the editor(s) disclaim responsibility for any injury to people or property resulting from any ideas, methods, instructions or products referred to in the content.

MDPI AG
Grosspeteranlage 5
4052 Basel
Switzerland
Tel.: +41 61 683 77 34

Cancers Editorial Office
E-mail: cancers@mdpi.com
www.mdpi.com/journal/cancers

Disclaimer/Publisher's Note: The statements, opinions and data contained in all publications are solely those of the individual author(s) and contributor(s) and not of MDPI and/or the editor(s). MDPI and/or the editor(s) disclaim responsibility for any injury to people or property resulting from any ideas, methods, instructions or products referred to in the content.

www.ingramcontent.com/pod-product-compliance
Lightning Source LLC
LaVergne TN
LVHW072354090526
838202LV00019B/2547